Hearing Conservation in Industry, Schools, and the Military

Hearing Conservation in Industry, Schools, and the Military

Edited by

David M. Lipscomb, Ph.D.

Consulting Audiologist
Correct Service, Inc.
Stanwood, Washington

SINGULAR PUBLISHING GROUP, INC
San Diego, California

NOTICE TO THE READER

Publisher does not warrant or guarantee any of the products described herein or perform any independent analysis in connection with any of the product information contained herein. Publisher does not assume, and expressly disclaims, any obligation to obtain and include information other than that provided to it by the manufacturer.

The reader is expressly warned to consider and adopt all safety precautions that might be indicated by the activities herein and to avoid all potential hazards. By following the instructions contained herein, the reader willingly assumes all risks in connection with such instructions.

The Publisher makes no representation or warranties of any kind, including but not limited to, the warranties of fitness for particular purpose or merchantability, nor are any such representations implied with respect to the material set forth herein, and the publisher takes no responsibility with respect to such material. The publisher shall not be liable for any special, consequential, or exemplary damages resulting, in whole or part, from the readers' use of, or reliance upon, this material.

COPYRIGHT © 1994
Singular Publishing Group is a division of Thomson Learning. The Thomson Learning logo is a registered trademark used herein under license.

Printed in Canada
3 4 5 6 7 8 9 10 XXX 05 04 03 02 01 00

For more information, contact Singular Publishing Group, 401 West "A" Street, Suite 325 San Diego, CA 92101-7904; or find us on the World Wide Web at http://www.singpub.com

Library of Congress Cataloging-in-Publication Data
Hearing conservation in industry, schools, and military / edited
   by David M. Lipscomb.
      p.   cm.
   Includes bibliographical references and indexes.
   ISBN 1-56593-380-X
     1. Deafness, Noise induced—Prevention. 2. Noise control.
  I. Lipscomb, David M.
  RF293.7.H43 1994
  363.7'46—dc20   94-26675
                                                    CIP

Contents

■ *Preface*

A textbook on the subject of hearing conservation principles and practices in industry, schools, and the military is sorely needed. Other references exist on hearing conservation, but they are not particularly suitable as textbooks. It is our intention to find the middle ground between sophisticated engineer/acoustician-oriented material and basic information encountered in an introductory acoustics course. This text is also an attempt to provide the comprehensive information necessary to help the reader become a competent professional in a hearing conservation program. In this light, the book serves a second important function, that being to provide an authoritative reference for the professional hearing conservationist.

Information herein originates from a number of writers, since a single author can scarcely be expected to competently present all aspects of this multifaceted special application of hearing science. The contributors are some of the best people in the area of hearing conservation. Most of the authors are teachers, either in the academic classroom, or on the lecture circuit. By virtue of their continuing educational duties, they are up to date on the topic of their chapters. Therefore, the reader benefits from the quality that has made these writers desirable contributors.

The book begins with a historical accounting of the evolution of the hearing conservation guidelines promulgated by the U. S. Department of Labor. Then, the text turns to "nuts-and-bolts" chapters that provide pertinent facts and materials to use in establishing and maintaining effective hearing conservation programs. Finally, information is offered on the application of hearing conservation principles in two related arenas: school systems and environmental noise.

The need is acute for well-trained and skilled hearing conservationists. Completion of an advanced degree in audiology, medicine, engineering, or industrial hygiene is a good beginning. However, these credentials are no guarantee that you are suitably qualified or competent to manage a hearing conservation program. Nor does holding appropriate degrees certify that you are automatically endowed with adequate credentials to serve as a consultant to hearing conservation programs. In the next decade, educators will be challenged to develop an army of competent professional hearing conservationists so that standards of program performance become more lofty. As part of the evolutionary process, *Hearing Conservation in Industry, Schools, and the Military* presents essential and accurate information in order to offer a positive influence on the emergence of hearing conservation as a proud and viable occupation.

Contributors

Deborah A. Arthur, M.A.
Group Manager
Regulatory and Clinical Affairs
Smith and Nephew Richards
Memphis, Tennessee

John P. Balko, M.A.
President
Industrial Audiological Services
Sharpsville, Pennsylvania

Elliott H. Berger, M.S.
Manager, Acoustical Engineering
Cabot Safety Corporation
Indianapolis, Indiana

Wayne G. Bodenhemier, Ph.D.
Advanced Hearing Care, Inc.
Denver, Colorado

Russell J. Fankhouser, M.S.
Director of Industrial Services
Watauga Hearing Conservation, Inc.
Johnson City, Tennessee

John R. Franks, Ph.D.
National Institute for
Occupational Safety and Health
Physical Agents Branch
Robert A Taft Labs
Cincinnati, Ohio

Donald C. Gasaway, M.A.
Hearing Conservation Consultant
Cabot Safety Corporation
Southbridge, Massachusetts

Andrew P. Stewart, M.A.
Director, Audiological Services
E.L.B. & Associates, Inc.
Chapel Hill, North Carolina

Alice H. Suter, Ph.D.
Alice Suter & Associates
Ashland, Oregon

Introductory Information

Any academic subject has as its foundation those concepts that have become integral with the subject. In the case of hearing conservation, that foundation includes diverse topics ranging from legislative action to the physics of sound. In this section, selected materials are presented to introduce the study and applications of hearing conservation. For those experienced in the field, this material will serve as good review. For the uninitiated, these chapters are essential to understanding later chapters on applications of the principles in specific activities.

Three Little Words

David M. Lipscomb

A BRIEF HISTORY

Since 1969, when the Walsh-Healy Public Contracts Act was revised, hearing conservation practices have been introduced into a sizable segment of the occupational scene (approximately 70,000 plants) with the force of federal law. Literally dozens of books, hundreds of articles, and millions of words have been applied to this single regulatory action. Two years later, in 1971, Congress passed the Williams-Steiger Occupational Health Act (OSHA), which included word-for-word the hearing conservation portion of the Walsh-Healy revision. OSHA required hearing conservation for virtually all employees in the American work place not already covered by the Walsh-Healy Act.

On March 8, 1983, finalized regulatory guidelines were published in the *Federal Register*. This ended a 12 year gestation period for the guidelines, a comprehensive interpretation and delineation of features originally mandated in the Walsh-Healy action many years before. The lengthy, sometimes stormy, often controversial proceedings that ultimately gave rise to the guidelines also led to some confusion in the minds of persons charged with developing and monitoring hearing conservation programs according to the OSHA guidelines. It is the purpose of this chapter to return to the basics of the original legislative intent and summarize the three fundamental principles on which all of the guidelines are based. The essence of the legislative action can be summarized in deceptively simple terms. I propose that we review the meaning of the original act in terms of "three little words": *qualify*; *abate*; and *protect*. It may seem presumptuous that one would dare to advance such simplistic information in a textbook that will number hundreds of pages dealing with the meanings and applications of these principles. Yet, at the outset, these three little words will serve as useful memory tools and initial concept formation for our further, and increasingly sophisticated, discussions of the application of these "words." They are further offered to serve as a reminder that "hearing conservation" is more than simply the use of earplugs and/or monitoring hearing testing. Hearing conservation in the truest sense includes all three elements symbolized by the three little words. Further, the order of these words has importance. It is common in legislative activity to organize actions according to priority. Priority is signified by the order in which the items are listed. That principle holds true in the federal instructions for hearing conservation activity. Thus, the order in which these components are discussed is also their order of priority.

QUALIFICATION

This first facet of the triphasic hearing conservation paradigm means that one must first evaluate sound exposures in representative work environments in order to determine if persons are exposed to sound in excess of the allowable amount according to the dictates of federal law. Exposure criteria are summarized briefly in Table 1–1. It should be noted that this table is an abbreviated version of Table G–16a of the 1983 OSHA Guidelines. (The complete text of these guidelines is contained in Appendix A at the end of this book.)

TABLE 1–1.
Abbreviated Schedule of Exposure for Purposes of the Occupational Safety and Health Act. (The complete table is found in Appendix A.)

Sound Level	Permissible Daily Exposure
90 dBA	8 hours
95 dBA	4 hours
100 dBA	2 hours
105 dBA	1 hour
110 dBA	30 minutes
115 dBA	15 minutes

Qualification of an area is accomplished primarily through the use of sound measurement and exposure analysis (see Chapter 4). Sound *measurement* establishes the levels of sound present in specific locations at various times during an activity period. The measurement procedure incorporates a sampling process with the assumption that the sampling is representative of sound conditions in that environment. Measured sound *level* is meaningless alone. Why should we care if the level (in dB) of some machine consistently reaches 140 dB—as long as no one is present to be *exposed*? The key word is *exposure*. In qualifying areas as to their potential for hearing impairment hazard, *exposure analysis* is essential. Here, we factor in the additional feature *duration*. That is, the length of time one experiences a given sound environment. Note that the information contained in Table 1–1 includes both level *and* duration of exposure. This is consistent with historically employed methods of analysis of exposure advanced by numerous hearing damage risk criteria that preceded this legislative action. Inspection of Table 1–1 shows that an increase in sound level within an environment is countered by a decrease in the allowable duration of exposure. This relationship (sometimes called *time/intensity trading*) is defined specifically for OSHA as being a 5 dB trade. This means that each time the sound level changes 5 dB, the allowable duration of exposure is altered by a factor of two. The 5 dB time/intensity trade is quite generous, due to the fact that acoustical principles dictate that the doubling of energy (or exposure duration) is expressed as 3 dB (see Chapter 2 for more details). Less strigent criteria were accepted for the purposes of the hearing conservation regulations because of the intermittent nature of most industrial noise exposures. Breaks, down-time for equipment, and variability in equipment usage punctuates the exposure conditions over the course of a work day. Chapter 3 offers some specifics on this compromise.

Qualification is essential to a hearing

conservation program. The first efforts must be to identify those settings and activities that exceed allowable sound analysis conditions. If none of the areas are found to be excessive with respect to hearing risk, the hearing conservation program ends at that point. It is the more common experience, however, to note that certain work stations are in excess of the allowable guidelines, thereby indicating the need to introduce those regions to the next step of the hearing conservation process.

ABATEMENT

Once environments have been qualified for hearing conservation action, the most effective treatment for that condition is the abatement of sound so that the noise exposure guidelines are not longer exceeded. Some hearing conservationists and members of industrial management would prefer to skip this step and proceed directly to the final segment, protection of personnel. Abatement can be costly, and at times, it is either technologically or economically infeasible to decrease the sound level sufficiently to achieve a safe work environment in terms of hearing risk. However, the U.S. Department of Labor (DOL), in its mandated efforts to enforce the law, insists that abatement be included in the regimen for hearing conservation. At one point, it was determined by the DOL that it was requisite to undertake abatement procedures regardless of the possibility that hearing protective devices would still be necessary. That position has softened somewhat, as stated in a directive to field enforcement officials of the DOL. This modification, advanced in November of 1984, allowed that hearing protection could be considered a final solution when steady-state noise in a work environment did not exceed 100 dBA. If the environment contained steady-state sound greater than 100 dB, abatement of the sound was required as part of the program.

Sound reduction can take many forms and will be discussed in Chapter 7. For purposes of introduction, the following brief descriptions are offered.

ENGINEERING CONTROLS. When put into effect by qualified individuals, modification or redesign of equipment can be expected to decrease the noise output of sound sources. This is the most effective form of noise control because sound is eliminated at the source. Other forms of this method of control include equipment maintenance (squeaky metal can raise an awful racket), or placement of noise shields such as baffles or enclosures.

ADMINISTRATIVE CONTROL. If, for some reason, it is not possible to reduce the noise output of equipment influencing the acoustical properties of an environment, it may be appropriate to reduce overall exposure by modifying a worker's schedule. According to Table 1–1, a person should not be exposed to an average sound level of 95 dBA at his or her work station for more than 4 hours each day. However, a person who is found to be located in an area that has this average level for 6 hours each day would be exposed to sound in excess of the allowable guidelines. Rotation of the worker out of the environment and into quieter environments for 4 hours of each workday will accomplish the goal of keeping exposure level and duration within allowable limits. Administrative controls are often overlooked, yet they can be useful in meeting sound exposure guidelines. At times, however, administrative controls cannot be invoked due to the need for certain skilled persons within the area; union/management agreements; seniority rules; etc.

Successful noise abatement procedures obviate the need for a hearing conservation program. Once abatement has been accomplished and repeated sound measures indicate that exposure conditions no longer exceed the guidelines, no further activity will be needed other than periodic reassessment of the sound environment to assure that the sound exposure conditiions have remained within allowable limits. If excessive exposure according to the guidelines continues to be part of the work experience, the third part of a hearing conservation program is put into operation.

PROTECTION

Worker protection takes two forms:

HEARING PROTECTION. Some type of protective device (earplug, earmuff, etc.) is fitted and dispensed to employees, the purpose being to shield a person's hearing mechanism from high-level sound. Chapter 10 provides specific details of the design, fitting, and use of hearing protective devices (HPDs).

MONITORING AUDIOMETRY. The periodic measurement of hearing provides an ongoing evaluation of the effectiveness of the hearing that may be associated with noise exposure. Further details of the monitoring audiometrics incorporated into hearing conservation programs will be found in Chapter 13.

CONCLUSION

And so there they are—three little words: *qualify, abate,* and *protect.* No *one* concept represented by these words can stand alone. The combination of these three elements comprises the essence of an effective hearing conservation program. And now, what follows will be elaborations on three fundamentals.

REFERENCES

Federal Register. (1969). Walsh-Healy Public Contracts Act. *34*(96): 7891–7954.

Federal Register. (1983, March 8). Occupational noise exposure; hearing conservation amendment; final rule. *48*(46), 9737–9785.

Noise Control Act of 1972. (1972, October 27). Public Law 92–574. 92 Congress, HR 11021.

Williams-Steiger: Occupational Safety and Health Act. (1970). Public Law 91–596 Title 29CFR Chapter 17. Washington, DC.

2

What Is This Thing Called "Noise"?

David M. Lipscomb

THE CONCEPTS

*I*n one version of their highly useful reference text, Peterson and Gross (1972) asked: "What noise annoys an oyster?" That rhythmically phrased question raises the specter of quantification. Nearly every person alive can describe one or more sounds he or she finds particularly irritating. Yet, it is not usually possible to describe those sounds in objectively based, quantified terms. For the sake of universality, it is useful to have some objective way to describe sound exposure experiences. For legislative and enforcement purposes, it is essential to have some means of objectifying sounds that come under the purview of regulatory officials.

Since the advent of electronic sound measurement and analysis equipment, literally hundreds of *descriptors* have been proposed, some of which have found common usage. The term *descriptor* identifies a family of single-number identifiers intended to delineate acoustic parameters in some specific context. Decibel (dB) values can be included in this list. With a dB description, one can obtain some degree of objective insight into an acoustical condition. In this chapter, selected descriptors utilized in hearing conservation and related work will be introduced. The beginning student is well advised to develop a thorough grasp of this information. Persons working in any field that mixes acoustical principles and human response must have facility with these descriptors. In the event that one's activities lead toward serving as an expert witness, working knowledge of the descriptors is even more essential.

Sound is a complex physical phenomenon. Instantaneous variation of all acoustic signal parameters results in compound interaction between those components at any given moment. Further, changes over a wide range occur from moment to moment. Therefore, it usually is not feasible to track the "upness" and "downness" of sound amplitude or frequency over a lengthy period. Our best strategy is to combine all that variability and the elements that contribute to the variability into a simplified but reliable descriptor. At times, it is appropriate to describe only the crests or valleys of sound levels (in dB.) Other times, averaging is more descriptive. Another means of objectifying sound conditions is to compare the existing sound environment with some "norm" to determine exceedance.

It stands to reason that the more elements packed into a particular descriptor, the more assumptions one must make about acoustic parameters involved. For example, one of the descriptors we will be detailing utilizes a "time-weighted energy average" of sound in the environment. In this approach, all of the sound is combined with respect to the amplitudes (dB levels) attained from moment to moment to reflect a single-number dB value that is equivalent in energy to the time-varying amplitude in the real world. To base standards for sound exposure on such a concept, the assumption must be made that the average level reflects the hazardous or disturbing qualities of the time-varying sound. As we shall see, not all sound exposure conditions allow that assumption.

PARAMETERS OF SOUND

If foundation material for topics covered in this section is not clear, it is advisable to seek out references that offer detailed instruction on the physical parameters of sound (e.g., Borden & Harris, 1980; Deutsch & Richards, 1979; Durrant & Lovrinic, 1984).

The physical attributes of sound and their psychological counterparts are listed in Table 2–1. Sound *amplitude*, whose unit is the dB, is perceived as *loudness*. *Frequency*, with its unit the Hertz (Hz), is described as pitch in terms of perceptual experience. The *complexity of a sound gives it some identifiable characteristics that musicians call timbre* (pronounced *tamber*). It is not difficult to distinguish an oboe from a violin when they

TABLE 2–1.
Physical and Psychological (Perceptual) Attributes of Sound with Their Most Common Units in Parentheses

Physical Attributes	Psychological Counterpart
Amplitude (dB)	Loudness (sone)
Frequency (Hz)	Pitch (mel)
Complexity	Timbre

are both sounding the same musical note because of their unique respective complexity resulting in audible differences in timbre. This third component of sound reflects its *spectral energy*, that is, the frequencies contained in the sound with respect to their relative amplitudes.

THE DECIBEL

The decibel (dB) is a form of shorthand notation describing the logarithm of a ratio of two sound pressures or powers. It is the most widely used single-number descriptor of sound environments. The dB scale is a *ratio scale* in that it expresses nonlinear progression throughout its range. Contrast this to a linear scale such as that found on a yardstick, where each unit—the inch—is identical to each preceding unit. Depending on location in the dB scale, one dB does not reflect the same amount of sound pressure or energy as a dB at another point along this scale. This nonlinearity was made necessary because very early psychoacoustical observation determined that *sensation increases as the log of the stimulus*. In English, this means that in order to make equal increases in sensation, one must continually multiply the value of the stimulus (Table 2–2). All ratio scales must be developed in concert with two requirements: an arbitrary *zero* (reference point) must be defined; and the physical parameter must be identified.

Zero dB does not mean "no sound." Rather, it stands for that level of sound that has been arbitrarily established as the starting point or the *reference* power or pressure

TABLE 2-2.

Comparison of Growth of Linear and Nonlinear Scales When the Nonlinear Scale is Constructed with Base 10

Linear	Nonlinear
1	1 (10^0)
2	10 (10^1)
3	100 (10^2)
4	1000 (10^3)
5	10000 (10^4)

upon which the scale is constructed. In the form of dB scaling common to hearing conservation applications, the zero reference is 10^{-16} watts/meter² for sound energy or power and 20 microPascal (μP) for sound pressure. Over the years, several reference pressures have been described. Although they appear to be different, all of these historically significant pressures express the same amount of force exerted by sound at the "zero" dB level: 0.0002 dyne/centimeter²; 0.0002 microbar (millionths of barometric pressure); and 20 microNewtons/meter². Although 20μP has been designated the internationally standardized reference zero unit, many of us refer back to 0.0002 microbar (μb) as being more descriptive. Thus, with apologies to the standard-bearers of international standardization, this section will utilize the μb notation in some examples.

Note that a careful distinction has been made between sound *power* and sound *pressure*. They are two distinctly different physical phenomena. Sound *power* contains energy that can accomplish *work* and that work is effected when the energy strikes a surface. In doing so, the energy is distributed over the area of contact in the form of pressure or force. So, although energy and pressure must be considered separately, they do relate in a predictable manner:

$$E \sim p^2. \tag{1}$$

Energy is proportional to the square of pressure. This relationship is reflected in the dB equations (#2–#5) covered later in this section.

Initially, it may be confusing that we must deal with two factors of sound amplitude (energy *and* pressure). This dualistic feature is required because the output of any sound source is in the form of *energy* but it can only be measured as *pressure*. The force exerted on the diaphragm of a microphone is translated by sound measurement equipment into dB values (see Chapter 4). It is not possible to measure sound energy directly.

Decibel Notation

By memorizing three simple relationships (Table 2–3), it is possible to utilize the dB scale quite fully without need for dB equations or scientific calculators. In the context of sound power, the column in Table 2–3 under the word "energy" must be utilized. For example, anytime sound power (intensity or energy also) is doubled, that increase is expressed as a 3 dB rise in sound level. In like manner, a 10-fold change in sound energy is reflected as a 10 dB change in sound level. Halving sound energy results in a reduction of 3 dB. When energy is changed by a factor of 3.16 (the square root of 10), there is a 5 dB alteration in sound level.

Looking at the "pressure" column of Table 2–3, it is immediately apparent that the dB values in that column are double those in the "energy" column. This is a reflection of the two-to-one relationship in energy and pressure exponents mentioned in equation #1. It is necessary to double and then redouble sound power in order to accomplish a 6 dB increase (doubling) in sound pressure. Using values found in Table 2–3,

TABLE 2-3.

Three Mathematical Relationships for Sound Energy (Power) and Pressure Which Can Be Used for Memory Applications of the Decibel (dB) Concept

Multiplier	Energy	Pressure
2 ×	3 dB	6 dB
3.16 ×	5 dB	10 dB
10 ×	10 dB	20 dB

dB notation for sound pressure can be accomplished without the aid of calculators in steps as small as 2 dB (1 dB steps with respect to sound power).

Functional use of these relationships can be demonstrated by developing a continuum of sound pressures (Figure 2–1). Sound levels in dB are located above the dividing line and sound pressures in μb appear below the line. Using the 0.0002 μb reference pressure, this sound pressure level (SPL) scaling indicates the actual pressures requisite to produce each dB level. Beginning at 0 dB, the sound pressure is defined (arbitrarily) as 0.0002 μb. Each 10-fold increase in pressure results in a 20 dB increase in level (0.0002 = 20 dB; 0.02 = 40 dB; etc.). Note that the sound pressure (in μb) for an 80 dB sound is 2 μb. By decreasing the sound pressure by a factor of 2 (halving), there is a decrease of 6 dB (80 dB to 74 dB). Thus, the sound pressure, the physical force exerted by a 74 dB sound, is 1 μb. Multiplying 1 μb by 3.16 would result in the pressure equivalent for an 84 dB SPL. Increasing the 74 dB sound pressure 10-fold, the level rises to 10 μb (94 dB). This schedule continues all the way to theoretical maximum for a pure tone, 194 dB. This is maximum possible force a sound can exert and retain the characteristic of a "pure tone." At 194 dB (SPL), sound pressure is one million μb (one atmosphere). A pure tone, by definition, requires balanced compressional and rarefaction conditions. At a force of one atmosphere, the rarefaction phase would be a total vacuum. Since it is not possible to have greater rarefaction than a total vacuum, this level is the "theoretical" maximum for a pure tone. In reality, transducer limitations make it impossible to accomplish this theoretical level for pure tones, although some high-intensity sounds such as rocket firings by NASA do create sound levels in excess of 200 dB, but the sound is not tonal in nature.

A magnification of one part of the range in Figure 2–1 is found in Figure 2–2. Note that by using only those three relationships found in Table 2–3, it is possible to divide the sound pressure range into two dB steps.

Recall that it is necessary to both define a zero (reference) point and to name the physical phenomenon involved. When a sound level is identified with "SPL," conventional usage has determined that those two criteria have been met in that the scaling found in Figure 2–1 is implied. Therefore, when a report or research article designates a sound level, for example, 121 dB (SPL), it should be interpreted to mean that the "zero" is 20 μP (0.0002 μb in our examples) and the physical phenomenon is sound pressure. One additional assumption is possible as well. With the previous designation (SPL), the inference is that the sound was measured without the use of weighted filters. (See the following sections and Chapter 4.)

For energy, the same procedures can be followed using watts rather than μb and by employing the "energy" column of Table 2–3 (Figure 2–3).

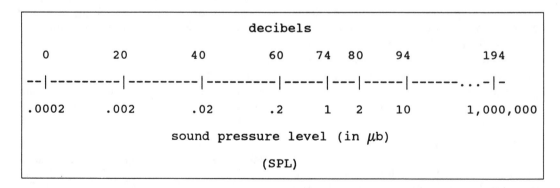

Figure 2–1. Example of the dB scale using pressure values ranging from 0.0002 μb to theoretical maximum for pure tones, one atmosphere.

decibels										
76	78	80	82	84	86	88	90	92	94	where: (rounded)
-|----|----|----|----|----|----|----|----|----|-										1.3 = 4/3.16
1.3	1.6	2	2.5	3.16	4	5	6.3	8	10	1.6 = 3.16/2
			sound pressure (in μb)							2.5 = 2(4/3.16)
			(SPL)							5.0 = 2(2.5)

Figure 2-2. Expansion of one portion of the scale shown in Figure 2-1 illustrating 2-dB steps possible with the use only of information contained in Table 2-3.

Figures 2-1, 2-2, and 2-3 illustrate an important feature of dB scaling, regardless of the physical phenomenon being used: note that each increment in the dB values represents a logarithmic increase in the amount of pressure or energy required to produce that increment. In Figure 2-1, at the low pressure end (left side), it took a change of only a tiny fraction of one μb to produce a 20 dB increase in SPL. However, at the upper end, a 20 dB increase required the addition of 900,000 μb! That is a graphic illustration of the fact that the dB scale is nonlinear.

For greater precision, use of the two available dB equations is necessary. The original equations for the Bel were:

For sound energy:

$$dB = \log E_1/E_0 \qquad (2)$$

where E_1 is the energy experienced; and
E_0 is the reference energy level

For sound pressure:

$$dB = 2 \log P_1/P_0 \qquad (3)$$

where P_1 is the pressure present; and
P_0 is the reference pressure.

The step sizes obtained in these equations were too large (1 Bel = 10 dB). Thus, the *deci*bel scale was developed to divide the steps into 10 parts. When one multiplies one side of an equation by 10, it is necessary to perform the same operation on the other side of the equation. The resulting equations for the *decibel* are identical to those given previously except that the multiplier is 10 times greater:

For sound energy:

$$dB = 10 \log E_1/E_0 \qquad (4)$$

where E_1 is the energy experienced; and
E_0 is the reference energy level

For sound pressure:

$$dB = 20 \log P_1/P_0 \qquad (5)$$

where P_1 is the pressure present; and
P_0 is the reference pressure.

decibels											
0	10	20	30	40	50	60	70	80	90	100	140
|----|----|----|----|----|----|----|----|----|---...-|-											
-16	-15	-14	-13	-12	-11	-10	-9	-8	-7	-6	-2
			sound energy in watts/m^2 (powers of 10)								

Figure 2-3. Example of the dB scale using power values ranging from the power reference point (10^{-16}) to the commonly cited auditory "threshold of pain," 140 dB.

The two equations bear distinct resemblance to each other with the exception of the multiplier. That two-to-one relationship refers back to the initial distinction between energy and pressure—that energy is proportional to the square of pressure (equation #1).

Combining Decibel Values

Using the data given in Table 2–3, it is possible to add dB values *providing* that the values are identical. That is, combining two sound sources to determine the total sound energy, one would simply add 3 dB to the output of the two sound sources. For example, if a turbine sound level is 81 dB, turning on a second, identical (81 dB) turbine would result in a *theoretical* increase of 3 dB (doubling of sound sources results in a 3 dB increase in the sound level). This is in theory because the acoustical properties of the environment and the phase relationships between the sound sources will govern the degree to which the two sources are identical. Another simple application of these concepts is commonly used in hearing conservation consulting activities. Realizing that the OSHA regulation has established a 90 dB baseline (90 dB exposure for an 8 hour work day), many plant engineers are being challenged to take the necessary steps to reduce sound in work environments to 90 dB. In the case of a manufacturer who has a cluster of machines, each individually outputting a sound level of 80 dB, he may ask how many of these machines he can put in a single room without *exceeding* 90 dB. This can be determined in the blink of an eye if one has Table 2–3 committed to memory. Recall that combining sound sources *always* employs the "energy" concept, for it is the energy output that is combined, rather than the sound pressure. If each machine is creating sound at 80 dB, that means that one must consider how many additional machines will result in a 10 dB greater sound. Looking in the "energy" column of Table 2–3, we see that a 10-fold increase is expressed as 10 dB. Therefore, the answer to the client's inquiry is that he can safely include *10* 80 dB machines in the room without exceeding 90 dB.

But it is frequently desirable to combine dB values that are not equal. To accomplish this, it is possible to utilize an equation:

$$dB_{(t)} = 10 \log \frac{A^2 + B^2 + C^2 \ldots N^2}{(0.0002)^2} \quad (6)$$

where: A, B, C, etc. are the sound pressures (in μP) of each dB value being combined.

Thus, to combine the dB values of 90, 96, and 88 dB, the sound pressure for each must be determined, squared, and then inserted into equation #6, resulting in a combined value that will be somewhat greater than the highest value being combined. Computer programs can be written quite easily that will accommodate this equation. However, if we do not have such programs available, a surprisingly accurate chart can perform the same task with very little loss of precision (Figure 2–4). To use the figure, one must combine two levels at a time. First, find the *difference* between the levels (in dB). In the previous example, the difference between 90 and 96 is 6. Find that point on the curved portion of the chart (on the right side of the chart). Using a straight-edge, note the point on the ordinate (vertical axis) that is parallel to the "6" on the curve. That designates the amount to be added to the

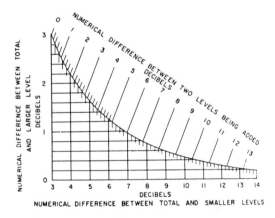

Figure 2–4. Chart for combining levels of uncorrelated noise signals.

highest of the two dB values being combined. In this example, the value to be added is approximately 1.0 dB, resulting in a combined level of 97 dB. Then, combine that level with the remaining level, 88 (the difference being 9 dB). Finding 9 dB on the curve points us to an approximate level of 0.5 dB on the ordinate. This raises the total level to 97.5 dB. Thus, the combination of 90, 96, and 88 dB sound levels results in a total sound level of 97.5. Using equation #6 and a programmed calculator, the level was calculated to be 97.49, virtually identical to the total obtained using the chart in Figure 2–4. Thus, Figure 2–4 can be used with the confidence that combining dB values will be accomplished with accuracy and considerable speed.

Solving dB Problems

A more detailed discussion of this topic will be found in Lipscomb, Diefendorf, and Bargelt (in press). Since it is sufficiently important to know how to solve dB problems, some information will be included here as well.

A four-step procedure will set the format for solving any dB question that might arise:

Step 1: *Choose the proper equation.* Is sound energy or sound pressure the physical phenomenon under consideration? An error at this point will result in a dB value that will be incorrect by a factor of 2. Remember, if the problem involves the combination of sound sources, the energy equation is *always* employed.

Step 2: *Insert data into the selected equation.* Many of us are not mathematicians, thus, initially it may not be easy to determine the proper points of insertion. However, experience and complete grasp of the parameters involved will augment correct completion of this step. If the given information is a dB value, that number is *always* inserted on the left side of the equation and it is necessary to solve for the numerator of the ratio. If a ratio or the experienced sound pressure (in μP) or sound energy (in watts) is given, this is inserted into the numerator

and one must solve for the dB value. Remember, calculation is only possible if the denominator is known. If a ratio is given, the denominator is always "1." If watts or μP are given, the denominator would be either 10^{-16} watts or $20 \mu P$, respectively. Note: the $20 \mu P$ value is used here in view of the international standardization of that pressure value. It may be easier to use 0.0002 μb instead. Either will result in a correct dB calculation.

Step 3: *Calculate.* Most of us have trouble balancing a check book, therefore, it stands to reason that errors can creep into this step with unpredictable results.

Step 4: *Answer the question.* By mentally or physically writing out an answer to the problem, one can often discover errors because the answer seems strange. This step, then, provides an informal check of the calculations.

Example 1: An engineering consulting firm is invited to bid on a noise abatement project. This firm proposes that they will reduce the noise energy by a factor of 78:1 and the cost will be $21,000. *Question:* In dB how much reduction will be accomplished and what will be the cost per dB?

Step 1: The energy equation (#4) will be used because the bid stated reduction in sound *energy* would be attempted.

Step 2: The data inserted into the equation:

$$dB = 10 \log 78/1$$

Step 3: It calculates as follows:

$$dB = 10 \log 78$$
$$= (10) (1.8921)$$
$$= 18.9 \text{ dB}$$

Note: the value is rounded to the nearest tenth

Step 4: *Answer:* The bid indicates that the noise reduction will be 18.9 dB which will cost $1055.28 per dB.

Example 2: One weaving loom creates sound measured to be 79.5 dB. *Question:* What would be the predicted sound level (in dB) if there are 100 identical looms in the loom room?

Step 1: Since the problem involves calculating combined sound levels, the solution will be based on the *energy* equation.

Step 2: Inserting the given data into the equation:

$$dB = 10 \log 100/1$$

Step 3: The calculation is as follows:

$$dB = 10 \log 100$$
$$= (10)(2.0)$$
$$= 20 \text{ dB}$$

Step 4: *Answer:* With looms that each generate 79.5 dB, the total sound level of 100 looms operating simultaneously will be 20 dB greater than the output of any one loom, or 99.5 dB. Note that the answer contained a two-step feature itself, which proves the value of providing a specific answer to the question. Had the answer been left at 20 dB, it would have been wrong. It is always necessary to return to the substance of the original question to be sure that all of the information is factored into the problem-solving process.

Example 3: This exercise will demonstrate the calculation for the numerator of the dB equation. *Question:* What is the sound pressure (in μb) for a 98.5 dB (SPL) sound?

Step 1: The question dealt with a sound pressure, thus, the pressure equation will be selected.

Step 2: Inserting the given data into the equation:

$98.5 = 20 \log P_1/0.0002$ (Note: dB values are *always* placed on the left side of the equation.)

Step 3: Calculation as follows:

$4.9250 = \log P_1/0.0002$ (obtained by dividing both sides of the equation by 20)

$84139 = P_1/0.0002$ (obtained by finding the antilog of 4.9250)

$P_1 = 16.8 \mu$b (obtained by multiplying 84139 by 0.0002)

Step 4: *Answer:* The sound pressure (physical force) exerted by a sound of 98.5 dB (SPL) is 16.8 μb.

OTHER DESCRIPTORS

The great English philosopher, Alfred North Whitehead, suggested that we seek simplicity, but he cautioned us not to trust it. This warning by the great man of letters should be taken to heart by all those who use single-number sound analysis and sound exposure descriptors. The ideal descriptor would comprehensively provide a description of the exposure condition; it would be totally reliable; it would be easily obtained, either by calculation or through the use of specialized equipment; it would be easily understood; it would be free of contaminating influences; and it would provide a single number representing the magnitude of the sound under analysis. Unfortunately, some of these criteria are mutually exclusive. So, single-number descriptors should be identified as simplifications—sometimes oversimplifications—of highly complex interactions of numerous time-varying physical parameters. Some information may be lost and the limitations of a descriptor brackets its utilization. This is not to deprecate the efforts that have resulted in such descriptors because many of them are being effectively utilized in several types of applications. One must always know and understand the assumptions underlying the development of a descriptor in order to utilize it appropriately.

Weighting Scales

One of the early attempts to resolve variability in spectral features of sound was to use "weighting scales" in adjusting the frequency response of sound measurement and analysis equipment. Three weighting scales are now standardized (Figure 2–5). Several versions of a fourth scale have been proposed, but none have been standardized at the time of this writing. Essentially, this

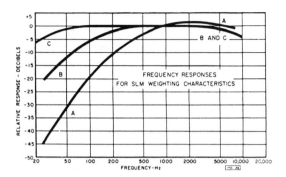

Figure 2–5. Frequency response characteristics (weighting scales) in the American National Standard Specification for Sound-Level Meters. ANSI–S1.4–1971.

fourth scale, the "D" weighting scale, follows the "B" scale for the lower frequencies and has an 11 dB "hump" at 2000 Hz concurrent with human sensitivity to that frequency range as compared to lower and higher frequencies.

"C" weighting is practically linear in that there is virtually no weighting of a spectrum in the greatest part of the audible range. The "B" scale is a rough approximation of the inverse of the 70 phon line, one of the equal loudness curves (see Figure 2–6). 'A' weighting is based on a smoothed inverse of the 40 phon curve. Both of the latter two weighting curves discriminate against low frequency sound energy—as does the ear. They are attempts to adjust sound measuring devices to approximate human experience at two levels of stimulation, 40 and 70 dB levels.

The down side of these single-number descriptors is that they do not reflect the sound spectrum as accurately as one might like. Considerable information is lost due to the automatic filtering of some frequencies. Weighted sound measurement values are not always comparable. That is, a 95 dBA environment at one location is not necessarily equivalent either in sound quality or in potential hearing risk to a 95 dBA environment elsewhere. As will be elaborated in Chapter 4, sound measures using weighting filters can be used to note "peak,"

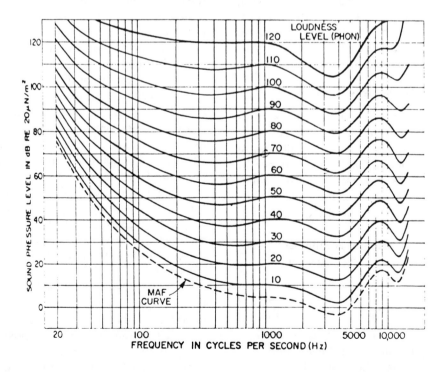

Figure 2–6. Equal loudness curves (phon) obtained by asking trained listeners to equate the loudness of sound at various frequencies to the comparison tone of 1 kHz.

"midpoint," or even "valley" sound levels without distinction. Further, the weighting curves are only valid for the phon levels from which they were derived. This means that the commonly used "A" weighting is theoretically appropriate for use only with relatively low level sound (40 phons). It is not correct to employ this scale for levels commonly encountered in the work place. However, the use of A-weighting has become so conventional, that this truth is now overlooked. In actuality, the "D" curve is more rational for high level applications since it is the smoothed inverse of the 90 phon line. In this case, theory gives way to one of the practicalities of life—too much has gone on with the use of A-weighting to attempt reversing the trend.

There is no firm agreement as to how one should designate measures obtained with equipment that has used weighting filters. It is conventional to represent A-weighted measurement results as "dBA." However, some users prefer dB(A), and still others maintain that the only correct notation is AdB or (A)dB. This text, while giving intellectual assent to the correctness of proponents of other notation styles, will use the "dBA" notation due to the sheer weight of conventional usage.

Daily Noise Dose

Rather than rely on a single peak, midpoint, or valley level, single-number compound descriptors such as "daily noise dose" incorporate a *time-weighted average*. In the case of daily noise dose, the time-weighting is according to the 5 dB time/intensity trade dictated by the OSHA regulation.

Calculation of *dose* is accomplished with the use of the following equation:

$$\text{Dose} = 100 \left(\frac{(C_1 + C_2 + C_3 \ldots C_n)}{(T_1 + T_2 + T_3 \ldots T_n)} \right) \quad (7)$$

where: C is the duration of exposure at a particular sound level; and

T is the allowable duration of exposure for that level according to the OSHA guidelines

An example: An industrial worker is exposed on a daily basis to 92 dBA sound for 3 hours, to 95 dBA sound for 2 hours, and to 97 dBA sound for 30 minutes. The remaining time on his shift the sound exposure falls to levels below 90 dBA and, thus, is not considered in this calculation. The equation would be filled in as follows:

$$\text{Dose} = 100 \left(\frac{3}{6} + \frac{2}{4} + \frac{.5}{3} \right)$$
$$= 100 (0.5 + 0.5 + 0.17)$$
$$= 100(1.17)$$
$$= 117\% \text{ of the}$$
allowable exposure.

In this example, the worker is exposed to more sound than is permissible.

Equivalent Levels

The composite sound exposure descriptor that has been favored by the federal government since 1975 is the *equivalent level* (L_{eq}). Although the L_{eq} does not meet all of the qualifications for a "perfect" descriptor, it has been found to be satisfactory in a range of applications. Essentially, L_{eq} is a time-weighted energy average that represents the total sound energy experienced over a given period of time as if the sound was unvarying. For example, $L_{eq(1)} = 75$ means that all of the sound integrated over a 1-hour period presented the same energy as an unvarying 75 dB sound. Lost in the use of this descriptor is the range over which the sound varied.

As a predictor of auditory damage risk, the L_{eq} is gaining popularity because, unlike the 5 dB time/intensity trade fostered by OSHA, it uses a 3 dB trading relationship. That feature renders the L_{eq} more conservative than the exposure scales used by the U.S. Department of Labor (DOL). Using the information under the "energy" heading in Table 2–3, it is possible to mentally determine equivalent exposures for many situations. For example, if one encounters a 100 dB sound for 1 minute, a 103 dB exposure for 30 seconds would provide the same sound energy to that person's ears. Or, a 110 dB sound exposure for 6 seconds (1/10 minute) would also be equivalent in the amount of sound energy experienced.

Another useful quality of this descriptor is that it is possible to compare widely different sound exposures in some sort of "context analysis." In seeking answers to the relative hazard of sounds, the L_{eq} can be utilized to compare the total amount of energy experienced in each sound exposure condition and allows an analysis of the relative hazard posed by each.

Calculation of L_{eq} values is based on an equation derived from an integral calculus equation. The derivation of the functional equation along with details on background and application of L_{eq} can be found elsewhere (Taylor & Lipscomb, 1978). For the purposes of this chapter, discussion will be limited to use of the functional equation to a select number of applications.

Conventional use of the L_{eq} concept utilizes sound measures taken using the "A" weighting scale (see previous section on weighting scales).

The L_{eq} equation resembles the dB equation for energy:

$$L_{eq} = L_i + 10 \log x_i \qquad (8)$$

where L_i is the level experienced for a period of time and x_i is the proportion of time L_i occurs with respect to the total time.

Example 1: A worker is exposed for 1 hour to an average sound level of 97 dBA. *Question:* What would be the equivalent exposure if the energy were distributed over 8 hours ($L_{eq(8)}$)?

$$
\begin{aligned}
L_{eq(8)} &= L_i + 10 \log x_i \\
&= 97 + 10 \log 1/8 \\
&= 97 + 10 \log 0.125 \\
&= 97 + (10)(-0.9030) \\
&= 97 + (-9.0) \\
&= 88 \text{ dB}
\end{aligned}
$$

Answer: A 1-hour exposure to 97 dBA sound is equivalent to exposure to 88 dBA over an 8-hour period.

Example 2: Question: What is the 24 hour equivalent level of a 97 dBA sound that occurred for 1 hour?

$$
\begin{aligned}
L_{eq(24)} &= L_i + 10 \log x_i \\
&= 97 + 10 \log 1/24 \\
&= 97 + 10 \log 0.04 \\
&= 97 + 10(-1.3802) \\
&= 97 + (-13.8) \\
&= 83.2 \text{ dB}
\end{aligned}
$$

Answer: The 24 hour equivalent level of a 1-hour exposure to 97 dBA is 83.2 dB. (Note: the relationship between 8 hour and 24 hour equivalent levels is *always* that the 24 hour value is less than the 8 hour value by 4.8 dB—usually it is sufficiently accurate to round to 5 dB).

Example 3: A person receives a high-level sound exposure of 132 dBA for 0.5 second. *Question:* What are the 1 hour, 8 hour, and 24 hour equivalents to that exposure?

$$
\begin{aligned}
L_{eq(1)} &= L_i + 10 \log x_i \\
&= 132 + 10 \log 0.5/3600 \\
&\quad (3600 \text{ sec/hour}) \\
&= 132 + 10 \log (0.000139) \\
&= 132 + 10(-3.8573) \\
&= 132 + (-38.6) \\
&= 93.4 \text{ dB} \\
L_{eq(8)} &= L_i + 10 \log x_i \\
&= 132 + 10 \log 0.5/28800 \\
&\quad (28800 \text{ sec/8 hours}) \\
&= 132 + 10 \log 0.00001736 \\
&= 132 + (10)(-4.760) \\
&= 132 + (-47.6) \\
&= 84.4 \text{ dB} \\
L_{eq(24)} &= L_i + 10 \log x_i \\
&= 132 + 10 \log 0.5/86400 \\
&\quad (86400 \text{ sec/24 hours}) \\
&= 132 + 10 \log 0.000005787 \\
&= 132 + 10(-5.2375) \\
&= 132 + (-52.4) \\
&= 79.6 \text{ dB}
\end{aligned}
$$

Answer: The equivalent levels for an exposure of 132 dB for 0.5 second are 93.4 dB, 84.5 dB, and 79.6 dB respectively for 1 hour, 8 hour, and 24 hour durations.

This example offers a means whereby uncommon exposure conditions can be put into context with other exposure levels in order to determine the potentially hazardous nature of an exposure or to assist in evaluating whether an alleged hearing impairment could have resulted from a given exposure condition. More will be made of this application in the section on forensic applications.

One final point should be made regarding this descriptor—it is "peak sensitive." Since the L_{eq} is based on the geometrically progressive scale identical to that of the dB, sound sources with high acoustic energy will influence the time-weighted average energy level considerably more than will lower level sounds. Two relationships can be described for use in exemplifying that fact:

1. A sound whose level exceeds an L_{eq} target level by 50 dB will "use up" the allowable amount of sound energy for a 24 hour period in 1 second.

2. A sound exceeding the target level by 40 dB will consume the allowable energy in 10 seconds.

Example: A community has a noise code that sets a 24 hour L_{eq} maximum of 60 dBA. An unmuffled motorcycle with an expansion chamber roars past a given monitoring location producing for 1 second a sound level of 110 dBA. *Question:* What is the 24 hour equivalent level for that single sound source?

$$
\begin{aligned}
L_{eq(24)} &= L_i + 10 \log x_i \\
&= 110 + 10 \log 1/86400 \\
&= 110 + 10 \log (0.00001157) \\
&= 110 + 10 (-4.937) \\
&= 110 + (-49.4) \\
&= 60.6 \text{ dB}
\end{aligned}
$$

Answer: In 1 second, the unmuffled motorcycle generates more sound than allowed in the community for a 24 hour period.

Day/Night Equivalent Level L_{dn}

A sister descriptor to L_{eq} is the L_{dn}. It is predicated on the observation that late night and early morning environmental sound is generally lower in sound level than during the busier parts of the day. Therefore, any intrusive sound is more noticeable during these hours than during daytime. To account for this altered intrusiveness, the L_{dn} descriptor is calculated by adding a 10 dB constant to any sound that occurs

betweeen the hours of 2200 (10 P.M.) and 0700 (7 A.M.). Thus, a truck that passes a house at noon emitting a sound of 90 dB will be considered to be a 100 dB sound source at midnight. It should be emphasized that this descriptor is not appropriate for evaluating sound exposure with respect to risk of hearing impairment. It is to be applied to land-use planning and community noise environment assessment only. The calculation of L_{dn} proceeds exactly as with L_{eq} once the 10 dB "penalty" has been added to nighttime sounds.

Example: The following average sound levels have been noted at a boundary between commercial and residential zones:

```
0700 – 1200 = 68 dBA
1200 – 1700 = 71 dBA
1700 – 2200 = 62 dBA
2200 – 0700 = 55 dBA (65 dBA in the
                           calculation)
```

Question: What is the L_{dn} for this community environment?

This problem is worked out in several steps:

Step 1: Calculate the partial L_{eq} for each time period. Using equation #8 and procedures cited previously, partial L_{eq} values in this setting were found to be 61.2, 64.2, 55.2, and 60.7 dB, respectively.

Step 2: The combined equivalent level (with 10 dB added for the nighttime experience) is attained by combining the four dB values either using equation #6 or by using the chart in Figure 2–4.

Step 3: *Answer:* The L_{dn} for the described environment is 66.4 dB.

Exceedance Levels

Many state highway planning authorities utilize exceedance levels to describe the acoustical environmental impact of developments such as new, expanded, or rerouted highways. On occasion, projections are made for exceedance levels to determine whether certain properties adjacent to that development should be condemned for certain applications. The general designation for an exceedance level is:

$$L_x$$

where L is the level in dB and
    x is the percent of time that level is
    exceeded

Thus, an L_{10} of 75 dB signifies that sound in a particular environment exceeds 75 dB 10 percent of the time. An L_{90} of 45 means that the sound level is greater than 45 dB 90 percent of the time. These two exceedance levels are the most commonly used ones, but others (e.g., L_1, L_5, L_{50}) can and have been used. L_{90} is normally considered an expression of "ambient" and L_{10} is often used to describe the maximum acoustic condition allowable under certain circumstances. For example, some highway departments consider that a projected L_{10} of 70 dB warrants condemnation of property for residential use. Again, these descriptors are of no value in determining auditory damage risk from noise exposure, but, like L_{eq}, they have been utilized successfully in land-use planning applications.

CONCLUSION

This chapter is an attempt to survey a representative group of common concepts applied to hearing conservation and noise analysis activity. The descriptors cited here are important to those who desire to become involved in consulting activities in industry and in the legal arena. It cannot be over-emphasized how important it is to have a strong grasp of all of these descriptors.

With the advent of microchip-based equipment, which allows digitization of the acoustic parameters of sound environments, much of the labor necessary to generate descriptors has been relegated to our equipment. However, that does not absolve us from understanding the principles on which those measures and analyses are based in order to make them clearly understandable to those with whom we will be consulting.

ACKNOWLEDGMENT

The decibel concept became clear to me only after receiving instruction from Jack Mowry, publisher of *Sound & Vibration*. Much of the discussion in this chapter dealing with the decibel concept springs from those early contacts with this outstanding teacher.

REFERENCES

Borden, G. J., & Harris, K. S. (1980). *Speech science primer*. Baltimore: Williams & Wilkins Company.

Deutsch, L. L., & Richards, A. M. (1979). *Elementary hearing science*. Baltimore: University Park Press.

Durrant, J. D., & Lovrinic, J. H. (1984). *Bases of hearing science* (2nd ed.). Baltimore: Williams & Wilkins Company.

Lipscomb, D. M., Diefendorf, A. O., & Bargelt, J. (in press). *Introduction to speech & hearing science*. San Diego: College-Hill Press.

Peterson, A. P. G., & Gross, E. E. (1972). *Handbook of noise measurement* (70th ed.). Concord, MA: GenRad Corporation.

Taylor, A. T., Jr., & Lipscomb, D. M. (1978). A primer on the use of L_{eq} and L_{dn}. In D. M. Lipscomb & A. T. Taylor, Jr. (Eds.), *Noise control, handbook of principles and practices*. New York: Van Nostrand, Reinhold.

3

Basic Principles of Sound Measurement

David M. Lipscomb

*T*he duffer golfer who spends a great deal of money on a new set of clubs will not suddenly realize pro–tour–quality golf with the upgraded equipment. He will still be a duffer because he doesn't have the skills to play better. In like manner, a person cannot be expected to conduct high-level sound measurement and analysis without adequate knowledge of techniques and procedures, even if he or she is equipped with the world's finest and most advanced sound measurement devices.

Recall from Chapter 1 that the first stage of a hearing conservation program is to "qualify" various areas to their potential for causing a noise-induced permanent threshold shift (NIPTS) due to a combination of noise levels and exposure durations.

BACKGROUND OF SOUND MEASUREMENT PRINCIPLES

Sound measurement technique has been the subject of full texts and many references, so this material should be regarded as introductory. The purpose of this chapter is to prepare the reader to more fully comprehend the sophisticated levels of presentation found in those other materials.

A bit of philosophy: Strategy adjusts to the purpose. As the intent for sound measurement becomes clear, strategy will develop so one can choose appropriate courses of action during sound measurement activity. For those readers who are audiologists or audiologists-in-training, this format should be familiar. It is comparable to the development of a "clinical feel" for selection of appropriate hearing tests based upon information derived as the clinical hearing evaluation progresses. In clinical audiology, there are literally hundreds of tests and procedures available to the clinician. He or she must select those that fit the needs of the patient. Sound measurement protocol must also include only appropriate, intelligently selected measurement and analysis procedures in order to obtain adequate information in a minimum of time.

In hearing conservation efforts, the

fundamental purpose of sound measurement and analysis is to establish whether or not representative work place locations or environmental areas exceed applicable limits. In general, the work place standards to which one will relate sound measures are those administered and enforced by the U.S. Department of Labor (DOL). Other guidelines and/or levels can come into consideration for applications of sound measurement and analysis regarding land-use planning, acoustic considerations for human habitation of confined spaces, or assessment of product liability by measuring sound output of devices available in the market place.

SOUND MEASUREMENT PROCEDURES

Instrumentation

As will be seen in Chapter 17, it is possible to spend considerable sums of money to equip a hearing conservation program. Some computer-based sound measurement and analysis devices conduct detailed studies of acoustic parameters in real-time, and can cost well into the six-figure range. Fortunately, for the needs of most industrial hearing conservation programs, hand-held sound level meters with some basic capabilities will be sufficient.

Sound level meters are manufactured according to specifications set forth by the American National Standards Institute ([ANSI], 1983). There are three grades of meters: Type 1 meters are precision units meeting the most rigorous specifications (ANSI S1.4–1983); Type 2 meters are general purpose sound level meters and are adequate for most industrial sound measurement activities; and Type 3 meters are sound survey meters, devices intended for nonprofessional, hobby-type use. The Occupational Safety and Health Act (OSHA) accepts use of Type 2 meters for assessment of overall A-weighted sound conditions in occupational environments. Consultants should obtain at least one Type 1 meter for measures that might be reported or discussed

in courts of law. It is always preferable to state that measures were obtained using a precision sound level meter. That notwithstanding, in the hands of a capable user, A-weighted sound measures will be virtually identical whether Type 1 or Type 2 equipment is utilized. Type 1 equipment offers a wider range of capabilities, therefore, greater versatility in selecting types of measurement.

Type 2 Sound Level Meters

The basic components of a sound level meter consist of a microphone, a filter set, an amplifier, and some sort of display (Figure 3–1). As the level of sophistication of sound measuring devices increases, the

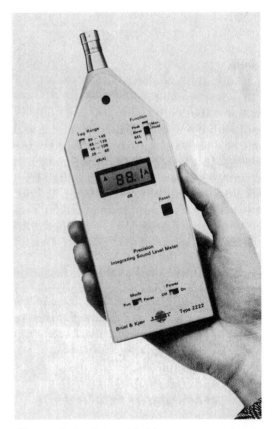

Figure 3–1. A hand-held Type 2 sound level meter. Note that the indicators are a stacked series of LED lights rather than the meter movement. Photo courtesy of Bruel and Kjaer.

type, quality, and number of components increases commensurately. Most equipment manufacturers supply basic units, such as the unit illustrated in Figure 3-1. These meters serve the purpose of monitoring A-weighted sound levels in various environments. The so-called idiot-proof devices are stripped to the barest essentials so that one does not need high levels of understanding to utilize the meter in order to note whether noise levels are changing over time. The simplified sound level meter may even have the display entirely on the face of the meter in order to circumvent the need for a range multiplier switch which might be confusing or misused by an uninitiated user.

Most basic meters provide a choice for type of indicating meter response. OSHA guidelines dictate that measures be taken using A-weighting and the ''slow'' meter speed. When set for ''slow'' meter response, the circuit integrates sound received by the microphone over a 1 second period (time window). Therefore, an needle-type indicating meter makes relatively slow changes from one dB value to another with the change of measured sound level. ''Fast'' meter speed, which is usually also available on these units, integrates each 0.2 second. This means that the needle will be dancing more rapidly with changes in sound level.

Additional features can be added to the sound measurement armamentarium for greater utility—at additional cost. These include a calibrator, octave-band measurement circuitry, impulse noise response circuitry, time-weighted averaging circuits, and input or output devices. All but one item in this component list are optional.

CALIBRATOR. It is imperative that some reliable method be used for maintaining the accuracy of the sound measurement equipment. Most devices have an internal electrical calibration signal which is useful to provide quick checks of the condition of its electronics. However, internal calibration circuitry does not take into account any changes that might have occurred in the response characteristics of the microphone. The transducer is the weakest point in the

system. Therefore, it is necessary to have a known sound source that one can periodically present to the equipment in order to check its accuracy. Such a known sound source (calibrator) offers the advantage of checking the accuracy of the microphone as well as all of the associated electronic circuitry.

Manufacturers provide calibrators intended for use with their sound level measuring instrumentation. Regardless of the amount of money one spends for sound measuring equipment, without a calibrator its value is nil. Room must always be left in the purchase budget for a calibrator even though that device might nearly double the expenditure.

It is not good practice to use a calibrator on a sound level meter if that calibrator was manufactured for use with another product. Acoustical coupling between the mouth of the calibrator and the microphone can be sufficiently different between brands of equipment to influence the accuracy of the calibration process.

OCTAVE-BAND ANALYZERS. After overall (full spectrum or weighted) measures are completed, it is often desirable to learn more about the acoustic environment with respect to the relative energy levels contained within its spectrum. Octave-band analysis electronically ''dissects'' the environment into a series of ''bands'' which have been specifically defined (Figure 3-2). This procedure allows one to identify those portions of the spectrum that contain the greatest amount of sound energy. As will be elaborated in Chapter 7, noise control strategies are often predicated upon the findings of octave-band analysis.

In acoustical parlance, an *octave* represents a band of frequencies where the highest frequency in the band is twice the frequency of the lowest. Thus, the interval between 500 Hz and 1000 Hz is one octave. In Figure 3-2, the standard specifications (ANSI S1.4–1983) for an octave band are illustrated. It is these designations that manufacturers of sound measurement equipment meet when they design and

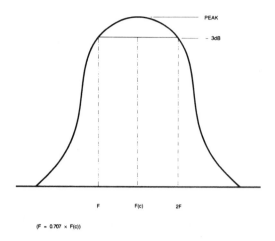

(F = 0.707 × F(c))

Figure 3-2. An octave-band showing standard parameters.

build octave-band analyzers. Until quite recently, it was not possible to provide filtering of sound with vertical slopes on either side of the octave band. Digital filtering can now do that. However, most equipment still relies on other electronic circuitry which allows some slope to the band on either side. Thus, it was necessary to specifically define the octave-band width. Figure 3-2 shows that the low- and high-frequency designations for band width are placed at the "half power point" (the level on the filter curve that is 3 dB below the peak). Vertical lines drawn from the half power point to the baseline should intersect with defined frequency locations. Specifically, the low "knee" for a filter slope should be 0.707 of the center frequency (f_c) for the octave band. The high knee will be double the frequency of the low knee. For example, an octave band whose center frequency is 1000 Hz would encompass a range of frequencies between 707 Hz (0.707 times 1000) and 1414 Hz. Usually, the numbers are rounded, so we would say the 1k Hz center-frequency octave band includes 700 to 1400 Hz. Knowing this range for the 1k Hz band, one can easily obtain the band widths for other octave bands. The width of the band centered on 500 Hz would be 350 Hz and 700 Hz. The 250 Hz octave band would

include frequencies between 175 Hz and 350 Hz and so on.

Results of the analysis can be recorded in various ways. Figure 3-3 shows one such method. The advantage of using a form which includes other measure, in addition to octave-band data, is that one can tell at a glance how all of the figures relate. This form offers space for sketching the environmental layout so repeated measures at some later time can be reliably undertaken. Another common format for reporting results of octave-band analysis is shown in Figure 3-4. Using this form, it is possible to graphically illustrate octave-band sound levels measured in an environment at one time as compared to sound measured there at some other time in order to demonstrate the effect of certain noise control activities. Or, using the format shown in Figure 3-4, measured octave band levels can be compared to reference levels for determination of exceedance.

This section has concentrated on the use of octave bands because they are the most commonly utilized in the practice of taking sound measures. Smaller band widths are available. For more refined determination of the frequency regions containing sound energy, 1/3 octave bands can be used. There are also available 1/2 octave-band measuring devices and even 1/10 octave-band equipment. Related units, although quite different in their function, are frequency analyzers which monitor sound energy in even smaller band widths to plot out the frequency spectrum of sound under consideration.

Plotting the results of band-width measures can cause some confusion. Forms, such as the one shown in Figure 3-4, use a *contiguous* scale. Each data point is independent of other data points. For convenience, we often connect those data points, but one must remember it is not valid to interpret the *slope* of those connecting lines. All of the energy in a band width is reported as one data point. In actuality, the energy might be concentrated at only one small portion of that band. A *continuous* scale reports the energy levels at all data

Date:

Instrument(s):

Location:

By:

Notes:

MEASURES TAKEN (in dB):

Center Frequency (Hz)	A	B	C	D	E	F	G	H
63	---	---	---	---	---	---	---	---
125	---	---	---	---	---	---	---	---
250	---	---	---	---	---	---	---	---
500	---	---	---	---	---	---	---	---
1000	---	---	---	---	---	---	---	---
2000	---	---	---	---	---	---	---	---
4000	---	---	---	---	---	---	---	---
8000	---	---	---	---	---	---	---	---
dBA	---	---	---	---	---	---	---	---
Leq(60s)	---	---	---	---	---	---	---	---
Impulse	---	---	---	---	---	---	---	---
Repetitions	---	---	---	---	---	---	---	---

Comments:

Figure 3-3. An octave-band data recording form.

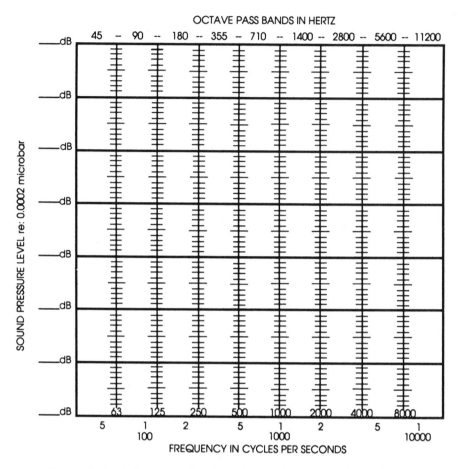

Figure 3-4. A data recording form for graphing octave-band data.

points, thus, slope of the line can be indicative of the relative energy levels along the spectrum being sampled. Only the frequency analyzer provides a *continuous* scale. Be sure to keep this distinction in mind in reporting and in interpreting band-width measurements.

IMPULSE/IMPACT NOISE MEASURING EQUIPMENT. It is becoming increasingly clear that steady-state sound and impulse sound are distinctly separate acoustical phenomena. They should be treated and measured according to the parameters of the nature of each sound classification. The major distinction between steady-state and impulse sound is the *rise time* of the sound. The sudden attack of sound energy occasioned by impulsive

sounds creates unique psychophysical reactions in listeners. It also creates specific injury in the ear if the exposure is excessive. Because of the unique qualities of impulse noise, specialized equipment must be utilized to "capture" the peak energy emitted by the impulse sound source.

Consider the "rise time" (first quarter wave) of a pure tone of 1000 Hz. The period of a tone is calculated using the equation:

$$p = 1/f \tag{1}$$

where:

p = period
f = frequency

Therefore, the total period of a 1k Hz tone is 1/1000 or 1 millisecond (ms). One-fourth of that would be 0.25 ms or 250 microseconds

(μs). The initial energy increment of such a tone is quite slow when compared to the report of a cap pistol whose rise time may be as brief as 5 μs. Ordinary sound measuring equipment is incapable of responding to sound pressure changes that occur that rapidly. Most Type 2 sound level meters have a frequency response ranging between about 20 Hz and 20k Hz, the commonly described audible range for human hearing. The 5 μs rise time mentioned previously is the duration of a 50k Hz tone. Thus, to be responsive to impulse sounds, frequency response characteristics of measuring devices must be extended well past the audible range for human hearing. Otherwise, the level of an impulse sound would be recorded to be at some point below its peak.

It is axiomatic that impulse noise measuring equipment will be required for thorough evaluation of a work environment which contains impulsive sound. This usually means that Type 1 (precision) equipment should be available.

INTEGRATING SOUND LEVEL METERS. Chapter 2 contained information on compound sound exposure descriptors such as L_{eq}. With the proliferation of microchip technology and its incursion into the field of acoustical measurement and analysis, equipment has become available with the capability of making large numbers of measures, integrating them, and producing a time-weighted average. Most of these integrating devices sample the sound environment four times per second or more. Some units are so small they can fit into a pocket, and can simultaneously report 15 or more types of measures (Figure 3-5).

The value of equipment with integrating capability is immense. No longer is it necessary to guess at the range of sound levels or to undertake arduous repetitive measures by hand. These little gadgets are happy to do all of that for us. In fact, in a day's time, using one of the integrating meters, it is probably possible to take more sound measures than have ever been taken individually by all persons who have taken

sound measures in the history of science.

Depending on the purpose of the measures, choice of equipment and equipment settings will be adjusted. If American work place sound exposure evaluation is to be undertaken, the integrating meter should be capable of so-called L_{osha} integration, which uses the 5 dB time/intensity trade. General environmental measures or European work place analysis would more appropriately utilize the L_{eq} approach, which utilizes the 3 dB time/intensity trade.

One subclass of integrating meters is *noise exposure dosimeters*. These are small cases equipped with a microphone that usually is attached to the case by a wire allowing the mike to be placed on the collar or hat of a subject worker. Both to conserve power and to maintain secrecy, there are no "readout" devices on the dosimeter. After a period of sampling, the unit is brought to a companion readout (or printout) device that will report the accumulated sound exposure condition experienced by a worker during the sampling period. Figure 3-6 shows hard-copy printouts from two types of dosimeters: statistical distribution and time history. In the former, the total sound exposure conditions are summarized statistically in the context of the total measurement period. The time history printout summarizes discrete minute-by-minute measurement results with a summary (in L_{eq}) at the end of the printout.

Dosimetry is a desirable concept. However, it is fraught with numerous problems of which the use should be aware:

1. It is sometimes difficult to obtain valid measures due to placement of the microphone.

2. Results can be influenced by horseplay (e.g., the wearer who shouts into the mike at close range, beats it against the wall, or muffles the microphone).

3. The presence of impulse noise can contaminate dosimetry indications, resulting in an inaccurate statement of noise exposure conditions.

4. Certain dosimetry equipment is set up on a pass/fail basis in that it simply

Figure 3-5. A multiple-function integrating community noise meter. Photo courtesy Quest Electronics.

indicates whether sound accumulated during the sampling period is over or under the predetermined allowable limit. Any inadvertent or intentional contaminating spurious sound can result in an overstatement of risk.

5. Dosimetry reflects only the sound reaching the microphone during the sampling period. Other periods on other days can yield quite different results, depending on the operations undertaken by the subject worker at that time.

INPUT/OUTPUT CAPABILITY. In addition to the indicators that are normally integral with the measuring device, it is possible in many cases to direct information to peripheral devices as well. For the purpose of capturing certain sound conditions for replay, tape recorder output is available on some meters. Hard copy can be produced by attaching a graphic level recorder or recording attenuator to the sound level meter. Relative sound levels can be measured by playing a tape through some devices in order to take advantage of metering indicators on sound level meters. Some limited sound level meters can be made more valuable by adding external peripherals such as octave-band filter sets or integrating circuits which

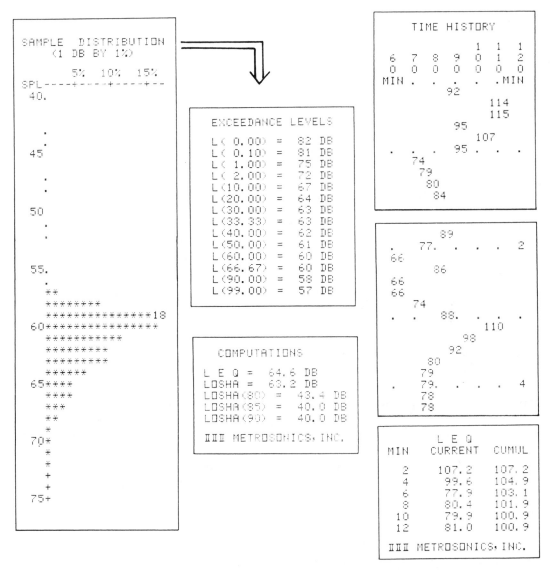

Figure 3–6. Two forms of dosimeter printouts. Information courtesy Metrosonics, Inc.

then interact with the electronics of the basic meter to provide additional useful information.

Procedures

Descriptions of procedural steps in taking sound measures are given with the understanding that the order and inclusion of these steps will be adjusted as conditions dictate. The emphasis of this text is on hearing conservation work in the industrial arena, therefore, the tone of this section will reflect that emphasis. Similar sound measurement protocol may be used in the schools, military, or other environments. Still other sound measurement activities such as community noise measurement will require adjustment of the following suggestions to meet the requirements for that setting.

Once a request has been made to evaluate the sound levels and exposure conditions of personnel at a manufacturing plant or other place of business, the following procedures are suggested.

Equipment Selection

The original selection and purchase of sound measuring equipment is discussed in Chapter 7. This discussion assumes that the hearing conservation service has a selection of sound measurement and analysis devices from which one can choose for particular assignments. Upon receiving an invitation to conduct sound measures at a work place, it is advisable to inquire about the environment or, if possible, make a preliminary visit to the setting to determine the types of noise to be included in the measures. It is probable that a Type 2 sound level meter will be needed for assessment of A-weighted sound levels at representative locations. The need for octave-band analysis will be almost as certain. Thus, it is the rare sound measurement activity that does not require these two forms of equipment. If the environment contains impulse sound, it will be requisite that an impulse sound level meter be included. Finally, if the request is for the determination of exposure conditions, it is useful to have time-weighted averaging available. Lest one feel that a 16-wheeler will be required to haul the measurement gear to the site, it should be noted that all of these capabilities are now available in single units from numerous manufacturers of sound measurement equipment. That is one of the advantages of springing for the Type 1 (precision) units at the outset.

Site Inspection

If a preliminary visit has not been made, it is advisable to first undertake a "walk through" to learn those areas of the plant that are considered potentially problematic regarding noise levels. The plant manager, plant engineer, or one of the supervisors will accompany any visitor for the sake of security and safety. They are the personnel who can answer pertinent questions regarding period of "up-time" and "down-time" for specific areas and processing machinery,

break schedules, lunch schedule sequencing, etc. It is doubtful one will ever encounter a measurement assignment wherein all sections of a plant are in full operation on a given day. Return visits may be necessary to fill in gaps left by nonfunctional equipment. This introduction to the layout and operations of the plant will offer a chance to customize a strategy for sound measurements to the plant which will be efficient and will reflect accurately the sound levels experienced in various locations.

Calibration

No attempts should be made to assess sound levels in any environment before appropriate calibration procedures are completed. It is recommended that acoustical calibrators be used for all equipment that will be utilized in the field study in order to check out the accuracy of all components of the equipment. Periodic calibration should be undertaken during the measurement period in the unlikely event that some change has taken place in the the function of equipment during sound measurement activity. At the end of the exercise, complete acoustical calibration must again be completed to assure the accuracy of the day's measures. This final step points up the need for periodic calibration checks. It would be sad indeed if one completed a full day of sound measurement work and the final calibration check showed one of the often-used measurement units to be inaccurate by 5 or more. It is unlikely that one would know when that change happened, so the entire day is lost—as well as the confidence of the people who requested the work.

Microphone Placement and Orientation

The primary intent of sound measures is to provide an indication of noise levels in the vicinity of work stations staffed by employees. Therefore, it is usually advisable to place the measuring equipment microphone at a location normally occupied by the ear of a worker. This location dictates that the worker should be moved from that location for the brief period measures are underway. It is not good practice to hold a microphone to the side of a worker's face near the ear for making determination of sound levels in that region. Head reflection and the head/body shadow effect can contaminate the measures by as much as 10 to 15 dB. Depending upon the type of microphone used, its orientation should be appropriate to the sound source being measured. Some microphones require that one point the long axis directly toward the sound source (0° or perpendicular incidence). A grazing incidence (90°) is recommended for other mikes wherein the long axis of the device is oriented to a plane of 90° from the

source so the sound "grazes" the surface of the microphone. Still other microphones are constructed so that a 70° incidence is required. Be sure that the required orientation is satisfied for the equipment under use.

Measurement Protocol

Consistency is one of the qualities one should strive to achieve in taking sound measures. By developing a consistent pattern of measurement steps, one can efficiently sample sound environments in work places. A flow chart (Figure 3–7) is offered to suggest successive steps and options for

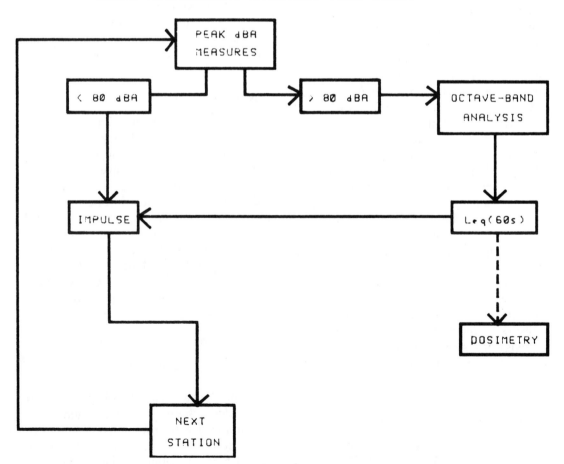

Figure 3–7. Sound level measurement activity flow chart.

each point of measurement. Peak overall readings (in dBA) are first completed at a work station. Depending on the outcome of those measures, subsequent steps follow logically. If the readings exceed 80 dBA, then octave-band measures at that location are recommended. These data are of little or no use for purposes of noise hazard assessment, but are of interest to the plant engineer or other persons interested in noise control actions. The next step would be to take brief time-weighted average measures (either L_{eq} or L_{osha}). Some devices sample an environment for 1 minute and provide a 60 second L_{eq} or L_{osha} which can often be used to generalize to the overall exposure at the work station under consideration. It is surprising to find that a great deal of industrial noise recycles within a 1 minute period. This can be checked by taking several repeated 1 minute measures. If the results vary widely, then longer-term dosimetry is advised. Next, if there is impulse noise in the area, an impulse sound level meter should be used to assess the peak impulse levels experienced. If impulse peaks exceed 100 dB, and if the repetition rate is high, it is necessary to count the number of repetitions in a 1 minute time period in order to later determine the possible hazard posed by the impulse noise. After completion of impulse measures, the steps shown in Figure 3–7 have been satisfied and it is time to move on to the next work station. This protocol is then followed throughout the plant, sampling all the appropriate work stations.

Note the alternate (left) "track" of Figure 3–7. In the event the first measure (dBA peak) did not exceed 80 dBA, impulse noise assessment should be conducted. If there is impulse noise in the environment, then go to the next work station.

Calibration Check

At the end of the measurement period for the day, conduct calibration measures before turning off the equipment.

Confer with the Appropriate Plant Personnel

No reasonable person will expect a full report at this juncture. But it is possible to offer an initial assessment of the measurement results to the plant manager or other operative for the hearing conservation program. A few minutes spent in this activity will lessen the shock of a later report that might outline numerous unsuspected potentially hazardous areas. Also, this mini-conference will serve as a preliminary introduction to the information to be contained in the report, and will ensure that the report will be better understood when it is received. In the conference, it is usually possible to summarize by indicating those areas of the plant that appear to "qualify" for a hearing conservation program and to identify, in general, those workers whose exposure exceeds the "action level" (see Chapters 1 and 13). In essence, it is good diplomacy, hence, good business, to provide this information as soon as possible.

Report Preparation

This is the item the industry is really purchasing. Time spent in gathering data comprises the bulk of the invoice one will submit, but a comprehensive, intelligently and logically prepared report must be provided. Help is available in one respect of report preparation in that there is a standard format to be followed (ANSI S1.13–1971). All raw data should be included in the report, either in the body along with text or as attachments at the end of the report. See Chapter 6 for details regarding report preparation.

CONCLUSION

Sound measurement and analysis is an acknowledged essential component in a comprehensive hearing conservation program. Initial sound measurement results establish the baseline for noise exposure control strategies. The measurement results are a point of comparison for evaluation of future control measures after they are completed. The importance of appropriate sound measures and the analysis of the measures to a successful hearing conservation program cannot be overstated. The results obtained form foundation information for developing and maintaining a hearing conservation program. If the foundation is shaky, the program cannot be otherwise.

The equipment and measurement protocol discussed herein are introductory in nature. Once this first exposure to sound measurement is completed, one should seek out further details from more advanced treatises. Some are cited at the end of this chapter.

REFERENCES AND SUGGESTED READINGS

American National Standards Institute. (1971). *American national standard methods for measurement of sound pressure levels.* (ANSI S1.13–1971 [R1976]). New York: American National Standards Institute.

American National Standards Institute. (1983). *American national standard specification for sound level meters.* (ANSI S1.14–1983). New York: American National Standards Institute.

Durrant, J. D., & Lovrinic, J. H. (1984). *Bases of hearing science* (2nd ed.). Baltimore: Williams & Wilkins.

Goldstein, J. (1978). Fundamental concepts in sound measurements. In D. M. Lipscomb (Ed.), *Noise and audiology* (p. 3). Baltimore: University Park Press.

Nabelek, I. V. (1985). Noise measurement and engineering controls. In A. S. Feldman & C. T. Grimes (Eds.), *Hearing conservation in industry* (p. 27). Baltimore: Williams & Wilkins.

Peterson, A. P. G. (1980). *Handbook of noise measurement* (8th ed.). Concord, MA: GenRad.

Williams, K. (1978). An introduction to the assessment and measurement of sound. In D. M. Lipscomb & A. C. Taylor, Jr. (Eds.). *Noise control, handbook of principles and practices.* New York: Van Nostrand Reinhold.

Determination of Noise Exposure

David M. Lipscomb ∎

Unwanted sound can be a problem at two levels:

1. *Hearing damage.* With high sound levels, if the duration of exposure is sufficiently long, hearing impairment can result.

2. *Other auditory effects.* This is the class of noise-related effects that, although not causing injury to the hearing mechanism, negatively impacts any of several functions. Examples are speech communication interference, sleep disturbance, interruption of concentration, and degradation of task performance.

HEARING DAMAGE RISK

Exposure to noise must be assessed in light of available criteria. In the context of potential for injury to the hearing mechanism, *hearing damage risk criteria* (DRC) have been developed. These criteria employ time/intensity relationships. Some early attempts added a second dimension for time—the overall duration a person encounters sound with potentially dangerous regularity. In these criteria, consideration was given for x duration of exposure to sound at y level of intensity over z years during a work life. Adding that third dimension provided a complexity factor so great that it has been dropped in the more recent attempts. the L_{eq} descriptor reinstates that concept because the integration period is flexible (see Chapter 2). Total duration of noise exposure can be extended to many years and incorporates such time factors as a work life for projecting hearing damage risk.

One must consider the *overall* exposure conditions in the context of currently applied DRC. It is doubtful that a person would lose any hearing by spending 5 hours in a 95 dBA industrial setting for a single day, although OSHA regulations limit the time of exposure to 4 hours at that sound level. Repeated exposures can, of course, ultimately result in hearing impairment. Two DRC systems are in frequent use at the present time: (1) the Occupational Health and Safety Act (OSHA), a U.S. regulation whose guidelines are printed in Appendix A; and (2) the Internation Organization for Standards (ISO) document 1999–1975 (primarily used in European countries). One feature these two quite different DRC guidelines have in common is the implication that the exposure gradients listed in each relate to long-term, occupational-type exposure conditions. Neither DRC is set forth as posing maximum levels of exposure for short-term or single-event accidental conditions. Occasionally, one will hear the statement: "The U.S. government says that *no one* should *ever* be exposed to sound over 115 dBA." In the context of legislative intent for OSHA, that statement is not consistent. Any noise exposure limitations inserted by OSHA relate to daily (or at least frequent) exposures and re-exposures to sound over many years throughout a work life. That same precept is advanced in the preamble of ISO 1999–1975 by referring to exposure during the years of a person's work experience.

The foregoing is not intended to speak for uncontrolled brief exposure experiences, but to set forth the limitations we now have regarding DRC guidelines available to us. At present, there are no similar guidelines available to us. At present, there are no similar guidelines specific to those accidental exposures one might encounter once in a lifetime at work or in the nonoccupational setting. Thus, the use of DRC values should be related to repeated exposure, generally in the context of occupational activities. They can be applied to other frequent exposure conditions, however, if the sound exposures are consistently experienced over several years in one's nonoccupational activities, for example, the hobbyist motorcycle racer. In such applications, the DRC guidelines do not have the force of law, but they can be used to provide a means for evaluating the exposure condition with respect to current DRC.

There are two important differences between the OSHA DRC and ISO DRC:

(1) the time/intensity trade for OSHA is 5 dB, whereas ISO uses the 3 dB exchange; and (2) OSHA designates a maximum allowable sound level for exposure for any period of time (115 dBA), whereas ISO does not. It should be noted that the example table provided in the ISO document does not express time/intensity features above 120 dBA, but there is no statement that 120 dBA is the maximum allowable sound level. One can extend that table, if necessary, by using the 3 dB time/intensity trade values.

One question that may arise is, Which of these two prominent DRC systems should one use? The answer can be complex; however, some general guidelines are offered:

1. If the noise exposure is ongoing and occurs in a U.S. work place, OSHA time/intensity levels apply with the force of law.

2. If one is constantly exposed to nonoccupational noise conditions, the more conservative ISO guidelines are recommended.

3. Single-event (e.g., accidental) or infrequent exposures to high-level noise are not appropriately addressed by either set of DRC. For context analysis of the exposure condition, it is possible one would benefit from consideration of the ISO guidelines with the awareness that these DRC assume repeated, rather than isolated, exposure conditions.

To further complicate matters, many persons whose occupational noise exposure levels are regulated according to OSHA also engage in nonoccupationally related noisy activities. In many states, farmers have taken employment at industrial plants and then they continue their farming activities during "off-hours." Other workers have noise exposure in nonoccupational hobby activities. Some participate in National Guard drills on specified weekends. Many use power equipment around their house after work. Still others moonlight at a second, perhaps noisy, job.

Thus, the analysis of a person's sound exposure condition is not always simple and straightforward, or even limited to the 8 hour work day. Hearing conservation is a 24 hour concern. Often, one does not obtain requisite information about an employee's total noise exposure until he or she is found to have a hearing impairment. Then it is too late to avoid some injury to hearing, but early signs of noise-induced permanent threshold shift (NIPTS), if heeded, can identify persons who are at risk due to some complex combination of exposure conditions, both occupational and nonoccupational. This points up the value of industrial hearing testing, including periodic testing of persons who are not considered to be overexposed to sound in their work stations. The combination of exposure at work and off-the-job noise conditions can result in a hearing impairment for which many persons would attempt to hold the employer responsible.

Application of DRC

Regardless of the DRC to which the noise exposure conditions will ultimately relate, there are some useful procedures one can follow to gain insight and gather data pertinent to individual exposure conditions.

Using only peak dBA level readings will generally overstate the potential risk of a sound environment because the low sound level swings in amplitude are not considered. Average dBA levels (midpoint of meter swings) tend to understate the risk because sound is "peak loaded" due to the logarithmic basis for the dB scale (see Chapter 2). So, in light of currently available technology, it is best to obtain some type of time-weighted average of a person's noise exposure during representative activity. Some forms of "noise dosimetry" is useful in the pursuit of quantifying a person's "true" sound exposure conditions.

Long-term dosimetry on every employee of even a modestly-sized manufacturing plant is a practical impossibility. Therefore, it is necessary to develop methods to short-cut the use of dosimetry without undue loss of accuracy and predictability with respect to risk. As indicated in the preceding chapter, short-term L_{eq} or L_{osha} samples are accurate in many industrial settings. If one

samples the sound level at a work station for 60 seconds repeatedly throughout a work day, it is not uncommon to find that the averaged measures do not vary more than 1 or 2 dB. Many industrial noise environments repeat their high-low sound amplitude cycles within the span of 1 minute. In such settings, a brief sampling of the sound environment will offer a reasonably accurate approximation of the total exposure conditions throughout the work period.

For those work stations where sound levels vary widely, representative short-term samples can be taken for each level of sound encountered, their proportion can be gauged, and a daily noise dose (Chapter 2) can be obtained in far less time that it would take for long-term noise dosimetry.

Example 1: A worker in a mobile home manufacturing plant assembles kitchen cabinets. About half of the time, this person is engaged in the assembly process, which includes operating a small saw, power screwdriver, and air jets to join and clean off the pre-cut pieces. The remaining half of the time is spent gathering materials to be assembled, and sound levels at the worker's ears fall below 85 dBA. During the assembly period, the time-weighted average (TWA) sound level to which this employee is exposed is 92 dBA. *Question:* Without undertaking lengthy noise dosimetry, what is the estimated exposure condition for this worker? In the context of OSHA guidelines (5 dB doubling rate—or, in this case, halving), one can estimate that the worker's total exposure each day is 87 dBA. This estimate was based on the fact that half of the exposure time was in low noise conditions. Therefore, it is appropriate to subtract 5 dB from the TWA level obtained while measuring only the high noise activity period. *BEWARE:* When using this technique, it must be clearly explained and justified in the report presented to the plant. Some enforcement officers are not particularly clear on the theoretical bases for hearing conservation activities and might suspect that a 5 dB "correction" has been arbitrarily used to lower the apparent exposure condition of employees. This exposure evaluation procedure is solidly based on the principles underlying sound measurement and exposure analysis protocol in the work place. The projection obtained by the short-cut method would be confirmed by long-term dosimetry.

Example 2: A worker in a machine shop spends 2 hours of a work day at a milling machine where the sampled short-term TWA registers to be 92 dBA. For 3 hours, he operates a turret lathe with a TWA of 95 dBA. For 2 hours he runs a welder (97 dBA) and the remaining hour is spent in non-noisy activities. Calculating daily noise dose (using the equation in Chapter 2) yields a total exposure index of 175 percent of the allowable exposure according to OSHA guidelines. This again offers an indication of total exposure without time-consuming and expensive dosimetry. *A word of caution:* Remember this technique is valid *only* when the sound environment being sampled is relatively unvarying in level or recycles during the sampling period.

There is no question that any method of sound exposure sampling is inexact at best. Day-to-day variability of work activities, equipment changes, maintenance problems, and background noise variability all signify that it is necessary to view these noise exposure analyses as being less than 100 percent accurate. That inadequacy can be compensated, at least in part, by the monitoring audiometry program which must also be part of the total hearing conservation package. If the total exposure estimate for a given worker is understated, any hearing shift will be detected by the annual hearing test, thereby identifying this person as one whose sound exposure should be more carefully assessed—often to include consideration of nonoccupational noise exposure.

Impulse Noise Exposure Assessment

Since impulse or impact noise is quite frequently incorporated into the total exposure of a subject worker, attention must be given to that condition as well. Most dosimeters are not designed according to

the specifications requisite upon an impulse sound level meter. Therefore, although impulse noise contaminates dosimetry, it is not accurately represented in the exposure reflected by dosimeters. Since impulse peaks are not accurately measured by dosimetry, impulse noise must be considered separately.

To the extent possible, repetition rates for impulses should be determined and the total number of repetitions per work day can then be estimated. According to the 1981 version of the OSHA guidelines, that schedule would be as shown in Table 4–1. Using this schedule, one can determine the risk factor of employee exposure to impulse noise.

Combining Steady-State and Impulse Noise Exposure

The work place may contain a combination of sounds, both steady-state and impulse. In that case, it is not possible with a single sampling to obtain an accurate indication of the total exposure resulting from the combined effects of sound types. One should consider steady-state noise exposure conditions and impulse noise exposure separately, then combine them to determine if a worker's daily noise dose is within allowable limits. The following example is a description of one method for making such a determination. Although it is a practical application of exposure analysis principles, it is not an "official" OSHA-sanctioned procedure. Therefore, the

TABLE 4–1.

Allowable Repetitions of Impulse Noise Exposure According to Peak Level of Impulses (in dB)

Range of Impulse Peaks (dB — SPL)	Allowable Repetitions/Day
>140	0
140	100
130–139	1000
120–129	10,000
110–119	100,000
100–109	1 million

rationale for using the procedure may require elaboration to enforcement officials on occasion.

Example: A printing press operator works a full shift (8 hours) in a location with steady-state noise averaging 95 dBA for 3 hours of the shift. His press prints computer forms with tractor-feed perforations which are punched as the printing process ensues. There are 75,000 repetitions of the punching sound each shift with a peak level measured to reach 118 dB (SPL). The steady-state noise exposure condition for this worker is within the allowable OSHA guidelines for steady state exposure (95 dBA exposure is allowed for 4 hours), and his impulse noise exposure is within the allowable amount according to the previous table (75 percent of the allowable number of repetitions). However, using the dose equation (Chapter 2), wherein the two exposures are summed and multiplied by 100:

$$
\begin{aligned}
\text{dose} &= 100(3/4 + 75{,}000/100{,}000) \\
&= 100(0.75 + 0.75) \\
&= 150\% \text{ of the allowable total} \\
&\quad\ \text{exposure}
\end{aligned}
$$

By combining the steady-state and impulse exposures, it is estimated that the worker's environment places him at risk for hearing impairment. The advantage of this procedure is that it utilizes *accurate* measures of both sound types to which the worker is exposed.

OTHER AUDITORY FACTORS

Discussions in this section will concern the impact of noise exposure on individuals, not with respect to hearing impairment, but in the context of its intrusiveness. One who is involved in industrial sound level assessment may be asked to determine whether some noise in an environment constitutes an excessive intrusion of some sort. Two major aspects of life are often considered: speech communication and sleep. Nearly everyone has had the experience of having to stop speaking until some type of noise source was quieted (e.g., an aircraft flying over at low altitude); and it is not uncommon to have one's sleep interrupted by some sound in the environment. If a sound source causes one or both of these experiences on a repetitive basis, complaints are likely, and sometimes legal action is initiated. On occasion, the sound source emanates from a manufacturing plant or from ancillary equipment such as trucks entering and leaving the premises. As part of sound analysis for hearing conservation, it is possible one will be asked to assess the impact plant noise is having on the neighborhood. In other cases, legal action may be brought against the source of offending noise and sound measurement is requested to determine the impact a noise source has on the complainant.

For those communities with legislation for the control of excess environmental noise, exceedance is often defined in the legal code. If a given sound source exceeds the legally allowable amount, the complaint is easily verified. However, some communities with noise codes do a less-than-adequate job of enforcing them. One nearby town has a chamber of commerce that would like to project an image of being "the playground of the southeast." Recently, this community passed a quite comprehensive noise code. But, when recreational facilities created noise in excess of the legal limits, neighboring residential inhabitants were unable to have the community noise code enforced. They found it necessary to take the case to court on their own. It is the apparent policy of that small city that the economy would suffer if noise-producing businesses were brought under the law.

In cases where there is either no noise code, or the existing code is not adequately enforced, two types of sound analyses are useful:

Boundary Noise Assessment

In general, what goes on within one's property with respect to noise or other environmental contaminant output is generally considered to be the concern of only the property owner. But, when effluent in the form of noise, dust, smoke, odor, etc., escapes over the boundary line into neighboring property, controls should be brought to bear. Community noise codes often have "boundary noise limits" which establish the maximum level of noise allowed at the boundary line. Most commonly, these codes relate to the property interfaces between industrial and residential applications. Some communities have a number of these noise limits according to the zoning in question. Figure 4–1 illustrates a form used to apply the boundary noise code of a representative county. Note that the limits are expressed in a series of octave-band levels, generally reflecting the nonlinearity of human hearing by allowing higher noise levels in low frequencies. In using the boundary noise limits, exceedance is demonstrated if any *one* of the octave bands is measured to contain sound in excess of the allowable amount. Often, multiple measures are used to sample several locations along a boundary. This is necessary in order to obtain a clear picture of the relative impact the noise source has on the surrounding community.

Sleep Environment Assessment

In those settings close to some noise source that continues for 24 hours each day, complaints are often made that the noise interferes with sleep. This extremely subjective complaint can be given a degree of objectivity with the employment of Noise Criterion (NC) Curves (Beranek, 1960). The essential purpose of this family of frequency/intensity curves based on octave-band measures is to provide acoustic context for several applications of spaces. Depending upon the use for an area, one of the NC Curves can be applied (Figure 4–2). For example, a sleeping environment most often requires NC = 25 to NC = 35. These equate to approximately 35 to 45 dBA. It is generally good practice to apply the NC = 30 NC Curve when evaluating sleep interference.

Example: Interpretation of community noise measures can be objectified within the limitations of the NC Curve applications as shown in Figure 4–3. These data were obtained in a case wherein large air conditioning units for a manufacturing plant were relocated at the back of the plant, directly across the boundary line from a residence. Comparative measures were taken when the compressors were operating and silent. Figure 4–3 shows definitively that the sound from the air conditioner compressors penetrated into the home and exceeded the "sleep curve." It was possible to conclude that the noise should be expected to create a disturbance of sleep, substantiating the allegations of the resident. The case went to court. In the trial, an illustration similar to Figure 4–3 was shown to graphically demonstrate that the noise was a problem. Ultimately, the judge ruled that the manufacturing plant must relocate the air conditioning units to reduce acoustical intrusion into the neighboring residence.

NC Curves can be applied to a wide range of space utilization (Tables 4–2 and 4–3). With accurate sound level measures and intelligent use of the comparative "appropriate" sound levels signified by the NC Curves, one can determine the adequacy of a sound environment for a projected purpose for the space.

CONCLUSION

Analysis of *exposure* to sound includes consideration for both hearing damage risk and interference of some human activity. The materials presented in this chapter summarize the most frequently used applications in the experience of the author. There are other comparative indicators one should have in his or her armamentarium in order to effectively respond to requests for sound exposure analyses in a variety of settings and for many different purposes. They can be found in some of the more advanced references cited at the end of this chapter and elsewhere in this text.

Date:

Instrument(s):

Location:

By:

Notes:

MEASURES TAKEN (in dB):

Center Frequency (Hz)	Permissible Level	A	B	C	D	E
63	79 dB	---	---	---	---	---
125	74 dB	---	---	---	---	---
250	66 dB	---	---	---	---	---
500	59 dB	---	---	---	---	---
1000	53 dB	---	---	---	---	---
2000	47 dB	---	---	---	---	---
4000	41 dB	---	---	---	---	---
8000	39 dB	---	---	---	---	---
dBA		---	---	---	---	---
Leq(60s)		---	---	---	---	---

Comments:

Figure 4-1. Boundary noise assessment form. Allowable limits are those currently in the Knox County, Tennessee code.

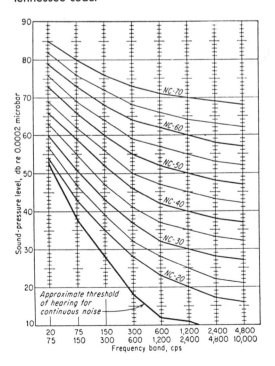

Figure 4-2. Noise Criterion Curves. Note: the octave-band boundaries have been changed to center frequency bands by ANSI since these curves were developed. The differences brought about by that change are, for these purposes, negligible. From Beranek (1960, p. 519), *Noise reduction*. Reprinted with permission from the author.

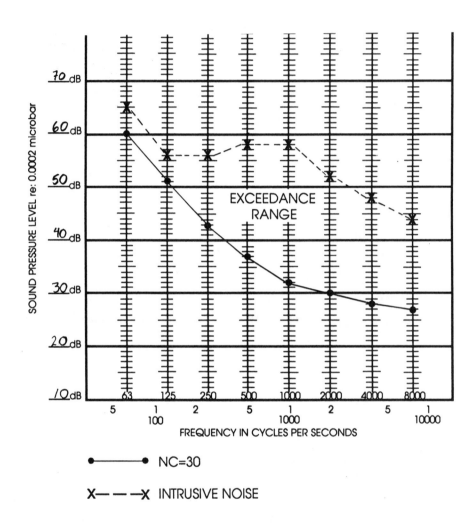

Figure 4–3. Example of one application of NC Curves in assessing the intrusiveness of noise into sleeping quarters.

TABLE 4-2.
Recommended Noise Criteria for Offices

NC Units	Communication Environment	Typical Applications
20–30	Very quiet office—telephone use satisfactory—suitable for large conferences	Executive offices and conference rooms for 50 people
30–35	"Quiet" office; satisfactory for conferences at a 15-ft table; normal voice 10 to 30 ft; telephone use satisfactory	Private or semiprivate offices, reception rooms and small conference rooms for 20 people
35–40	Satisfactory for conferences at a 6- to 8-ft table; telephone use satisfactory; normal voice 6 to 12 ft	Medium-sized offices and industrial business offices
40–50	Satisfactory for conferences at a 4- to 5-ft table; telephone use occasionally slightly difficult; normal voice 3 to 6 ft; raised voice 6 to 12 ft	Large engineering and drafting rooms, etc.
50–55	Unsatisfactory for conferences of more than two or three people; telephone use slightly difficult; normal voice 1 to 2 ft; raised voice 3 to 6 ft	Secretarial areas (typing), accounting areas (business machines), blueprint rooms, etc.
Above 55	"Very noisy"; office environment unsatisfactory; telephone use difficult	Not recommended for any type of office

From Beranek, L. L. (1960). *Noise reduction* (p. 521). New York: McGraw-Hill Book Co. Reprinted with permission from the author.

TABLE 4-3.
Recommended Noise Criteria for Rooms

Type of Space	NC Units	Computed Equivalent SLM Readings (Weighting Scale A, dBA)
Broadcast studios	15–20	25–30
Concert halls	15–20	25–30
Legitimate theaters (500 seats, no amplification)	20–25	30–35
Musicrooms	25	35
Schoolrooms (no amplification)	25	35
Television studios	25	35
Apartments and hotels	25–30	35–40
Assembly halls (amplification)	25–35	35–40
Homes (sleeping areas)	25–35†	35–45†
Motion-picture theaters	30	40
Hospitals	30	40
Churches (no amplification)	25	35
Courtrooms (no amplification)	25	30–35
Libraries	30	40–45
Restaurants	45	55
Coliseums for sports only (amplification)	50	60

From Beranek, L. L. (1960). *Noise reduction* (p. 523). New York: McGraw-Hill Book Co. Reprinted with permission from the author.

REFERENCES AND SUGGESTED READINGS

Beranek, L. L. (1960). *Noise reduction*. New York: McGraw-Hill Book Co.

Federal Register. (1981, January 16). Occupational Noise Exposure; Hearing Conservation Amendment; Final Rule. *46*(11), 4078–4179.

Federal Register. (1983, March 8). Occupational Noise Exposure; Hearing Conservation Amendment; Final Rule. *48*(46), 9737–9785.

Harris, C. M. (Ed.). (1979). *Handbook of noise control*. (2nd ed.). New York: McGraw-Hill Book Co.

International Organization for Standardization. *Acoustics—Assessment of occupational noise exposure for hearing conservation purposes*. ISO 1999–1975.

The Development of Federal Noise Standards and Damage Risk Criteria

CHAPTER

5

Alice H. Suter

THE NEED FOR NOISE REGULATION

Noise is the most prevalent health hazard in the manufacturing industries. It is easily recognized and annoying to workers, so that it is one of the most common sources of complaints to the Occupational Safety and Health Administration (OSHA). Because of its interest in all areas of noise exposure, the Environmental Protection Agency (EPA) commissioned a report entitled "Noise in America" (EPA, 1981). The authors of this report estimated that 9.27 million Americans are exposed to daily average noise levels of 85 dBA and above in the six major categories of noisy occupations. This number represents nearly 25 percent of the people employed in these occupations. Table 5–1 shows the breakdown of noise-exposed workers by occupational category.

On the basis of similar data, OSHA (1981) has estimated that approximately 2.9

American workers are exposed to daily average noise levels above 90 dBA and 5.2 million above 85 dBA. These exposures occur in more than 300,000 work places throughout the nation. Table 5–2 shows the percentage of workers exposed to daily average noise levels above 85 and 90 dBA according to type of industry. One can see that certain industries, such as lumber and wood, textiles, petroleum and coal, and primary metals are the source of serious noise problems, although all of the industries listed in this table have their noisy operations.

Not coincidentally, a 1965 U.S. Public Health Survey reported that 8.4 percent of the U.S. adult population had average hearing levels of 25 dB or greater at 500, 1000 and 2000 Hz (U.S. Department of Health, Education, and Welfare, 1965). If this estimate applies to today's population, it would amount to more that 20 million people. Although many of these people are affected by etiologies other than noise, it is

TABLE 5–1.
Summary of U.S. Population Exposed to Daily Average Noise Levels of 85 dBA and Above

Employment Area	Total Employed	Number Exposed to 85 dBA and Above
Agriculture	3,600,000	323,000
Mining	957,000	400,000
Construction	4,644,000	513,000
Manufacturing and Utilities	21,781,000	5,124,000
Transportation	4,345,000	1,934,000
Military	3,019,000	976,000
Totals	38,346,000	9,270,000

From: Environmental Protection Agency, Office of Noise Abatement and Control. (1981). *Noise in America: The extent of the noise problem.* (EPA Report No. 550/9–81–101).

reasonable to assume that a large portion of these losses are caused by occupational noise exposure.

Unfortunately, the free market system does not provide for adequate protection against the harmful effects of noise. Many noise-exposed workers do not have the option of shopping around for a quieter job, especially one that pays as well. After they have worked in high noise levels for a while, there is a tendency for workers to feel that noise is a necessary component of an industrial job, and noise-induced hearing loss is the inevitable consequence. Even if they are aware that noise is controllable, they are reluctant to complain for fear of jeopardizing their jobs. As a result, today's employers do not experience much pressure from workers, or even from labor unions, to control excessive noise.

Although the benefits of reducing noise are many, they are not always readily apparent to employers. They are somewhat difficult to quantify, and even more difficult to give a dollar value to. Worker compensation payments are often viewed as a major incentive for protecting employees, when, at least in the private sector, the current reality is that these expenditures are quite small in relation to the costs of noise abatement and hearing conservation programs. According to Shampan and Ginnold (1982), the total amount paid to private sector workers in 1977 was approximately $13 million, which amount to only about $3,000 per claim. Payments to civilian federal workers was somewhat more generous, due to the absence of various filing restrictions imposed by the states. Military personnel also receive more generous compensation through the Veterans Administration, which should provide added incentive for protecting these categories of employees.

There are some additional benefits to controlling occupational noise that are less tangible but more important than worker compensation costs in the long run. Preventing noise-induced hearing loss preserves the vital ability to communicate for thousands, if not millions, of Americans. It also preserves the quality of life in terms of enjoying music and the sounds of nature, as well as the capacity to hear warning signals and other sounds necessary for one's safety. In addition, controlling noise levels to within the range considered safe for hearing sensitivity reduces, and very possibly eliminates, the risk of noise-induced extra-auditory health effects such as cardiovascular, digestive, and neurological disorders. Controlling occupational noise reduces the risk of job-related accidents and injuries due to failed communication, and there is also some evidence that lower noise exposures result in fewer employee absences (Cohen, 1973).

With the exception of hearing loss, these benefits can be elusive on the surface, and

TABLE 5–2.

Percentage of Workers Exposed to Noise in the Manufacturing and Utilities Industries

Standard Industrial Classification (SIC) Code	Industry	Percentage of Workers Exposed to Daily Average Level	
		>90 dBA	>85 dBA
20	Food	16 %	28 %
21	Tobacco	6.6%	9.7%
22	Textiles	52 %	75 %
23	Apparel	0 %	1 %
24	Lumber and Wood	72 %	94 %
25	Furniture and Fixtures	12 %	30 %
26	Paper	21 %	40 %
27	Printing and Publishing	19 %	45 %
28	Chemicals	20 %	37 %
29	Petroleum and Coal	52 %	76 %
30	Rubber and Plastics	8.9%	20 %
31	Leather	0 %	1 %
32	Stone, Clay, and Glass	4.8%	16 %
33	Primary Metals	38 %	63 %
34	Fabricated Metals	19 %	34 %
35	Machinery (Except Electrical)	13 %	26 %
36	Electrical Machinery	2.5%	7 %
37	Transportation Equipment	13 %	23 %
49	Utilities	30 %	74 %
Total in Manufacturing Industries		19.3%	34.4%*

From: Occupational Safety and Health Administration. (1981). Occupational noise exposure: Hearing conservation amendment. *Federal Register, 46,* 4078–4179.

*The percentage exposed above certain levels will vary slightly according to the dates on which the employment data were gathered. These data are based on employment statistics that are somewhat more recent than those appearing in Table 5–1, and reflect a slightly smaller workforce.

most employers are not aware of them. Even hearing loss is insidious, in that it usually takes years to develop, has no visible manifestations, and results in no obvious lost workdays. So it is safe to assume that occupational noise is controlled primarily because of the great impetus provided by government regulation. Undoubtedly, concern for workers' health and the desire to avoid compensation payments may provide additional motivation, but the primary motivation is concern for compliance with government regulations.

HISTORY OF NOISE STANDARDS

Before embarking on a discussion of the development of noise standards, it would be useful to explain the similarities and distinctions among some commonly used terms: *standard, law,* and *regulation.* A *standard* is usually a codified set of rules or guidelines, and often is used interchangeably with the term *regulation.* Standards may be developed by consensus groups, such as the American National Standards Institute (ANSI) or the International Organization for Standardization (ISO), or by government agencies. A *regulation* is a rule order prescribed by an authority (the government, for example), usually a rule or set of rules that is somewhat more formal than a standard. A government regulation is published in an official document such as the *Federal Register,* and may include appendices, compliance procedures, and effective dates in addition to the basic standard. *Laws* are also prescribed by authorities, and for purposes of this discussion, they are enacted by our elected representatives. To put it succinctly, Congress legislates (makes laws), and the administration regulates. Congress enacts laws that very often give the administration certain duties, such as the making of regulations. Agencies in the administration, such as the Department of Labor's Occupational Safety and Health Administration (OSHA) promulgate and enforce regulations. Sometimes they use consensus standards in the formulation of their regulations.

Early Noise Standards

Perhaps the first standard or regulation for protection against noise was Air Force Regulation 160-3, which was issued in 1956 (U.S. Air Force, 1956). The regulation specified the use of hearing protection above 95 dB in four octave bands: 300-600 Hz, 600-1200 Hz, 1200-2400 Hz, and 2400-4800 Hz. Hearing protection was recommended for noise exposures exceeding 85 dB in those bands, and the regulation also called for audiometric monitoring.

A few years later the ISO proposed a standard suggesting a limit of 85 dB in the octave bands centered at 500, 1000, and 2000 Hz for daily exposures of 5 hours or more (ISO, 1961). This recommendation was based on a suggested permissible temporary threshold shift (TTS) of 12 dB at the 2000-hz audiometric frequency. Exposure to intermittent noise was to be assessed according to a complicated scheme involving the on-time of each burst, the off-time between bursts, the sound level, and the number of cycles per day.

In 1966, another consensus organization, the Committee on Hearing, Bioacoustics, and Biomechanics (CHABA) of the National Academy of Sciences/Nation Research Council published a report entitled "Hazardous Exposure to Intermittent and Steady-State Noise" (Kryter, Ward, Miller, & Eldredge, 1966). The report consisted of a series of graphs, containing curves representing tolerable levels and durations of octave and third-octave bands of noise for a range of approximately 85 to 135 dB. Higher levels were permitted as durations

become shorter. The rationale for CHABA's curve was again based on maximum permissible TTS: in this case, a median TTS not exceeding 10 dB at 1000 Hz, 15 dB at 2000 Hz, and 20 dB at 3000 Hz or higher frequencies. The working group that formulated the CHABA report referred to the curves as "damage-risk criteria," meaning that they represent the risk of hearing damage expected to occur as a result of exposure to various levels and durations of noise. Although the CHABA report did not constitute a standard or even a set of recommendations, it has often been referred to as the "CHABA standard."

The "Intersociety Committee," a consensus group composed of two members from each of five technical committees having an interest in the area of noise and hearing conservation, published guidelines for controlling noise exposure in 1967. The recommended curves were quite similar to those of the 1961 ISO standard, but the Committee also stated that these curves "may be approximated by the simple rule that for each halving of daily exposure time, the noise levels may be increased by 5 db [sic] up to a maximum of 115 db" (Intersociety Committee, 1967, p. 421).

Yet another consensus body, the American Conference of Governmental Industrial Hygienists (ACGIH), published "threshold limit values" for noise in 1969. The ACGIH recommended a maximum A-weighted sound level of 90 dB for an 8 hour exposure, with an allowable increase of 5 dB for each halving of exposure duration, up to a maximum of 115 dB or 15 minutes. They also recommended a peak sound pressure level of 140 dB as a limit for impulsive noise (ACGIH, 1969).

Federal Regulation

By this time, enough background work had occurred to pave the way for government regulation. Some years earlier, Congress had passed the Walsh-Healey Public Contracts Act, giving the Labor Department the authority to regulate certain benefits for workers whose employers had contracts with the federal government. Consequently, officials in the Bureau of Labor Standards decided to regulate these workers' exposure to noise. Late in 1968, the Labor Department proposed a regulation to limit exposure to steady noise to a level of 85 dBA (U.S. Department of Labor, 1968). Exposure to nonsteady noise was to be assessed over a weekly period, according to a large table indicating noise level and burst duration. This version was promulgated (made final) early in 1969, and the Labor Department quickly retracted it, issuing instead a final regulation virtually identical to the ACGIH version (U.S. Department of Labor, 1969). The regulation is commonly known as the Walsh-Healey noise standard, and it is still in effect today. It is reprinted in Appendix B.

Briefly, the standard requires employers to limit workers' exposures to daily average noise levels of 90 dBA for 8 hour exposures. For durations less than 8 hours, the exposure level can be increased by 5 dB for every halving of exposure duration, as shown in Table 5–3. This relationship between allowable level and duration is now called the "exchange rate." Some of the earlier consensus standards used a 3 dB exchange rate, which is commonly used in Europe, or a variable exchange rate depending on the noise level and duration (usually a small exchange rate, such as 2 or 3 dB for long duration exposures, and a large one such as 6 or 7 dB for short duration, infrequent exposures). The simplicity of the constant exchange rate was appealing to those who drafted the ACGIH and Walsh-Healey standards. Evidently, 5 dB was selected over 3 dB to allow for intermittencies during the typical work day.

The Walsh-Healey noise standard requires employers to use feasible engineering or administrative measures to keep employees' exposures to the levels specified in Table 5–3. Whenever these levels are exceeded (because engineering or administrative controls are infeasible or while these controls are being implemented), employers must issue, and workers must wear, hearing protective devices. Thus, the first responsibility is on the employer to control the noise

TABLE 5–3.

Allowable Levels and Durations of Noise Exposure According to the Walsh-Healey Noise Standard Table G-16

Duration/Day	Level— Slow Response
8 hr	90 dBA
6 hr	92 dBA
4 hr	95 dBA
3 hr	97 dBA
2 hr	100 dBA
1.5 hr	102 dBA
1 hr	105 dBA
0.5 hr	110 dBA
≥ 0.25 hr	115 dBA

From: U.S. Department of Labor, Bureau of Labor Standards. (1969). *Occupational noise exposure. Federal Register, 34,* 7946ff.

whenever possible, but workers also bear the responsibility of wearing hearing protection when, for whatever reason, employers are unable to control the noise.

Two other provisions of the Walsh-Healey noise standard deserves mention. The standard states that impulse noise should not exceed a peak sound pressure level of 140 dB. Because the standard uses the word "should" instead of "shall," this provision has been considered advisory, and has never been enforced. The 140 dB level has, however, been widely recognized by professionals as an appropriate maximum, at least for industrial impulses. The other provision is the section that called for "a continuing, effective hearing conservation program" whenever the levels in the table were exceeded. Although the Labor Department issued guidelines explaining this statement, enforcement was always difficult because the explanation did not carry the official weight of regulation. For example, when the Labor Department tried to require audiometric testing as part of the hearing conservation program, certain companies refused to comply, stating that audiometric testing was not actually specified in the regulation.

Shortly after the Walsh-Healey noise standard appeared, Congress enacted a monumental piece of legislation, known as the Occupational Safety and Health Act of 1970. The Act delegated important powers and responsibilities to the executive branch of the government. To the Department of Health, Education, and Welfare (now Health and Human Services), it gave the responsibility of funding and conducting research on occupational health and safety hazards, and the duty to develop criteria and recommend standards. These activities were to be carried out in a newly developed organizational body called the National Institute for Occupational Safety and Health (NIOSH). To the Labor Department, the act gave the duties of regulating occupational health and safety hazards, enforcing these regulations, assisting the states and other federal agencies in their occupational safety and health programs, and educating the public. These duties were to be carried out in a newly designated section of the Labor Department called the Occupational Safety and Health Administration (OSHA).

The Occupational Safety and Health Act also created a third governmental body known as the Occupational Safety and Health Review Commission (OSHRC, or sometimes called just the Review Commission). The Review Commission's job is to adjudicate disputes that arise between OSHA and employers in the process of enforcing health and safety regulations.

In addition to these general responsibilities, each of the three agencies created by the Occupational Safety and Health Act has a number of specific duties. One of OSHA's first duties was to make various existing governmental and consensus standards into OSHA standards, so that they would have the force of regulation and would apply to virtually all of American industry. The act directed OSHA to perform this rule-making (development and promulgation of the regulation) all at once rather than to have a separate rule-making for each standard. Consequently, along with numerous other standards, the Walsh-Healey noise standard became an OSHA standard in 1971, and instead of applying only to employers who contracted with the government, it applied to any employer engaged in interstate commerce, which involves most noisy industries.

Revising the OSHA Noise Standard

Not long after the noise standard became an official OSHA standard, the process of revision began. OSHA had two principal reasons for revising the noise standard. First, it needed more specific requirements for hearing conservation programs because its policy guidelines did not carry the force of regulation. Second, there was mounting evidence that the 90 dBA permissible exposure limit (PEL) was not sufficiently protective of the exposed population. In the meantime, NIOSH had conducted a large-scale study of the effects of noise on hearing, and had developed a detailed recommendation for a standard in the form of a "criteria document" (NIOSH, 1972). NIOSH delivered the criteria document to OSHA in 1972, which was the first step in one of the lengthiest rule-making processes in OSHA's history.

In its criteria document, NIOSH recommended that the 90 dBA permissible exposure limit be lowered to 85 dBA after the Labor Department had conducted a study to determine the technical feasibility of such a reduction. The recommended standard also included requirements for audiometry testing, warning signs, hearing protectors, employee notification, and recordkeeping. Although these requirements have been added to and modified in subsequent versions of the standard, they effectively formed the backbone of what later became the hearing conservation amendment.

The next step in the regulatory process was for OSHA to convene a Standards Advisory Committee. The Committee's composition, which is mandated by the Occupational Safety and Health Act, included representatives from the affected industries, organized labor, federal and state government, and professionals representing the concerned public. The Standards Advisory Committee on Noise discussed the

NIOSH recommendation and deliberated for many months. Finally, the committee recommended to the Secretary of Labor that the PEL remain at 90 dBA, but that audiometric testing and certain other hearing conservation measures be initiated at an average exposure level of 85 dBA. The committee accepted most of the other NIOSH recommendations and added a few of its own (Standard Advisory Committee on Noise, 1973). Although the subject of feasibility received little mention during the committee's deliberations, economic and political considerations undoubtedly played a part. The committee members knew that engineering controls were extremely costly, and many of them refused to accept the evidence that pointed toward hearing damage at exposure levels below 90 dBA.

During this time, OSHA contracted with an acoustical consulting firm, Bolt, Beranek and Newman (BBN), to study the economic and technical feasibility of controlling noise, either to the current 90 dBA level or to the 85 dBA level proposed by NIOSH. BBN found that the technology existed to control noise to the 85-dB(A) exposure level for virtually the entire workforce, but that such controls would require expenditures amounting to billions of dollars. In the first report, BBN estimated that employers would need to spend $13 billion just to achieve compliance at the current 90-dB(A) exposure level, and $31 billion for compliance with an 85-dB(A) standard (Bolt, Beranek and Newman, 1973).

Naturally, the BBN study was quite controversial. A few years later, OSHA commissioned BBN to perform another study, this time with more extensive and rigorous sampling procedures. In 1976, BBN issued another report, estimating the national cost of controlling to 90 dBA would be approximately $10 billion and to 85 dBA an estimated $18 billion. To put these estimates into some kind of economic perspective, the reader should be aware that the gross national product (GNP) in 1973 was $1294.9 billion, which means that industrial noise control would constitute about 1 to 2 percent of the GNP (Bruce, 1979). Also, the reader

should be aware that the average cost per exposed worker was less that $1,000 to comply with 90 dBA, and about $1,600 to comply with the 85 dBA level (see Bruce, 1979, Table 4, p. 605).

In accordance with the rule-making procedures spelled out in the Occupational Safety and Health Act, OSHA published a proposed regulation in the *Federal Register* on October 24, 1974 (OSHA, 1974). In developing the proposed regulation, OSHA personnel considered the recommendation of NIOSH and the Standards Advisory Committee, along with comments, data, and information submitted by the general public to the "record" or "docket" of the noise standard. In its proposal, OSHA retained the 90 dBA PEL to be achieved by engineering or administrative controls, and required hearing conservation programs to be initiated at 85 dBA. Most of the provisions were similar to those recommended by NIOSH and the Advisory Committee, with the addition of a new requirement for exposure to impulse noise. The peak sound pressure level remained at 140 dB, but the maximum allowable level decreased by 10 dB with each increase in the number of impulses above 100. In other words, a maximum of 100 impulses would be allowed at 140 dB, 1,000 at 130 dB, and 10,000 at 120 dB.

Also in accordance with procedures mandated by the Occupational Safety and Health Act, OSHA scheduled hearings, giving the public a chance to comment on the proposed regulation and on BBN's economic impact studies. The first hearings lasted for 5 weeks, while industry and trade association representatives, labor union officials, and independent professionals debated the details of the standard and the economic and technical feasibility reports. The key issues were: (1) the PEL, whether it should be 85 or 90 dBA; (2) the exchange rate, whether it should be 5 dB or the more conservative 3 dB used by most European countries; and (3) the method of control, whether primacy should still be given to engineering and administrative controls or whether hearing protective devices should be allowed as an equally acceptable means of limiting noise exposure. Not surprisingly, most industry representatives favored 90 dBA with the 5 dB rule, to be achieved by hearing protection on an equal footing with engineering controls. Most labor union representatives favored 85 dBA, the 3 dB rule, and the primacy of engineering controls, as did a panel of well-known professionals in audiology and acoustics assembled by the EPA.

Actually, two sets of public hearings took place. During the first set in 1975, the proposal and the first economic impact study by BBN were discussed. The second round of hearings, which occurred in 1976, was scheduled in response to the revised economic impact analysis. By the end of the proceeding, the record of the noise standard contained some 30,000 pages of testimony, comment, and other types of information, such as research reports, proceedings of conferences, and even books. OSHA staff began analyzing all of this material, studying the impact of various regulatory scenarios, and devising a strategy. Finally, it became clear that requiring engineering controls to an 85 dB level was politically infeasible, due to the economic climate of the time. Dr. Eula Bingham, who was then OSHA Director, agreed to leave the Walsh-Healey noise standard essentially unchanged, and to amend the standard with detailed requirements for hearing conservation programs.

The Hearing Conservation Amendment

The Hearing Conservation Amendment (HCA) was nearly a year in preparation. The standard itself was fairly similar to the original NIOSH recommendation and subsequent versions, in the form of the Advisory Committee recommendation and the 1974 proposed regulation. The standard did contain more detailed requirements for noise exposure monitoring, audiometric testing, and worker training and education, and a lengthy preamble containing a thorough justification of OSHA's action, a description of the costs and benefits inherent in compliance with the HCA, and

an explanation of the HCA's various provisions (OSHA, 1981). On January 16, 1981, the HCA was published in the *Federal Register* as a final standard, nearly a decade after the process of revision had been initiated.

Because it is sometimes difficult for employers to comply immediately with a new regulation, OSHA usually gives them a short grace period before enforcing the regulation. In this case OSHA set the effective date, the date on which the standard was to become enforceable, for April 15th, 1981. In the meantime, however, President Reagan had assumed office, and the entire upper echelon in the executive branch had changed. Even before the inauguration, David Stockman, who later became director of the Office of Management and Budget, had "targeted" the hearing conservation amendment because of its cost to American business. The result was that the new administration "stayed" or delayed the HCA's effective date in order to scrutinize the standard's requirements and its economic impact.

A lengthy exposé of the events of the next 2 years is unnecessary for purposes of this chapter. (The interested reader is referred to Suter, [1984].) It is sufficient to say that after one of the most powerful labor unions filed suit against OSHA for illegally delaying the amendment's effective date, certain portions of the amendment were allowed to become effective in August, 1981. Other provisions continued to be stayed, and public hearings were held on these provisions in 1982. Finally, on March 8, 1983, the entire revised HCA was published in the *Federal Register* and became effective shortly thereafter (OSHA, 1983). OSHA labeled the 1983 version the "final" standard, but this was a misnomer because the 1981 version was already a final standard. A more appropriate label for the 1983 version would be the "revised" standard.

Although there were numerous changes in the 1983 version of the HCA, most of the changes were not sufficiently serious to degrade the quality of the resulting hearing conservation programs. A few changes, however, may have an adverse impact.

Nearly all of the specifications for noise exposure monitoring were deleted, allowing employers to use either area or personal monitoring, and to use whatever techniques they choose. If employers or their consultants are not knowledgeable in proper noise monitoring techniques, or if they are not conscientious, some employees may not be identified as being noise exposed and may not receive the benefits of the program. Other major changes involve relaxing the requirement for baseline (initial) audiograms, relaxing the requirements for maximum noise levels in audiometric testing rooms, and simplifying the definition of significant now-"standard" threshold shift in such a way as to reduce its protectiveness. All of these changes may result in less protective hearing conservation practices, but a conscientious professional may (and should) improve on them. It is important to remember that OSHA standards usually represent the *minimum* standard of health practice, and employers and their consultants should be encouraged to improve upon them whenever it is possible and appropriate to do so.

The 1983 version of the HCA is the current standard with which employers must comply. The standard itself is reprinted as Appendix A in this book, and the entire regulation, which includes its preamble, may be found in the *Federal Register* (48 Fed. Reg. 9738–9785), or obtained from the OSHA national, regional, or area offices. Every professional working in the hearing conservation area should have a copy of the standard. The 1981 HCA is not the latest version, and therefore, it is not stocked in OSHA publications offices. Its preamble, however, is extremely useful, in that it gives a thorough explanation of every one of the standard's provisions, most of which are still current. It also presents a very lucid discussion of the effects of noise on hearing, and the expected costs and benefits of the amendment. The reader is urged to locate this portion of the *Federal Register* (46 Fed. Reg. 4078–4179) in libraries or at the headquarters of the *Federal Register* in Washington D.C.

The Amendment's Requirements

To summarize the major provisions of the current HCA, hearing conservation programs must be available to all employees whose 8 hour time-weighted average exposure levels (TWAs) equal or exceed 85 dBA. Employers must monitor at least once the noise exposures of workers whose TWAs are 85 dBA or greater. Remonitoring is necessary with a change of equipment or work process that causes a significant increase in exposure level. All continuous, intermittent, and impulsive noise between the levels of 80 and 130 dBA must be included in the exposure assessment. Area monitoring is permitted, but employers must use personal exposure monitoring when there is considerable variation of noise level over time. Workers must be be allowed to observe the monitoring procedures and must be told about their exposures.

Employers must provide baseline audiograms within the first year of an employee's exposure to 85 dBA and above, and annual audiograms thereafter. The tests must be conducted by trained and competent personnel and supervised by an audiologist or physician. Tests must be carried out in rooms that meet or exceed the 1969 ANSI criteria for background levels, and equipment must be calibrated according to specific schedules. Workers who experience significant (or in the final version "standard") threshold shifts are notified in writing, counseled as to the fitting and use of hearing protection, and referred to a specialist if necessary. A "standard" threshold shift is defined as an average shift from baseline levels of 10 dB or more at 2000, 3000, and 4000 Hz.

Hearing protection must be worn by all workers exposed to a TWA of 90 dBA and above. Employers must offer hearing protectors to workers exposed above 85 dBA, and all must be given a variety of protectors from which to choose. Employers must provide protectors that are suitable for the specific noise environments in which they are to be worn. OSHA allows employers to use any of three methods for assessing the adequacy of hearing protector attenuation. The standard recommends using the Noise Reduction Rating (NRR), which now appears on the protector package (see Chapter 10). To estimate the noise level under the protector, the employer subtracts the NRR from the worker's C-weighted exposure level. If C-weighted levels are not available, 7 dB must be subtracted from the NRR to obtain the A-weighted sound level at the ear.

Training and education sessions must be given at least annually to workers exposed above 85 dBA. These sessions must include information on the effects of noise on hearing, the purpose and procedures of audiometric tests, and the proper selection, fitting, use, and care of hearing protectors.

Last, employers need to keep records of noise measurements, audiograms, audiometer calibrations, and background levels in audiometric test rooms. These records must be given to employees or their representatives on request.

Any reader who intends to practice industrial hearing conservation should turn to Appendix A and become thoroughly familiar with the HCA's provisions, including all of its appendices.

The Court Challenge

Once the HCA had been revised and sanctioned by the Reagan administration, there was considerable effort on the part of noisy industry to comply. Employers set about training their nurses and other personnel, buying equipment, and hiring consultants. The market for hearing conservation consultants boomed. Everyone thought that the regulatory process had finally ended, but there was a final, somewhat surprising, step.

There is a tradition over the history of OSHA's regulatory activities for legal action the moment an OSHA standard is promulgated. The unions sue on the grounds that the standard is too lenient, and the affected industry sues because they believe the standard is too stringent. Each group tries to select the court that is likely to favor its position (theoretically, such groups must

sue in the court closest to their headquarters), and the court that actually hears the case is the one where the plaintiff filed the earliest. This practice results in the closest possible scrutiny of OSHA's regulatory timetable, and once the *Federal Register* "posts" the OSHA standard, there is a headlong rush to the courts.

The 1983 version of the HCA, however, prompted no such rush to litigate. Because the issues had been so exhaustively debated and because most of the controversial ones had been deleted or watered down, few people expected a challenge in court. Despite this, the Forging Industry Association (FIA) filed suit in the Fourth Circuit Court of Appeals in Richmond, Virginia. On April 5, 1984, a three-judge panel heard arguments from attorneys for the plaintiff (FIA) and the defendant (OSHA), and in November of that year, issued a decision striking down the amendment for the entire nation (U.S. Court of Appeals, 1984). The court's rationale was that nonoccupational noise can cause hearing damage, that the results of occupational and nonoccupational exposures are virtually identical on the audiogram, and that employers should not have to take action for hearing losses that they may not have caused. One of the three judges voted to uphold the HCA on the grounds that the amendment was perfectly reasonable, and that occupational exposures clearly dominated the occasional nonoccupational ones.

At this point, much of the newly initiated hearing conservation activity ceased while employers waited to see what OSHA would do. After numerous delays and much equivocation, Robert Rowland, OSHA director at that time, finally appealed the decision to the Fourth Circuit Court, requesting the entire court to hear OSHA's arguments. The full court did agree to rehear the case, and in a move without precedent in OSHA's history, voted *unanimously* to reverse the earlier decision of the three-judge panel (U.S. Court of Appeals, 1985). The Court found that the argument for nullifying the standard on account of nonoccupational exposure was without merit. In addition, it denied every one of FIA's petitions, and in doing so greatly strengthened OSHA's position on many of the standard's specific requirements, as well as the regulation as a whole. Although the FIA could have appealed its case to the U.S. Supreme Court, it chose not to, and the HCA has been safely in effect ever since. This does not mean that the standard cannot be revised at some future date, but the decision put an end to legal efforts aimed at eliminating any or all of the HCA.

THE ANALYSIS OF DAMAGE RISK CRITERIA

For any government agency, or, for that matter, for any consensus organization to set a standard for protection against harmful effects of noise, those who develop the standard must analyze the available damage risk criteria. That means that the regulators or consensus committee must examine the available research to determine the levels, durations, and types of noise that produce various amounts of hearing loss in various segments of the exposed population. Sometimes experts will convene and simply agree, or vote to accept a certain level and duration as safe and another combination as harmful. But until the group examines all of the possible evidence in an open-minded way, their consensus is meaningless.

In developing criteria for noise-induced hearing loss, as in the development of any kind of occupational health criteria, three issues should be examined initially:

1. The amount of hearing to be preserved
2. The determination of an acceptable degree of risk
3. The best way to present the information to potential users

The Amount of Hearing to be Preserved

Different groups have had vastly different ideas about what constitutes a hearing impairment, or conversely, how much hearing should be preserved. Most would agree that hearing should not be lost if it would jeopardize the understanding of speech. For many years the American Academy of Ophthalmology and Otolaryngology (AAOO) advocated preserving the ability to understand speech by preventing average hearing levels at 500, 1000, and 2000 Hz from exceeding 25 dB (e.g., American Academy of Ophthalmology and Otolaryngology, 1973). The Department of Labor used this criterion in drafting the Walsh-Healey noise standard (U.S. Department of Labor, 1970). The AAOO had determined that individuals needed this minimum amount of hearing sensitivity to understand everyday speech, which they defined as simple sentences perceived in a quiet background (American Academy of Ophthalmology and Otolaryngology, 1959).

The NIOSH Criteria Document stated that hearing no better than 25 dB at 500, 1000, and 2000 Hz was inadequate for understanding speech under less than optimal conditions (e.g., some distortions or a noisy background), and recommended a slightly more conservative criterion: 25 dB at 1000, 2000, and 3000 Hz (National Institute for Occupational Safety and Health, 1972). The EPA submitted testimony to OSHA during the public hearings, recommending a 25 dB average hearing level at the frequencies 1000, 2000, and 4000 Hz (Environmental Protection Agency, 1975). Later, the EPA submitted another report to OSHA recommending the same frequencies at a slightly lower hearing level—approximately 22 dB (Suter, 1978). The basis for these recommendations was the importance of high frequency hearing sensitivity for the recognition of speech in noisy backgrounds. According to Suter (1978), individuals with average hearing threshold levels greater than approximately 22 dB at 1000, 2000, and 4000 Hz were less able to recognize speech than their normally hearing counterparts. Because of these kinds of research results, and because the American Academy of Otolaryngology (AAO) acknowledged that everyday speech communication often takes place in noisy surroundings, the AAO changed its formula to include 300 Hz, while stil maintaining the 25 dB average hearing level, or "low fence" (American Academy of Otolaryngology, 1979).

The Occupational Safety and Health Act directs OSHA to promulgate standards that will ensure "that no employee will suffer a material impairment of health or functional capacity even if such employee has regular exposure to the hazard dealt with by such a standard for the period of his working life" (1970, Sec. 6(b)(5)). The definition of "material impairment" is likely to differ according to various hazards and their effects on workers. With respect to noise-induced hearing loss, OSHA studied all of the recommendations submitted to the record of the noise standard, and decided that a material impairment of hearing is loss exceeding 25 dB at the frequencies 1000, 2000, and 3000 Hz. This then became the protection goal.

One other criterion deserves mention. In the Noise Control Act of 1972, Congress told the EPA to identify levels of environmental noise that would be safe for the entire exposed population, and then told the EPA to add a margin of safety to its calculations (Noise Control Act, 1972, Sec. 5(a)(2)). The EPA needed to decide what kind of loss to protect people against in the identification of its safe level. Because the congressional mandate was even more conservative in this instance (including an added margin of safety), the EPA decided upon a criterion of no more than 5 dB hearing loss at 4000 Hz (Environmental Protection Agency, 1974b). The EPA selected the 5 dB value because it is the smallest generally measured audiometric increment, and the 4000 Hz frequency because it is usually the most sensitive to noise exposure.

Determination of Acceptable Risk

The next step in the development of damage risk criteria is to determine the level of an acceptable risk. In other words, to address the issue as to the proportion of the exposed population that could be allowed to become materially impaired. In the best

of all possible worlds, no one should become materially impaired by an occupational hazard. However, in the real world, economics and politics play a subtle part in this decision, as do the relative effects of the hazard in question. Undoubtedly there will be greater efforts toward minimizing the risk of a carcinogen than toward minimizing the risk of noise-induced hearing loss.

The EPA selected a criterion of 4 percent, meaning that 96 percent of the population exposed to its identified "safe" level would not exceed a 5 dB hearing loss at 4000 Hz. Other criteria have been suggested, such as 5 percent (Committee on Hearing, Bioacoustics, and Biomechanics, 1968), and even 20 percent (International Organization for Standardization, 1971). In 1970, the Department of Labor admitted that the 90 dBA PEL would "not produce disabling loss of hearing in more than 20% of the exposed population" (U.S. Department of Labor, 1970).

In the preamble to the 1981 version of the HCA, OSHA published a table of risk estimates developed by the ISO, EPA, and NIOSH, showing the percentage of the population at risk as a function of daily average noise exposure experienced over a 40 year work life. These estimates had been compiled and published by the EPA (1974a), and they are reprinted as Table 5-4. The ISO's risk estimates are based on a study by Baughn (1973), the NIOSH estimates originate from the Institute's own data (National Institute for Occupational Safety and Health, 1972 & 1974), and the EPA's estimates reflect a compilation of the data of Baughn (1973), Passchier-Vermeer (1968), and Burns and Robinson (1970). The reader can see that the risk of incurring material impairment of hearing is 21 to 29 percent at the 90 dBA PEL, 10 to 15 percent from exposure to 85 dBA, and 0 to 5 percent at 80 dBA.

OSHA has never tackled this issue openly by specifying an acceptable level of risk for noise-induced hearing loss. Because the PEL remains at 90 dBA for actual noise control, one might say that OSHA still

TABLE 5–4.

Estimated Percentages of the Population at Risk of Exceeding an Average Hearing Threshold Level of 25 dB at 500, 1000, and 2000 Hz* as a Function of Average Noise Exposure Level for 40 Years

Organization	Noise Exposure in dBA	Risk %
ISO	90	21
	85	10
	80	0
EPA	90	22
	85	12
	80	5
NIOSH	90	29
	85	15
	80	3

From: Occupational Safety and Health Administration. (1981). Occupational noise exposure: Hearing conservation amendment. *Federal Register*, *46*, 4078–4179; and Environmental Protection Agency. (1974). Occupational noise exposure regulation: Request for review and report. *Federal Register*, *39*, 43802–43809.

*Because these data are based on a 25 dB level of impairment at 500, 1000, and 2000 Hz, the risk figures could be expected to be slightly higher using OSHA's current definition of material impairment of hearing, which omits 500 Hz and adds 3000 Hz.

condones the 20 to 29 percent level of risk, which is extremely high for any health hazard. However, since the HCA requires audiometric monitoring and numerous other actions at 85 dBA, OSHA seems to imply that the 10 to 15 percent risk figure is more acceptable.

Presenting the Information to Users

The third step in developing damage risk criteria is to decide how best to present the hearing loss data to potential users. Consensus groups and federal agencies decided some years ago to discontinue the practice of using octave and third-octave bands for hearing risk criteria, in favor of single number schemes. Occupational noise exposure is almost always given for A-weighted levels, averaged throughout an 8 hour day. The question, then, is whether to present the data in terms of noise-induced

permanent threshold shift (NIPTS) or in terms of the percentage of exposed individuals "at risk" of incurring material impairment of hearing.

Before examining these two alternative methods of presenting data, it would be useful to discuss an important aspect of collecting hearing loss data, so that the reader may see how data collection methods can influence the resulting figures.

It is extremely important to match the experimental (noise-exposed) and control (non-noise-exposed) populations for such characteristics as age, nonoccupational noise exposure, and otologic abnormalities. If one group has more of these charcteristics than the other, the effects of noise exposure cannot be properly assessed. There are two basic approaches: one can use either "screened" or "unscreened" data. Screening is the process of removing from the study of any subjects who have had military noise exposure, who have noisy hobbies, who have pre-existing hearing loss from whatever cause, whose otologic exams show some sort of abnormality, or whose exposures are uncertain for any reason. The advantage of such screening is that one can be reasonably sure that any remaining hearing loss will be due to noise exposure, assuming that the data are properly adjusted or otherwise controlled for aging (presbycusis). The disadvantage is that by the time an investigator concludes the screening process, there may be hardly any subjects left.

The other alternative is not to screen either the experimental or the control populations. The advantages of this approach are that subjects are much more plentiful and the population being studied is more reflective of a true-to-life population. The disadvantages are that the resulting populations must be quite large so as to minimize the effects of bias in either group, and the investigator must be reasonably sure that the control population is closely related to the experimental population in every respect except occupational noise exposure. The two groups must still be well matched for age. Sometimes an investigator will report that the populations have been partially or moderately screened and the reader may only hope that the screening of both groups has been similar in degree and in the type of exposure that disqualifies the subject.

NIPTS

Noise-induced permanent threshold shift, or NIPTS, represents the amount of hearing threshold shift remaining after subtracting an amount for "normal" hearing from the hearing levels of a noise-exposed person or population. The components of normal hearing are presbycusis, hearing loss from aging; sociocusis, hearing loss from nonoccupational noise sources; and otologic abnormalities, covering, in effect, everything else. The amount of NIPTS will, of course, depend greatly upon the level and duration of noise exposure, but it will also depend on the amount that is subtracted for normal hearing. This value reflects the hearing levels of the control population. If the two populations are closely matched, this subtraction will be appropriate. If they are not closely matched or if the screening has been uneven, then the subtaction for normal hearing can give an erroneous picture of NIPTS.

Table 5–5 shows NIPTS at various audiometric frequencies for the 90th, median, and 10th percentiles of noise-exposed populations. NIPTS is given as a function of exposure level, from 75 to 100 dBA, and durations from 10 to 40 years. The figures reflect a compilation of the data of W. Burns and D. W. Robinson and W. Passchier-Vermeer. They were published as an EPA report (Johnson, 1978) and later in the preamble to the 1981 HCA (OSHA, 1981). The reader can see that NIPTS tables offer an enormous quantity of data, but it can be difficult to imagine the overall effects of noise with so many factors to consider. Another minor disadvantage with presenting the user with NIPTS data is that the hearing threshold shifts appear abnormally small. For example, the median NIPTS at 2000 Hz from exposure to 90 dBA for 30

years is only 5.4 dB. If, however, the values for "normal" hearing are added, this figure would be somewhat higher.

NIPTS data can also be given in graphic form, as in Figures 5–1 and 5–2. These figures show median NIPTS values for durations of 10 years (Figure 5–1) and 40 years (Figure 5–2). These data were developed by Passchier-Vermeer (1968 & 1971). It is interesting to note that after 10 years, 4000 Hz is the most vulnerable frequency at any exposure level, but that 3000 Hz follows it closely, especially at higher exposure levels. After 40 years, the threshold at 3000 Hz is actually worse than 4000 Hz at levels above 90 dB, and 2000 Hz is also severely affected at the higher levels.

Risk

The percentage of the exposed population at risk of developing material impairment of hearing is that portion of the population exceeding a certain amount of hearing loss (or "fence") at certain frequencies for a specific exposure duration, *after* subtracting the percentage of the population that would normally incur that amount of loss from nonoccupational causes. The most commonly used fence is 25 dB, the audiometric frequencies are usually the combinations 500, 1000, 2000 Hz or 1000, 2000, 3000 Hz, and the exposure duration is usually 30 to 40 years. Sometimes age is given instead of exposure duration when the population in question has worked in the same noisy conditions since early adulthood.

Figure 5–3, from Baughn (1973), shows the percent of the exposed population expected to develop material impairment of hearing (in this case, losses greater than 25 dB average at 500, 1000, and 2000 Hz) as a function of 8 hour exposure levels from 80 to 115 dBA. Also shown in the lower right hand portion of the graph is a curve representing the hearing levels of non-noise-exposed office workers, based on the data of Glorig and Nixon (1962). Because Baughn's curve for workers exposed to approximately 80 dBA so closely parallels the curve for non-noise-exposed office

TABLE 5-5.

Noise-Induced Permanent Threshold Shift (NIPTS) in Decibels as a Function of Audiometric Frequency, Resulting from 8 Hour Daily Average Noise Exposure Levels of 75 to 100 dBA. Data are Presented for the 90th, 50th, and 10th Population Centiles, and for Durations of 10 to 40 Years

Sound Level [dB]	Freq. [Hz]	10 yrs.			20 yrs.			30 yrs.			40 yrs.		
		.9	.5	.1	.9	.5	.1	.9	.5	.1	.9	.5	.1
75	500	0	0	0	0	0	0	0	0	0	0	0	0
80	500	0	.1	.6	.1	.1	.3	.4	.1	.4	.8	.2	.4
85	500	.1	.2	1.4	.1	.3	1.2	.5	.7	1.5	.9	.4	1.1
90	500	.2	.5	2.3	.2	.7	2.4	.6	.8	2.5	1.1	1.0	2.4
95	500	.3	.9	3.5	.5	1.3	4.0	.6	1.7	4.5	1.0	1.9	4.6
100	500	1.7	3.9	7.6	2.8	5.1	9.2	3.9	6.2	10.6	4.9	7.2	11.4
75	1000	0	0	0	0	0	0	0	0	0	0	0	0
80	1000	.1	.1	.7	.1	.2	.4	.4	.2	.6	.8	.2	.6
85	1000	.1	.3	2.0	.2	.5	1.9	.5	.6	1.9	.9	.7	1.7
90	1000	.3	.7	2.9	.4	1.0	3.3	.8	1.3	3.6	1.3	1.5	3.6
95	1000	.5	2.7	5.9	.8	3.7	7.2	3.3	4.6	8.4	4.3	5.3	8.9
100	1000	3.1	6.1	11.0	4.8	8.1	13.5	6.4	9.9	15.6	8.3	11.3	16.9
75	2000	0	0	0	0	0	0	0	0	0	0	0	0
80	2000	.1	.3	1.6	.1	.4	.9	1.1	.5	1.2	1.8	.6	1.3
85	2000	.3	.9	4.9	.7	1.3	4.8	1.8	1.8	4.6	2.8	2.1	3.9
90	2000	.6	2.4	8.0	1.6	3.9	9.3	3.6	5.4	11.8	5.3	6.6	10.6
95	2000	1.2	5.5	14.2	3.7	8.7	17.4	7.4	12.0	20.1	10.6	14.8	21.9
100	2000	2.3	9.2	21.5	6.5	14.6	26.6	12.0	19.9	35.9	16.7	24.1	33.9
75	3000	0	0	0	0	0	0	0	0	0	0	0	0
80	3000	.2	2.0	4.1	1.9	2.4	3.4	3.5	2.7	2.7	5.1	3.0	2.3
85	3000	1.6	4.4	7.7	3.6	5.3	7.8	5.5	6.2	7.6	7.4	6.7	7.2
90	3000	3.9	9.2	16.9	6.6	11.0	17.9	8.5	12.6	18.4	11.3	13.7	18.4
95	3000	8.1	16.0	26.8	11.6	18.9	29.1	14.9	21.4	30.1	17.8	23.1	30.5
100	3000	15.8	25.4	37.5	20.6	29.5	39.7	24.8	32.7	41.0	28.6	35.0	41.6
75	4000	0	0	0	0	0	0	0	0	0	0	0	0
80	4000	.3	3.1	1.9	.3	3.4	2.1	1.8	3.7	6.3	3.3	3.8	5.3
85	4000	1.6	6.7	12.3	3.4	7.4	12.2	5.0	8.0	11.9	6.7	8.3	11.0
90	4000	6.3	11.9	19.1	8.4	13.3	19.5	10.3	14.4	19.4	12.1	14.9	18.6
95	4000	13.7	20.4	28.2	16.4	22.5	28.7	18.7	23.9	28.5	20.7	24.6	27.6
100	4000	22.3	30.2	37.8	25.6	32.6	37.8	28.4	34.1	37.2	30.6	34.8	36.1
75	6000	0	0	0	0	0	0	0	0	0	0	0	0
80	6000	.3	2.0	4.0	1.8	2.2	3.1	3.2	2.4	2.2	4.6	2.5	2.1
85	6000	1.1	4.8	8.9	2.8	5.3	8.6	4.3	5.8	8.2	5.8	6.1	7.3
90	6000	1.9	8.5	15.6	3.8	9.5	16.0	5.6	10.4	16.0	7.3	10.9	15.3
95	6000	4.3	13.7	23.5	6.7	15.5	24.4	8.9	16.8	24.5	10.8	17.6	23.8
100	6000	9.2	20.3	32.2	13.3	23.7	33.9	17.0	26.5	34.9	20.2	28.4	35.1

From Johnson, D. L. (1978). *Derivation of presbycusis and noise-induced permanent threshold shift (NIPTS) to be used for the basis of a standard on the effects of noise on hearing* (U.S. Air Force Report AMRL-TR-78-128). Wright-Patterson AFB, OH; and Occupational Safety and Health Administration. (1981). Occupational noise exposure: Hearing conservation amendment. *Federal Register*, 46, 4078–4179; using the combined data of W. Burns and D. W. Robinson and W. Passchier-Vermeer.

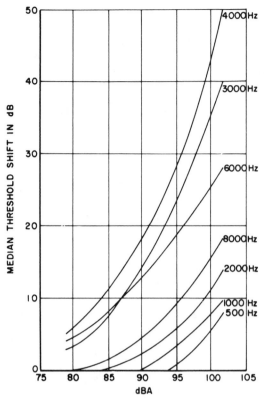

Figure 5-1. Median NIPTS for various audiometric frequencies as a function of A-weighted daily noise exposure level for 10 years. (From Passchier-Vermeer, 1971 & OSHA, 1981.)

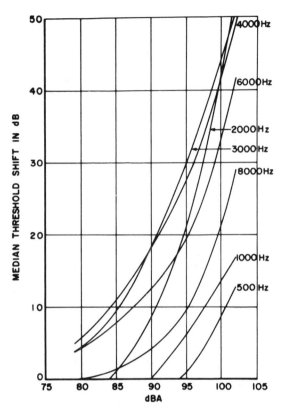

Figure 5-2. Median NIPTS for various audiometric frequencies as a function of A-weighted daily noise exposure level for 40 years. (From Passchier-Vermeer, 1971 & OSHA, 1981.)

workers, Baughn used the 80 dBA curve to represent "normal" hearing. To find the percentage of the population "at risk," one first locates the age and exposure level of interest, and identifies on the ordinate the percent of the exposed population expected to exceed the 25 dB fence. Then, one subtracts the percentage of a non-noise-exposed population that would exceed this fence from aging or other nonoccupational causes. The remainder is the population at risk. For example, Baughn's data indicate that 46 percent of the people exposed to daily average levels of 90 dBA would be expected to exceed the 25 dB hearing threshold level by age 55. An estimated 24 percent of these individuals would normally exceed the fence by this age, leaving the estimated population at risk as 22 percent.

Although this method of portraying data is attractive because it uses a single number and is relatively easy to understand, there are also some potential problems. Two quite reputable studies can yield different risk estimates for the same exposure levels and durations. Risk figures are easily influenced by the amount of screening and by the choice of frequencies and fence used to determine impairment (or material impairment). For example, Baughn's (1973) risk figures are somewhat higher than those of Burns and Robinson (1970), at least in part because Baughn's population was totally unscreened, whereas the population of Burns and Robinson was rigorously screened. In most cases, risk figures using higher frequencies, such as the 1000, 2000, 3000 Hz combination, will be greater than those using 500, 1000, and 2000 Hz. Despite

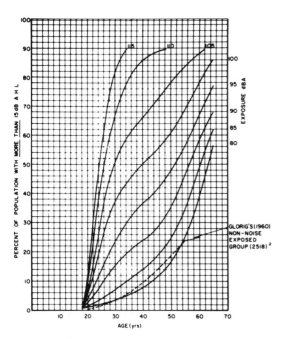

Figure 5-3. Percent of the population with average hearing levels exceeding 25 dB for the frequencies 500, 1000, and 2000 Hz as a function of A-weighted noise exposure level. (From Baughn, 1973 & OSHA, 1981.)

these disadvantages, percentage risk has remained a popular and informative way of estimating the effects of noise on large groups of people.

MAJOR HEARING LOSS STUDIES

Studies of the effect of noise on hearing can take many different forms. Laboratory studies can use either human or animal subjects. With human subjects, only temporary threshold shifts (TTS) can be induced, and these experiments must be performed very carefully so as not to cause any permanent damage. Animal models are used to study permanent threshold shift and anatomical changes due to noise exposure.

Damage risk criteria are generally developed from field studies today, although the results of TTS and animal experiments can add useful information. In the development of their criteria and standards, the EPA,

NIOSHA, and OSHA used the results of field studies—large studies of noise-exposed individuals who had worked in the same jobs for many years. Actually, there are two types of field studies: longitudinal and crossectional. Longitudinal study involve following the noise exposure levels and audiometric thresholds of a group of people over a number of years. They are rarely conducted, except in very moderate noise conditions, or unless subjects are wearing hearing protection. Longitudinal studies do have a very important use in the evaluation of hearing conservation programs. In the development of damage risk criteria, however, investigators have used crossectional studies, which constitute a "snap-shot" of a large industrial population. Trained personnel conduct careful noise measurements, perform audiometric tests, and interview the subjects to determine the length of exposure (usually in years) to noisy processes. Determining exposure duration involves the assumption that noise levels have not changed throughout a subject's exposure history. Although this assumption is sometimes risky, it is an integral component of this kind of study. The alternative would be to conduct a longitudinal study on workers in high noise levels without making hearing protection available, which is, of course, unethical.

In its analysis of the literature and subsequent development of the HCA, OSHA gave primary significance to four major hearing loss studies: Baughn (1973), Passchier-Vermeer (1968, 1971), Burns and Robinson (1970), and the National Institute for Occupational Safety and Health (1972, 1974).

Baughn

Dr. W. L. Baughn was medical director of the Guide Lamp Division of the General Motors Corporation. Although he and his colleagues collected the data between 1960 and 1965, the data were not published until 1973 (Baughn, 1973). The population consisted of 6,835 automobile workers exposed to noise levels of approximately 78, 86, and 92 dBA for varying numbers of years. Noise levels had been measured over a period of 14 years, and histories could be estimated for much longer periods because noise levels had remained relatively stable. Neither the experimental nor the control populations were otologically screened, although subjects were eliminated if their exposure histories were uncertain. Data are displayed in terms of the percentage of the population exceeding various fences (from 15 to 50 dB) at 500, 1000, 2000 Hz as a function of exposure level and duration. Figure 5–3 shows these curves for the 25 dB fence. Baughn and his colleagues used statistical procedures to ''smooth'' the data into these curves, which also allows for extrapolation to higher levels and longer durations.

Every major hearing loss study has certain drawbacks, and because studies such as Baughn's have formed the basis of standards, they have been the subject of some controversy. Baughn's study has been criticized because some of the audiometric testing was performed shortly after subjects were removed from noisy working conditions. This practice could have inflated thresholds to some extent, due to TTS. Also, the losses at 500 Hz were slightly greater than would have been expected, which may have been due to TTS, or more likely to background noise in the testing room. Another potential drawback is that Baughn used the 80 dBA exposure group as his controls, which means that his non-noise-exposed group may have had some noise exposure. If indeed the hearing levels of this group were inflated, they would cause the actual risk due to noise exposure to appear smaller. This being the case, these two potential problems would tend to offset each other.

Baughn's study has also been criticized because the populations were not otologically screened. Baughn believed that such screening would have drastically reduced his population, and that the resulting changes of hearing level would not have been sufficient to justify it. Besides, as we have indicated earlier, there are certain

advantages in data that are reflective of a real-life, industrial population. Despite these potential disadvantages, Baughn's study has been relied upon during recent decades as the largest and one of the strongest industrial studies of noise-induced hearing loss.

Passchier-Vermeer

Another study extensively relied upon by the EPA and OSHA is the work of the Dutch scientist, Dr. Wilhelmina Passchier-Vermeer (Passchier-Vermeer, 1968). The study was actually a compilation of 10 studies of noise exposure and hearing loss, which had been performed in England, Holland, Sweden, and the U.S. The total population exceeded 4,000 subjects, who were exposed to average noise levels of approximately 80 to 102 dBA. Passchier-Vermeer presented data for median and quartile NIPTS, that is, the 25th, 50th, and 75th centiles of susceptibility. Figures 5–1 and 5–2 show some of Passchier-Vermeer's median NIPTS curves. Passchier-Vermeer's data were also used by Johnson (1978) to compile Table 5–5.

A potential disadvantage of Passchier-Vermeer's study is the variation inherent in 10 studies. There must have been differences in screening techniques, and Passchier-Vermeer does not provide much information about subjects' exposure histories. The use of data from different investigators may, however, be seen as an advantage, in that the process of averaging large quantities of data should minimize any idiosyncratic effects.

Burns and Robinson

In 1970, Drs. William Burns and Douglas Robinson published the report of their crosssectional study of noise-exposed English factory workers (Burns & Robinson, 1970). The study's population was somewhat smaller than those of the two previously mentioned studies. In the words of the authors, the population was rigorously screened, eliminating all subjects whose exposure histories were not readily quantifiable, and those who had been exposed to gunfire, or showed signs of ear disease or abnormality. Out of 4,000 audiograms, the investigators ended up with 759 noise-exposed subjects and 97 controls. Daily average exposure levels (in octave bands) varied from 75 to 120 dB, and durations varied from 1 month to 50 years. After thoroughly analyzing their data, Burns and Robinson noted a consistent relationship between exposure level and duration in the effects on hearing. They found that certain relatively high levels and short durations produced the same amount of hearing loss as lower levels experienced for longer durations, and that a mathematical formula, known as the "hyperbolic tangent" described these relationships quite well. On the basis of this formula, the authors developed the concept of noise "immission" level, a single number that combines exposure level and duration. They also developed graphs and tables that can be used to predict NIPTS or hearing level in nearly any centile of the exposed population. Johnson (1978) used the data and prediction method of Burns and Robinson combined with those of Passchier-Vermeer to construct his elaborate NIPTS tables for the EPA (an example of which is shown as Table 5–5).

The study by Burns and Robinson has numerous advantages. The experimental procedures were carefully controlled and explained in great detail, and the data were thoroughly analyzed. The resulting prediction methods can be used with any population, and a hearing level can be estimated for almost any degree of susceptibility. The study has been the subject of some controversy, with detractors claiming that the prediction method may oversimplify the real-life picture, and that we are stretching its validity to use it for very high and very low population centiles. Also, the number of subjects was somewhat limited in the highest and lowest exposure categories, and there is the danger that too much screening caused the study to be less representative of a real-life population. Some of the possible disadvantages may have offset each

other, and the advantages have outweighed them to the extent that the Burns and Robinson data and method has been used extensively by the EPA and OSHA, as well as the British government, in the setting of standards.

NIOSH

The fourth major study on which OSHA relied in its analysis of damage risk criteria was conducted by the National Institute for Occupational Safety and Health between 1968 and 1972 (NIOSH, 1974). The Institute's team studied industrial workers exposed to noise levels of approximately 85, 90, and 95 dBA, and control subjects exposed to levels below 80 dBA. After screening out subjects who had been exposed to gunfire noise or who showed some signs of ear disease or audiometric irregularity, the investigators were left with 792 noise-exposed subjects and 380 controls. The NIOSH report presented hearing level data as a function of noise exposure level and duration for the 10th, 25th, 50th, 75th, and 90th centiles. Figure 5–4 gives an example of these data for noise-exposed workers and controls aged 43 to 51 years. Although the data are given for hearing levels, NIPTS can be calculated easily by subtracting control from noise-exposed thresholds.

The NIOSH study appears to have been well controlled, and the data are clearly and interestingly displayed. The only obvious disadvantage is that it did not include lower and higher exposure levels, but NIOSH may not have considered these levels necessary to its mission, which was to develop criteria and recommend a standard for OSHA. Although the control group was exposed to levels below 80 dBA, there is a possibility that some of these subjects may have incurred mild noise-induced hearing loss, which would result in smaller differences between noise-exposed and non-noise-exposed subjects. With respect to screening, the pros and cons discussed previously apply to the NIOSH study as well, although the screening was probably not as rigorous as it was in the Burns and Robinson study.

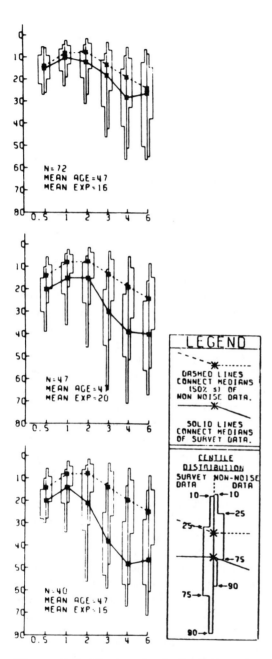

Figure 5-4. Hearing level distribution for workers aged 43 to 51 years exposed to daily average noise levels of 85, 90, 95 dBA, and for non-noise-exposed workers. (From NIOSH, 1974 & OSHA, 1981.)

Summary

OSHA did examine other data and research reports in its analysis of the effects of noise on hearing. The results of these studies, as well as the four major studies just discussed, were all in remarkably close agreement. These results gave OSHA ample justification for requiring the initiation of hearing conservation programs at a daily average noise level of 85 dBA. They would, in fact, give OSHA sufficient justification to lower the PEL to 85 dBA and the "action" level for initiation of hearing conservation programs to 80 dBA, if the economic and political climate would allow it. By examining Table 5–4, one can see that the risk of developing material impairment of hearing virtually doubles between 85 and 90 dBA, regardless of what body of data one chooses. The same is generally true of NIPTS, which can be seen in Table 5–5. Although these relationships will change according to audiometric frequency, exposure duration, and affected centile, the ratio is still approximately two to one between 90 and 85 dBA.

Some individuals are fond of saying that we should not set noise standards, especially at the 85 dBA level, because there is insufficient evidence to support this action. They will say that all of the hearing loss studies are flawed, and until the "perfect" study appears, we have no business regulating. But whenever optimistic investigators attempt a "perfect" study, they find that they must make a few compromises, or they will have no subjects. Such attempts can result in studies that are considerably more flawed than existing studies because rigid rules are broken, and experimental protocols are changed midway in an attempt to ensure an adequate sample size.

Another reason why we cannot wait for the perfect study is that hearing conservation programs are now required above an average level of 85 dBA. All workers exposed above 90 dBA and many above 85 dBA should be wearing hearing protectors. Once this has happened, their exposure histories become extremely complex, due to the varying effectiveness of hearing protectors, and we can no longer make safe assumptions about exposure durations.

The truth is that we have vast amounts of evidence about the harmful effects of noise on hearing, especially in the 85 to 95 dBA region. By examining the major bodies of evidence, as the EPA and OSHA have done, and by combining various bodies of data, along the lines of Passchier-Vermeer (1968, 1971) and Johnson (1973, 1978), certain conclusions become inescapable. These conclusions provide ample grounds for protecting people against the harmful effects of noise.

REFERENCES

American Academy of Otolaryngology Committee on Hearing and Equilibrium, and the American Council of Otolaryngology Committee on the Medical Aspects of Noise. (1979). Guide for the evaluation of hearing handicap. *J. Am. Med. Assoc.*, *241*, 2055–2059.

American Academy of Ophthalmology and Otolaryngology. (1973). Guide for Conservation of Hearing In Noise (rev. ed.). *Trans. Am. Acad. Ophthal. Otolaryngol.*, (Suppl.), Rochester, MN.

American Academy of Ophthalmology and Otolaryngology, Subcommittee on Noise of the AAOO Committee on Conservation of Hearing. (1959). Guide for the evaluation of hearing impairment. *Trans. Am. Acad. Ophthal. Otolaryngol.*, *63*, 236–238.

American Conference of Governmental Industrial Hygienists. (1969). *Threshold limit values for physical agents.* Cincinnati, Ohio.

Baughn, W. L. (1973). *Relation between daily noise exposure and hearing loss based on the evaluation of 6,835 industrial noise exposure cases.* (Joint EPA/USAF study, AMRL-TR-73-53). Wright-Patterson AFB, OH.

Bolt Beranek and Newman, Inc. (1973). *Impact of noise control at the workplace.* (Report 2671; also Exh. 7 in OSHA docket OSH–011).

Bolt Beranek and Newman, Inc. (1976). *Economic impact analysis of the proposed noise control regulation.* (Report 3246; also Exh. 192 in OSHA docket OSH–011).

Bruce, R. D. (1979). The economic impact of noise control. In R. W. Cantell (Ed.), *Symposium on noise: Its effects and control* (pp. 601–607). *The Otolaryngologic Clinics of North America, 12,* Philadelphia: W. B. Saunders Co.

Burns, W., & Robinson, D. W. (1970). *Hearing and noise in industry.* London: Her Majesty's Stationery Office.

Committee on Hearing, Bioacoustics, and Biomechanics, National Academy of Sciences/National Research Council, (1968). *Proposed damage-risk criterion for impulse noise (gunfire).* (Report of Working Group 57). Washington, DC.

Cohen, A. (1973). Industrial noise and medical, absence, and accident record data on exposed workers. In *Proceedings of the International Congress on Noise as a Public Health Problem* (U.S. EPA Report No. 550/9-73–088).

Environmental Protection Agency, Office of Noise Abatement and Control. (1981). *Noise in America: The extent of the noise problem.* (EPA Report No. 550/9–81–101).

Environmental Protection Agency. (1975). *Testimony of Alvin F. Meyer, Jr. at the public hearings on the proposed standard for occupational exposure to noise.* (Submitted to U.S. Dept. Labor, OSHA, as Exh. 57 in docket OSH–011).

Environmental Protection Agency. (1974a). Occupational noise exposure regulation: Request for review and report. *Federal Register, 39,* 43802–43809.

Environmental Protection Agency, Office of Noise Abatement and Control. (1974b). *Information on levels of environmental noise requisite to protect public health and welfare with an adequate margin of safety.* (EPA 550/9–74–004).

Glorig, A., & Nixon, J. (1962). Hearing loss as a function of age. *Laryngoscope, 72,* 1596–1610.

Intersociety Committee. (1967). Guidelines for noise exposure control. *Amer. Ind. Hyg. J.,* 418–424.

International Organization for Standardization. (1971). *Acoustic-Assessment of occupational noise exposure for hearing conservation purposes.* (R1999).

International Organization for Standardization. (1961). *Draft secretariat proposal for noise rating numbers with respect to conservation of hearing, speech communication and annoyance.* (ISO TC43, Acoustics).

Johnson, D. L. (1978). *Derivation of presbycusis and noise-induced permanent threshold shift (NIPTS) to be used for the basis of a standard on the effects of noise on hearing* (U.S. Air Force Report AMRL-TR-78-128). Wright-Patterson AFB, OH.

Johnson, D. L. (1973). *Prediction of NIPTS due to continuous noise exposure.* (Joint EPA/USAF Study. AMRL-TR-73-91. EPA-550/9-73-001-B).

Kryter, K. D., Ward, W. D., Miller, J. D., & Eldredge, D. H. (1966). Hazardous exposure to intermittent and steady-state noise. *J. Acoust. Soc. Amer., 39,* 451-464.

National Institute for Occupational Safety and Health. (1974). *Occupational noise and hearing: 1968-1972.* (NIOSH Pub. No. 74-116).

National Institute for Occupational Safety and Health. (1972). *Criteria for a recommended standard: Occupational exposure to noise.* Pub. No. HSM 73-11001.

Noise Control Act. (1972). PL 92-574.

Occupational Safety and Health Act. (1970). PL 91-596.

Occupational Safety and Health Administration. (1974). Occupational noise exposure: Proposed requirements and procedures. *Federal Register, 39,* 37773-37778.

Occupational Safety and Health Administration. (1981). Occupational noise exposure: Hearing conservation amendment. *Federal Register, 46,* 4078-4179.

Occupational Safety and Health Administration. (1983). Occupational noise exposure: Hearing conservation amendment; Final rule. *Federal Register, 48,* 9738-9785.

Passchier-Vermeer, W. (1971). Steady-state and fluctuating noise. Its effects on the hearing of people. In D. W. Robinson (Ed.), *Occupational hearing loss* (pp. 15-33). New York: Academic Press.

Passchier-Vermeer, W. (1968). *Hearing loss due to steady-state broadband noise.* (Report 35). Sound and Light Division, Research Institute for Public Health Engineering. Delft, Netherlands.

Shampan, J., & Ginnold, R. (1982). The status of workers' compensation programs for occupational hearing impairment. In M. B. Kramer & J. M. Armbruster (Eds.), *Forensic audiology* (pp. 283-324). Baltimore: University Park Press.

Standards Advisory Committee on Noise, U.S. Department of Labor, Occupational Safety and Health Administration (1973). Final draft recommendation, 1910-95: Occupational noise exposure. (Exh. 2(b) in docket OSH-011).

Suter, A. H. (1984). OSHA's hearing conservation amendment and the audiologist. *ASHA, 26,* 39-43.

Suter, A. H. (1978). *The ability of mildly hearing-impaired individuals to discriminate speech in noise.* U.S. Environmental Protection Agency and U.S. Air Force reports EPA 550/9-78-100 and AMRL-TR-78-4, Washington, DC.

U.S. Air Force, Office of the Surgeon General. (1956). A.F. Regulation 160-3.

U.S. Court of Appeals. (1984). *Forging Industry Association v. Secretary of Labor,* No. 83-1420, decision 11-7-84.

U.S. Court of Appeals. (1985). *Forging Industry Association v. Secretary of Labor,* No. 83-1420, decision 9-23-85.

U.S. Department of Health, Education and Welfare, Public Health Service, National Center for Health Statistics. (1965). *Hearing levels of adults by age and sex—United States 1960-1962,* Series 11, No. 11.

U.S. Department of Labor, Bureau of Labor Standards (1970). *Guidelines to the Department of Labor's occupational noise standards for federal supply contracts,* Bulletin 334.

U.S. Department of Labor, Bureau of Labor Standards (1969). Occupation noise exposure. *Federal Register, 34,* 7946ff.

U.S. Department of Labor, Bureau of Labor Standards (1968). Proposed rule making: Occupational noise exposure. *Federal Register, 33,* 14258-14260.

6

CHAPTER

Client Communication

Russell J. Fankhouser

*I*ndustrial hearing conservation programs can be delineated into three basic phases: qualify and quantify, noise abatement, and protection of the exposed personnel (see Chapter 1). Certainly, these phases are interrelated and are each critical to the value of the program. Of equal value is the communication relationship between the professional hearing conservationist and the industrial client. Effective client communication can dramatically enhance the success of a hearing conservation program. Communication is the instrumental basis for interfacing all the phases of the comprehensive program. Just as good communication serves to enhance the effort, poor communication can lead to misunderstandings and thus, result in minimal success for the program. The hearing conservation consultant is ethically obligated to provide his client with the highest standard of services. Quality client consultant communications help to achieve this obligation.

The intent of this chapter is to acquaint the reader with effective client communication techniques. The greatest emphasis is to present a fundamental format for adequately compiling and reporting sound survey results. In addition, other examples of client communication forms will be addressed and illustrated. The sound survey report completes the first phase of the hearing conservation program and provides a basis for the entire program.

REPORTING THE RESULTS OF A SOUND LEVEL SURVEY

Reporting sound survey information with the accompanying analysis is considered one of the most important forms of communication with the industrial client. A consultant retained by an industrial client to perform a sound survey has the responsibility of conducting the actual sound measurement procedure and reporting the results to the client in a logical and meaningful fashion. Clear communication between the consultant and the client is paramount to ensure the comprehension and initiation of any recommendations made.

The initial phase of the hearing conservation effort begins by qualifying and quantifying the industrial noise situation by means of a sound level survey. The obvious intent of the survey is to verify if a noise problem exists and, if so, to determine as accurately as possible, the extent to which it may be potentially harmful to the exposed employees. To meet the intent of the survey, the data must be judged comparatively against some prescribed or established criteria. The set of criteria assumed for use in this chapter is the Noise Standard and Hearing Conservation Amendment (CFR 1910.95) set forth by the Federal Occupational Safety and Health Administration (OSHA) (see Appendix A at the back of this book).

The sound survey report serves as a client communication medium that may address many issues. The importance of the survey report can be illustrated by listing several of these issues. A report may be structured to examine any one or a combination of all the following areas of concern:

- OSHA compliance requirements
- baseline sound levels in employee work areas
- record of employee noise exposure conditions
- administrative noise control procedures
- noise control engineering perspectives
- hearing protection device selection
- manufacturer's noise specifications in situ
- noise source impact on possible third parties (community)
- reference to compare past or future noise evaluations

The sound level survey and the accompanying report have the potential for serving as the basis of a hearing conservation program. The industrial client needs a precise assessment of production area sound levels. An analysis regarding this aspect of the plant environment may be crucial to protect employees from potential hazard and to protect the business from unnecessary expenditures. Accurate and thorough documentation is critical to the understanding and acceptance of virtually any aspect of hearing conservation. The client must be presented with a factual account of the sound measurement activities that leads to a rational analysis adequately describing his or her needs for hearing conservation. Communications regarding these issues should be conveyed objectively and clearly. This is the purpose of the sound level survey and analysis report.

The sound survey report may be used in legal proceedings and thus, may come under close scrutiny. Sound survey documentation is necessary in many industries to demonstrate compliance with OSHA requirements. For these reasons, results and descriptions of the sound survey must be obtained and reported by current and accepted methods. The accuracy of the report can be influenced by how sound measures are obtained and by what standards the data are judged. Current sound survey techniques and descriptions of regulations regarding hearing conservation standards are discussed in detail in Chapters 3 and 5.

A report structured to address OSHA hearing conservation compliance should, at a minimum, document essential employee participation in the hearing conservation program, provide baseline documentation of sound levels in the plant work areas, identify specific machinery that may be considered for engineering perspectives, and determine appropriate hearing protective devices for use in specific plant areas. The report should describe and examine these issues based on the measured sound levels. Basically, it should contain the sound measurement data, analysis of the data, conclusions, and recommendations.

The consultant must have a thorough knowledge of sound physics, sound measurement instrumentation, and the comparative criteria to be used. An accurate analysis cannot be expected otherwise. These topics are discussed more completely in other chapters of this text. Thorough preparation for the sound survey is essential to ensure that all of the critical issues are addressed completely.

Preparation for the Sound Survey and Report

A professional report is dependent upon a technique that incorporates the gathering of sound level data and all other pertinent data necessary for a complete analysis of the noise situation. Preparation is necessary to facilitate data collection and serves to enhance the accuracy and the validity of the report. Organized preparation in advance of the survey may help to decrease the time involved in preparing the report. Minimal turnaround time for the report augments the effectiveness of client communication.

Basic to the preparation process is a skillful sound survey technique. Even more basic in this process is an initial client introduction and consultation. Prior to undertaking the noise survey, the consultant should interview the client to clearly establish the intent of the sound level survey and to obtain other relevant information. The intent of the survey may affect the technique used for gathering sound data as well as the direction and indications of the report. The consultant needs to know what a client manufactures, the manufacturing processes (from raw materials through to the final product), suspected high noise areas, and the estimated number of employees thought to be involved.

The consultant should inquire if previous sound surveys have been performed. Noise reduction controls or any other hearing conservation efforts that have been previously attempted or that are currently in use should be noted. During the interview, the consultant should make notes regarding this information and request copies of previous and relevant reports.

The consultant should ask to be taken on a preliminary tour of the facility. In this manner, useful insight may be gained regarding the plant operations and the noise situations. Details of these inquiries and experiences will later be considered for incorporation in the report.

Preceding the actual sound survey, data collection forms should be designed to efficiently record information. The form may be either a standard form or one designed specifically for a particular client survey. Spaces should be provided to record the client company, locations, dates, times, consultant's name and qualifications, instrumentation, and actual noise data as it is obtained. An example of a completed sound survey form is shown in Figure 6-1.

This survey form was first generated on a word processor, then printed and stored in blank form. Copies of the blank form were used during the sound measurement survey to record data. The stored blank form was retrieved from the word processor memory as the final report was prepared. Raw data and other information were then inserted appropriately for the final report copy. The word processor serves as an invaluable tool to enhance the speed and professional preparation of the report.

The sound survey form contains a reserved area to sketch sound source locations within the test areas. A sketch of the sound measurement area should represent the relationships between the sound sources, the test area, the stationed employees, and as precisely as possible, locations where sound measurement data were obtained. Representations may be drawn to scale and the scale should be identified. If a scale is not used, distances from test microphones to sound sources should be documented as closely as possible. Original survey forms may later be converted to typewritten copy for the final report.

Other data record forms or notepads may be desired to record information at

WATAUGA HEARING CONSERVATION, INC.

SOUND SURVEY FORM

Date: 2/19/86

Company: S.C. Arnold Co.

Location: Temperline HT3

Instrumentation: Metrosonics db 301/306/352
Quest 215R/PH-35/OB-45 - B & K 2209/1613

Personnel: R. Fankhouser, M.S.

Notes: Measures taken 2:00-3:45PM

Blower Room SL's- below basement:
100 dBA/99 dBA L(osha)/111 dBA Max
Adjacent workarea - 86 dBA L(osha)

Refer to page 22 for octave band
analysis at locations.

For job/location descriptions see
page 32.

Representation of test area: letters refer to columes below

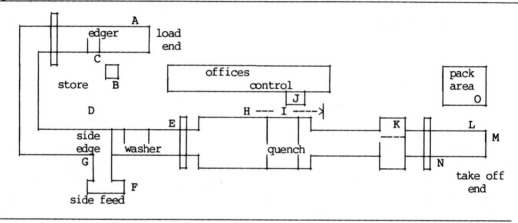

MEASURES TAKEN (in dB):

LOCATIONS	A	B	C	D	E	F	G	H	I	J	K	L	M	N	O
dBA	83 85	84	90 94	85 86	86 87	83 85	86 88	89 108	108	99 90	83 84	82 83	81	83 84	83 87
L(osha)	85		95		86	84	87	98	109	90	84	84		84	84
Leq(60)	86		96		86	85	87	97	108	90	85	83		85	86
Impulse								121A	118A		100A				
L(max)	88		97		91	86	89	110	115	96	88	90		88	96
Sample Time Min/Sec	2:28		11:01		6:27		5:01		10:23		7:17				2:45
						4:40		19:40		8:14		5:38	2:17		

Figure 6-1. Data form used in a sound survey. Note that each column represents sound measures taken at marked locations on the diagram.

employee work stations. Relevant information to note might include products manufactured, production techniques, and employee job cycles or job practices. An example of a word processor–created record form is included as Figure 6–2. The acoustic environment should be noted for general areas and for individual work areas. Environmental factors that may affect employee noise exposure should be recorded. Near and far field sound sources, ambient and background noise sources, as well as the temperature of the environment should be recorded. Space should be made available to describe noise measurement techniques for particular areas. These notes serve as an invaluable reference when compiling the actual report.

Format of the Report

The manuscript style employed in the report is left to the discretion of the consultant, however, the style presented in the Publication Manual of the American Psychological Association should be considered for use. The content of the report should be organized in a format according to that prescribed by the American National Standards Institute Publication S1.13–1971. In this format, each area of concentration is divided into the appropriate major topics and subtopics. These topics and subtopics shall be discussed at length in the next section of this chapter.

The logical progression of the reporting format should gradually build understanding and comprehension of the sound level data. Ultimately, the entire report should be structured and written so that the consultant clearly communicates the needs of the client. The content of the report should address the following prescribed topics: an introduction, the sound source(s) under test, the acoustic environment, the instrumentation employed, the sound measurement strategy, the sound measurement results, the analysis of results, conclusions, and recommendations. An appendix containing the raw measurement data, suggested hearing protection devices, and any additional information deemed necessary for inclusion should be attached.

Title and Introductory Sections

A title page should include the main title of the report, the submission date, the client's name and address, the consultant's name, qualifications, business name, telephone number, and address. A table of contents should then be provided so that the client may refer easily and directly to any of the content topics discussed.

The introduction of the report should establish the intent of the sound level survey and the report. If the survey was undertaken for purposes of OSHA compliance, then a statement regarding the noise standard and hearing conservation amendment should be made. The date(s) of the sound measurements to be reported should be noted and dates of any preliminary data or previous sound survey data obtained should also be mentioned. The introduction should also acquaint the client with the format of the text.

Sound Source Under Test

Descriptions of the sound sources under test should include the type of machine or device and the product(s) manufactured. The manufacturer of the equipment, model numbers, serial numbers, dimensions, installation location, mounting method, and operating conditions (load, speed, capacity, temperature, and others) should be stated. Details may be documented in the text or in an appendix. Employee job titles, number of stationed employees, and relative position(s) of the employees should be described. Employee work habits that may have an effect on the sound produced by the source should be detailed.

The descriptions of sound sources should advance source by source through the manufacturing processes as they were encountered during the sound survey. It should be noted if the plant or individual production areas were operating at typical

Appendix 2

Description of work activity by locations as measures were obtained. Please see attachments in appendix 1 for workarea reference maps.

Page 10 - Main Production Area Pipe Shop

A   Measures at trim saw - large diameter pipe (1-2 ft) is cut one end at a time to the appropriate length, deburred pneumatically and loaded onto a large transport dolly. Two workers here - one at the saw end and one at the opposite end. Chains from an overhead tow motor lift the pipe into position. Worker noise exposure is above the action level and is mostly near field.

B   Pipe length weld - two or more workers here. One welder and the other helpers. First the pipe is welded then the ends are "touched up". Higher sound levels are mainly due to the adjacent trim saw. Worker noise exposure is above the HC action level and is a mix of near and mostly far field conditions.

C   Natco break - three men here. Large metal plate is inserted inch by inch into the break - each time the break bends the metal lengthwise - slowly this makes the beginnings of a pipe. Ambient measures ranged from 76 to 84 dBA however there are significant far field conditions (saws & hammering) to consider. L(osha) measures reflect low and high activity periods. Worker noise exposure conditions are just at the HC action level.

D   Small diameter edge joiner - 11:20 am - not much b/g noise here. Worker noise exposure conditions would appear to be marginal in terms of the HC action level.

E   Bertsch roller - afternoon three men - uses tri-rollers and ½ ton ceiling mount tow motor with chains to roll metal into pipe form. Worker noise exposure was below the HC action level. Sound conditions do not reflect much b/g noise as the adjacents workareas were inactive.

F   Pipe straightening - afternoon two men - this process uses a tow motor hook-up to pull the pipe through guides that ultimately straightens the pipe. Most worker noise exposure is a mix of various far field operations and is above the HC action level. Higher measures reflect a grinder and the trim saw in operation. 1

G   Round up of large diameter pipe at 1:10 pm. Three men working. First measure (G)
H   is a short sample of only the small hammer. The second measure (H) reflects total load, set-up, small and sledge hammers and disc grind smooth - plus routine b/g noise. The process involves the measurement of the pipe's diameter and hammering out the "high" spots to ensure roundness within a certain percentage. The 10 minute measure indicates lesser noise exposure conditions than the first measure but also indicates worker exposure conditions to be above the HC action level.

I   Large tandem break presses - 1:33 pm - four men here - two to a break. Large (20'x 4') pieces of steel are pressed lengthwise by the break inch by inch - this curves the blank into a round pipe form. There is an operator end and helper end. The helper end did exhibit the higher sound exposure conditions. Measures at "I" & "J" were obtained when the "K" saw was operating - "L" was without the saw. As noted, the saw plays a significant role in the break workers exposure. The break creates a impulsive condition as the press action has intervals greater than one second in duration. Impulses to 118 dB during routine pressing and to 124 dB when the finished pipe was hammered off the press were noted.

Figure 6-2. Portion of a sound measurement report that describes job descriptions with respect to noise type and amount.

capacity. Changes in operating capacity should be noted, especially if changes occurred during the measurement period.

Acoustic Environment

An account should be given of the dimensions of the room, area, and building. The locations of the sound sources and the relative locations of the employees within the environment should also be stated. The acoustical properties of the walls, ceilings, and floors should be noted, including any sound treatment modifications that might have been installed. Any sound-reflecting objects or surfaces should be detailed, as well as evident noise reduction strategies such as source or receiver acoustical enclosures or barriers.

Sound transmission pathways from source to receiver should be distinguished as either air or structurally borne. The origin of noise reaching the employee work stations should be identified as either primarily near field, far field, or mixed sound field. The noise should be classified in terms of whether it is continuous, intermittent, impulsive, or cyclical in occurrence. The general spectrum and specific frequencies that were measured for each measurement location may be described as necessary.

Ambient or background noise sources such as air conditioners, air compressors, cooling fans, exhaust fans or blowers, fork truck traffic, tow motors, radios, intercom systems, or similar sources should be listed with their contributions to the overall noise condition differentiated, if possible. Background sound levels measured while primary noise sources were not operating should also be mentioned.

Instrumentation

The equipment used for obtaining the sound measurement data should be listed. This equipment should be delineated as to name, type (ANSI specifications), manufacturer, and serial number. Battery checks and the calibration technique employed should be reported in conjunction with the date and place of calibration. The frequency response and the time response characteristics of the equipment should be reported. Certification of the devices' calibration traceable to the National Bureau of Standards might also be given.

Any ancillary equipment used concurrently with the principal equipment, such as a remote or cable microphone extension, windscreen, remote graphic or magnetic tape recorder or mounting tripod, or others should be noted and described. If personal dosimeters were used, a statement describing instrument wearer attachment and microphone placement should be made. A general listing of the types of measures obtained by each specific instrument may be appropriate.

Measurement Strategy

The sound measurement strategy chosen for use should be discussed. This section of the report should state the rationale for the use of a particular measurement methodology. Compliance with the OSHA noise standard may influence the choice of strategy. The noise standard requires that measurements at employee work areas indicate a time-weighted average (TWA) noise exposure. OSHA accepts TWA noise exposures evaluated by either an area sampling method or by personal dosimetry.

The choice of methodologies is left to the consultant and client, but the chosen method must be shown to accurately identify employees who are to be included in the hearing conservation program. Decisive factors in the selection of the methodology should be presented. Characteristics of the sound sources under test, number of possible employees exposed to the noise, size of the work area, and cost efficiency may be used to substantiate the choice of measurement methods.

Measurement Results

Sound measurement data should be communicated in two basic modes: narrative descriptions in each area of concentration, and graphic presentation of raw data initially recorded during the sound measurement activity. The narrative measurement results section should first identify sound measurement locations. Reference should be made to raw data contained in the appendix. All actual data obtained for the individually sampled areas need not be included in the text. The client should be advised to refer to the appendix for specific measurement data. Each sound survey form for specific areas should be listed in their order of appearance in the appendix.

The following sound survey information should appear in either the text or appendix; the date(s) and time(s) that measures were obtained, locations of the collecting microphones, angles of microphone orientation in relation to the sound sources tested, microphone or other equipment corrections, all sound level data obtained, and the durations of the measurement periods.

The types of sound level measurements selected for use should be delineated. Reference should be made if special instrumentation was required to obtain the measures. Specific measures should be listed respectively for sampled areas.

Analysis of Results

An analysis of sound level data may be based on several types of comparisons. The number and type of comparisons should be determined by the scope and intent of the sound survey. The analysis should be prepared carefully and presented conclusively.

A comparison of the obtained data to other previously reported data should be considered. Comparisons in this regard can substantiate current findings or demonstrate changes in the noise exposure conditions. This kind of comparison may be valid only if sound level data were obtained identically under the same conditions at different times. Previous sound level data should be reviewed and used with prudence.

Comparisons for inter-test reliability may be made if two or more similar instruments were employed to collect the same or similar data at individual sample locations.

Confidence in such observations can be expected, depending on instrument calibrations and measurement techniques. A contrast of slow dBA measures versus L_{eq} measures may provide insight into the temporal characteristics of the noise under evaluation.

A comparison of the data to criteria established by the intent of the sound level survey should be made. Comparisons should be made individually for each measurement location. The significance of any resulting comparisons should be stated clearly. Collective tables may be presented, listing all measurement locations that meet, exceed, or fall below the criteria. Some locations may need to be individually discussed in greater detail.

A complete noise analysis should include a recommended method to select appropriate hearing protection devices for use in specific noise areas. A method or even several methods for this determination may be included in the appendix. To meet compliance with the OSHA guidelines, the client must document that chosen hearing protection devices are adequate to reduce employee noise exposure to specified levels. Selection methods and criteria for hearing protective devices are discussed in Chapter 10.

A worksheet to determine the desired Noise Reduction Rating (NRR) of a chosen hearing protection device for use in a known noise level is shown in Figure 6–3. By furnishing the client with a method of selection, he or she is able to narrow down and adequately select appropriate devices from an ever-increasing range of possible choices. The use of the NRR method is probably the easiest method for the client. The basic selection criteria is the desired NRR.

Correspondingly, a list of suitable protective devices may be detailed in the appendix for the client. A computer-generated list of specific hearing protection devices by area is shown in Figure 6–4. This list was generated by a computer program that compares manufacturer's published hearing protection device (HPD) attenuation data to the measured octave band sound levels in a specific noise environment. The resulting

WATAUGA HEARING CONSERVATION, INC.

Attachment

Noise Reduction Rating (NRR) Worksheet

The Noise Reduction Rating (NRR) of hearing protective devices is the amount of attenuation or reduction of noise that a properly inserted HPD will provide at the eardrum. HPD manufacturers are required by the Environmental Protection Agency to list the NRR of their HPD. OSHA requires that chosen HPD's for use in noise exposure conditions be evaluated in terms of their effectiveness for the chosen environment. This worksheet is not designed to determine the actual rating of a HPD but rather is designed to determine the desirable NRR in a specific noise environment.

For a HPD to be appropriate for a specific environment exposures should be brought down to below 90 dB, preferably 85 dB or better. A safety factor of 10 dB should be considered in the evaluation process because "real world" performance of a chosen HPD is often less than it's official rating.

The desired NRR for use in a known dBC or dBA sound level may be determined by using the following formulas.

Example For dBC measures:

Desired NRR = dBC measure + 10 dB (safety factor) - 85 dB (desired level)

NRR = 100 dBC + 10 dB - 85 dB

NRR = 25

Example For dBA measures:

Desired NRR = dBA measure + 7 dB (correction factor for "A" weighting)
            + 10 dB (safety factor) - 85 dB (desired level)

NRR = 95 dBA + 7 dB + 10 dB - 85 dB

NRR = 27

Figure 6-3. Noise Reduction Rating (NRR) worksheet for evaluating hearing protector adequacy. See Chapter 10 For details regarding NRR.

comparison yields specific HPDs that allow no more than 80 dB of sound pressure at the ear. A list can be prepared for all of the client's noise areas and provides solid choices for the client. This type of analysis and listing gives the client the documentation necessary for OSHA compliance.

Conclusions

Specific conclusions concerning the measurement results should be based on the analysis of the measures and should be directed to the intent of the sound survey. Conclusions may also be drawn from any other area of the report content. These conclusions should be strong, but should be stated in broad terms to form the basis for specific recommendations.

Recommendations

Recommendations should be direct and to the point. They should be organized and stated in order of importance to address the critical issues primary to the intent of the survey. Each recommendation should be supported by the factual data and judgment criteria presented in the preceding sections of the report. Conservatively, the consultant should provide any recommendations for improvement that may be deemed necessary.

```
                    WATAUGA HEARING CONSERVATION, INC.

                    HEARING PROTECTOR SELECTION PROGRAM


       This hearing protector use analysis was prepared specially for L.A. HIGGINS CORP.

Date of analysis: 6/22/86

Work location under consideration: MYERS REDUCER      LEVEL:  109  dBA

Octave-band levels at the work location:

   FREQUENCY BAND: 125Hz  250Hz  500Hz 1000Hz 2000Hz 4000Hz 8000Hz
                   -----  -----  ----- ------ ------ ------ ------
                    91     95     101   106    106    107    102
```

The following named hearing protective devices have been found to allow no more than 80 dB at the ear. The procedure used in this analysis was to reduce the specified attenuation (noise reduction) for each protector by two standard deviations. This insures that the protector(s) named below can be expected to provide adequate protection for 97.5% of the workers in the area for which the octave-band levels were listed above. This assumes, of course, that the hearing protective device has been appropriately fitted to each ear and that it is worn properly.

```
AMERICAN OPTICAL 1720 EARMUFFS
BILSOM SOFT EAR PLUGS
BILSOM PER-FIT EAR PLUGS
BILSOM UF-1 EARMUFFS
BILSOM VIKING EARMUFFS
CESCO E1-2 EAR INSERTS
E-A-R PLUGS
MINE SAFETY APPLIANCES MARK V EARMUFFS
NORTON SILENT PARTNER
NORTON COMFIT EAR PLUGS
```

There are 10 hearing protectors suitable for this work location:

The listed protectors are available from:

```
     SAFECO - Kingsport, TN (246-4164)
              - or -
     Safety Equipment Distributing Co. - Knoxville, TN (584-8663)
```

Figure 6-4. Computer printout showing hearing protectors selected according to the sound levels present in a given section of a plant.

Recommendations for instituting a hearing conservation program should be presented in detail and based on two broad premises: noise abatement and protection for the employees. Details should be included to ensure continued success with the hearing conservation program. The client should be able to refer to this section of the report and find all the basic information necessary to proceed with an effective, comprehensive hearing conservation program.

Appendices

Appendices to the report should include the sound survey data forms, and may include notes obtained through the interview with the client. In addition, the inclusion of notes taken during the sound measurements and selections for hearing protection devices would add strength to the report.

The Report Finalized

The acceptance and the level of confidence in the sound survey report or any professional report can be greatly influenced by its appearance. Use a quality paper stock that is white or reserved in color, unlined, and nonerasable. It is suggested that the material be prepared and printed with a word processor. If a word processor is unavailable, professional typing or printing should be considered. All documents should be proofread carefully before submission to the client. Errors in spelling and grammar do not enhance the appearance of the final copy. If possible, have the final copy bound, or present the report in an attractive cover.

The person to whom the report is submitted is usually a personnel director, safety officer, assistant manager, or plant manager. This person does not always have the authority to follow through with recommendations made in the report. Additionally, he or she may have little, if any, knowledge on the subject of noise and hearing damage. The client must be able to read and digest the information in the report so that he or she can present it logically and sensibly to upper management so that recommendations can be acted upon. A personal follow-up by the consultant with the client is necessary to assure a strong communication relationship.

In summary, the sound measurement report is probably one of the most important documents for an industrial hearing conservation program. Documentation of sound levels is required by OSHA to formally comply with the Noise Standard and Hearing Conservation Amendment. Certainly, the report establishes the need for hearing conservation efforts, as well as providing a continual basis for success of the program.

CLIENT COMMUNICATION FORMS AND INFORMATION PRESENTATIONS

Briefly, several other client communication documentation forms shall be mentioned and examples shown. All services provided to the client company must be documented. Typical communication forms may include audiograms, audiogram follow-up test forms, educational materials, audiometer calibration, and sound booth attenuation. The client communication forms serve to document all aspects of the hearing conservation effort as closely as possible. The OSHA hearing conservation mandate requires substantial record keeping and the consultant has the responsibility and expertise to assist the client in this manner.

Figure 6–5 shows an audiometer calibration and sound booth attenuation form. This form was designed on a word processor and is used on location for mobile test equipment calibrations. Audiometer calibration documentation is required by OSHA, and the client company should have this information on file. On this form, spaces are provided to record audiometer output levels and sound booth attenuation characteristics prior to and after mobile hearing testing. This type of calibration recording demonstrates the operational stability of the test equipment.

The professional hearing conservationist has the responsibility to monitor and update a client's program on a periodic basis. This will help to ensure continued success of the program. This process can only be accomplished by effective communications between the consultant and client. An excellent medium of client communication is a newsletter. An example of our newsletter cover page is presented as Figure 6–6.

The periodical mailing of information regarding hearing conservation and consultant services can serve to strengthen the relationship of client and consultant. OSHA requirement updates, new or additional services, and new hearing conservation

WATAUGA HEARING CONSERVATION, INC.
On Site Calibration Of Audiometer
and Mobile Sound Booth

Calibration Instrumentation: Precision Type 1

Sound Level Meter: Bruel & Kjaer 2209	S/N: 477741
Octave Band Filter Set: Bruel & Kjaer 1613	S/N: 480667
Artificial Ear: Bruel & Kjaer 4152	S/N: 490341
Microphones: 4144 audiometer calibration	S/N: 499223
4145 ambient noise levels	S/N: 486095
Pistonphone: Bruel & Kjaer 4220	S/N: 501508

Personnel: R. Fankhouser, M.S.

Calibration Check Before Testing: Date:

Audiometer: Model: S/N:

Standard: ANSI S-3.6 1969   Earphone: TDH-39
                                                     TDH-49
Sound Level Meter Reading dB: (70 dB audiometer dial)   TDH-50

	Right Earphone							Left Earphone						
Freq.	.5K	1K	2K	3K	4K	6K	8K	.5K	1K	2K	3K	4K	6K	8K
TDH-39	81.5	77	79	80	79.5	85.5	83	81.5	77	79	80	79.5	85.5	83
TDH49/50	83.5	77.4	81.1	79.7	80.7	83.4	83	83.5	77.4	81.1	79.7	80.7	83.4	83
Actual*														
Correct														

Linearity Check @ 1K&2K:YES ☐ NO ☐   Meets Calibration Standard: YES ☐ NO ☐

Calibration Check After Testing: Date:

Sound Level Meter Reading dB: (70 dB audiometer dial)

	Right Earphone							Left Earphone						
	.5K	1K	2K	3K	4K	6K	8K	.5K	1K	2K	3K	4K	6K	8K
Actual*														
Correct														

Linearity Check @ 1K&2K:YES ☐ NO ☐   Meets Calibration Standard: YES ☐ NO ☐

*Actual - includes mic correction factors in dB @ 3K,-.5; 4K,-1; 6K,-1; 8K,+1

Mobile Sound Booth: S/N: Date:

Location:

Standard: ANSI/OSHA   Meets Allowable Ambient Noise Standard   YES ☐ NO ☐

Octave Band Ambient Noise Levels dB :

Frequency Hertz (Hz)	.5K	1K	2K	3K	4K	6K	8K
OSHA Allowable Sound Pressure Levels at Octave Bands	40	40	47	---	57	---	62
Actual Sound Pressure Levels at Octave Bands Before Test							
Actual Sound Pressure Levels at Octave Bands After Test							

Figure 6-5. Audiometer calibration form which is completed on each unit and sent to each client for their records.

WATAUGA HEARING CONSERVATION, INC.

215 EAST WATAUGA AVENUE • JOHNSON CITY, TENNESSEE 37601 • TELEPHONES 615/928-1901 OR 928-6142

W. T. Mathes, M.D.
James F. Wood, M.D.
Arthur Harris, M.D.
Mark A. Howell, M.D.
Otolaryngologists

Laurie A. Higgins, M.S., CCC-A
Russell J. Fankhouser, M.S.
Audiologists

David M. Lipscomb, Ph.D.
Susan C. Mattingly, Ph.D.
Consulting Audiologists

```
HH    HH  EEEEEE  AAAAAA   RRRRRRR    NN    NN  OOOOO   WW      WW
HH    HH  EE      AA   AA  RR    RR   NNNN  NN  OO  OO  WW      WW
HH    HH  EE      AA   AA  RR    RR   NN    NN  OO  OO  WW      WW
HHHHHHH   EEEE    AAAAAAAA RRRRRRR    NN NN NN  OO  OO  WW      WW
HH    HH  EE      AA   AA  RR RR      NN    NN  OO  OO  WW      WW
HH    HH  EE      AA   AA  RR   RR    NN  NNNN  OO  OO  WW WW WW
HH    HH  EEEEEE  AA   AA  RR    RR   NN    NN  OOOOO      WW WW
```

SO   OTHERS   WILL   HEAR   LATER

Volume 5   Number 1   January 15, 1987

A Brief Newsletter

and

Hearing Conservation Update

Prepared at irregular intervals

(when the spirit moves)

Sent to you with our best wishes,

WATAUGA HEARING CONSERVATION, INC.

Figure 6-6. Cover page of an information newsletter sent several times each year to clients and prospects in the area.

techniques can be presented to all clientele efficiently through the newsletter. Strong communication between the hearing conservationist and the client can only enhance protection of industry's most important asset, the employee.

REFERENCE

American National Standards Institute. (1971). *American national standard methods for the measurement of sound pressure levels.* (ANSI S1.13–1971 [R1976]). New York: American National Standards Institute.

■ *Components of an Effective Hearing Conservation Program*

Although "three little words" summarize the substance of the intent of hearing conservation activities, many more words will be devoted to the "nuts and bolts" coverage of hearing conservation techniques and activities. This section provides that information requisite to designing, staffing, and conducting a hearing conservation program. The emphasis is on the industrial scene, but most of the principles discussed relate just as well to other environments such as military settings or in the schools. This section contains roughly 50 percent of the bulk of this text and comprises the "heart" of the message. Learning the materials presented herein and putting the concepts into practice will take one far toward becoming a successful contributor in the hearing conservation arena.

Principles of Noise Control

David M. Lipscomb

C H A P T E R

7

Noise abatement and control is best left to those professionals who are skilled in the practice. Engineers who concentrate on noise abatement procedures develop special techniques which allow them to (1) provide an accurate assessment of the noise control potential for environments; (2) consider specifications for noise reduction; (3) prepare an effective noise control strategy; (4) estimate costs; and (5) make requisite changes in sound sources to bring their sound output to more acceptable levels. The most effective of these professionals combine their knowledge and experience with ''good ol' common ordinary horse sense.'' There are limits of practicality in this endeavor that must be recognized and respected. The makers and enforcers of the hearing conservation legislation recognize that fact and allow for noise control within the bounds of ''technical and economic feasibility.''

Directors of hearing conservation programs are seldom engineers. Therefore, hearing conservation personnel are not trained specifically in noise control techniques and are thus unqualified to attempt any but the most superficial treatments. It is helpful, however, for the hearing conservationist to be conversant in noise control techniques in order to intelligently discuss control needs and to evaluate recommendations made by noise control engineers.

This chapter is presented as an introduction to some of the factors that underlie noise control. The discussion will include some foundation material, the knowledge of which will increase the value of the hearing conservationist to the program. It is not the purpose here to train noise control engineers. That will be left for other works in other sources. The goal is for the reader of this chapter to become adequately prepared to understand noise control strategies and to be sensitive to methods incorporated in noise reduction.

BACKGROUND

Sound Transmission

The elements of the propagation of unwanted sound are diagrammed in Figure 7–1A. This simplistic illustration shows that sound emanates from a source, and travels via a path to a receiver. That sequence of events is usually more complex in that there may be multiple sound sources, several paths (including reflected sound), and more than one receiver.

Once sound analysis is completed and excessively high level sound sources are identified, noise control strategies must be developed. The two elemental forms of noise reduction strategy are (1) decrease sound output of the source; or (2) interrupt the pathway of the sound. Engineering controls, proper maintenance, or equipment purchase specifications are included in strategy element (1), whereas the use of "sound barriers" or increasing the distance between the source and receiver are among the modes of choice for strategy element (2).

Concentration of Sound Energy

A major reason one should conduct octave-band sound analysis during a survey (see Chapter 4 and 6) is to provide requisite information for noise control procedures. The steps followed to quiet sound sources depend largely on the spectral content of the sound source. Predominantly low frequency output will be treated with the use of equipment modification or massive barriers. A sound source emitting high frequency energy is a likely candidate for approaches using sound absorption.

Barrier Penetration by Sound

Sound transmission through barriers varies according to the spectrum of the sound. This fact is directly related to the wave length (lambda - λ) of sound frequency content.

The wave length of a pure tone can be calculated according to the following equation:

$$\lambda = ss/f \qquad (1)$$

where: ss = the speed of sound (feet/sec or meters/sec) and
f = frequency of the tone

Low frequency sounds with long wave lengths penetrate barriers more readily than do short wave length, high frequency sounds. This principle is demonstrated each time an apartment dweller's neighbor cranks up the stereo. Most of the sounds that penetrate the walls separating apartments are the bass notes. High tones are reflected and absorbed.

Near and Far Field

Relative location of a sound source and receiver are of significance in considering sound treatment to deal with a particular exposure condition. If a person is located in the "near field," the only effective treatment of the sound path between the source and receiver is treatment of the source or the use of a barrier located directly between the sound source and any receivers. However, if the person is located several wave lengths away from the source (far field), acoustical treatment of the room may be effective in decreasing the amount of sound that reaches that person. Knowledge of these concepts will help to avoid unnecessary expenditures of money for acoustical treatments that are not effective. For example, it is sad to note that frequently, in the interest of reducing noise exposure in industry, plant owners invest in acoustical tile for ceilings and sometimes for walls, only to find that the attempt to reduce noise exposure was futile because the workers were actually in a "near field" exposure condition.

COMPOUND STRATEGIES FOR NOISE CONTROL

Specialists in drama maintain that there exist only a limited number of possible plots. According to that view, all dramatic offerings are simply variations on one or more of those fundamental plots. In a like manner,

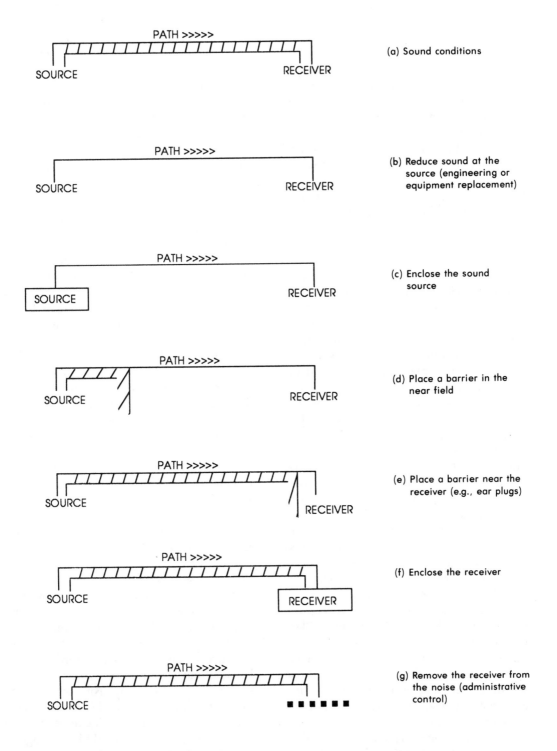

Figure 7-1. Conditions in noise transmission and basic noise control strategies.

there are a total of six possible strategies (''plots'') for noise reduction (Figure 7–1B through 7–1G). These complexities in noise control are due to the fact that each of these fundamental principles offers as many variations as there are noise sources that require attention. But, by reducing the strategies available for noise reduction to their lowest common demoninators, some of the mystery surrounding noise control techniques should evaporate.

Each of the approaches to noise control illustrated in Figure 7–1 can stand alone, but it is often desirable to combine two or more methods to achieve adequate reduction of sound level at the source. The strategies are:

1. REDUCE SOUND AT THE SOURCE (Figure 7–1B). This step is usually in the province of the noise control engineer because it involves modification of equipment. Some of the techniques used range from extensive redesign of one or more pieces of machinery to alteration of lubrication and maintenance procedures. The most drastic move is to remove or replace those units causing the noise problem.

If cutting edges are involved, more frequent sharpening, alteration of the cutting angle, or changing speed of cutter rotation may cause measurable decrease in sound output. Wood product manufacturers have found that changing the design of blades can reduce saw and planer noise as much as 20 dB with no loss in cutting efficiency or speed.

Although it will influence production, one approach to consider in reducing noise at the source is to reduce the speed of the equipment. This may place an economic hardship on the plant, however, if other departments have to slow down to await materials from the slowed equipment. Plant production schedules must be met and careful attention should be given to noise control attempts that may affect those schedules. Jobs may be at stake and it is not good policy to take away work positions in the name of hearing conservation.

Noise control at the source is the most effective means of noise abatement and control. However, technical and economic feasibility factors may dictate that other means of controlling noise be utilized.

2. ENCLOSE THE SOUND SOURCE (Figure 7–1C). Decisions on whether to attempt the use of an enclosing barrier and the composition of that barrier depend upon the spectral content of the sound emanating from the source (see previous section on barrier penetration). Generally, the effectiveness of sound isolation by an enclosure is a function of its mass. Lead has been a popular sound reducing material because of its high mass. However, the use of lead is now being regulated heavily due to the hazard it poses in the event of fire.

Placement of an enclosure by itself is of no value unless there is some companion treatment of the inside of the enclosure. The physical law of conservation of energy states that energy can neither be created nor destroyed. The only choice remaining is the conversion of energy from one form to another. Sound absorption is, in actuality, the reduction of acoustic energy by transforming it to heat, which is then dissipated. To simply place a solid enclosure over a sound source will be ineffective. The sides of the enclosure must contain sound-absorbing material as well.

Although enclosing a sound source is quite effective as a noise treatment, it also poses some problems. From a feasibility standpoint, some sound sources cannot adequately be enclosed. Their size or shape may not be amenable to enclosure. Many manufacturing devices are part of a production process which cannot be interrupted by an enclosure. Some equipment requires frequent modification and maintenance. Entering the enclosure repeatedly makes for delays that cost money in the form of lost production and down-time. One further consideration cannot be overlooked: an enclosure is a heat trap. For some equipment, this would render enclosures impracticable because the heat inside the enclosure may rise to a level that could be damaging to the machine. Heat dissipation must be considered in most instances where enclosures are contemplated.

The interpretation of the term *barrier* should be expanded to include placing the equipment in a separate room. On occasion, developing a room enclosure is the outgrowth of an attempt to create a barrier for sound.

3. PLACE A BARRIER IN THE NEAR FIELD (Figure 7–1D).

Barriers become less effective as the distance increases between a barrier and a sound source. In order to create a *sound shadow,* a barrier must be approximately 5 wave length (λ) high. If the sound contains high intensity low frequency sound, this physical law poses a problem. For example, the wave length of a 100 Hz tone is 11 feet. Therefore, in an open space, a wall higher than 50 feet would be required to cause an appreciable sound shadow for sounds containing that frequency. Therefore, many barriers actually become separate enclosures in order to reduce transmission of sound to one or more receivers.

Some examples of barriers are lead curtains, drapes, or solid shields which are placed close to a sound source. Such applications redirect sound energy away from potential receivers or absorb sound as it strikes the barrier. Some barriers both reflect and absorb sound. As a general rule, the use of barriers is restricted to marginal noise exposure conditions where slight reductions in sound level are needed. If the sound environment is more than a few dB over safe levels, other approaches than the use of barriers are warranted.

4. PLACE A BARRIER NEAR THE RECEIVER (Figure 7–1E).

The use of the term *barrier* in this context is misleading, because it is not very effective to locate curtains or shields near the receiver. However, hearing protectors are themselves barriers. Although it is not as desirable as reduction of the sound at the source, use of hearing protection devices (HPDs) is the second most effective form of noise reduction. See Chapter 10 for details on hearing protector use.

5. ENCLOSE THE RECEIVER (Figure 7–1F).

Often, it is more effective and practical to place workers in sound-treated control rooms or offices to reduce their noise exposure than it is to enclose the sound source. Printing press rooms have been modified by this technique with considerable success. Enclosing a large Webb press is impractical because of its size and since material must be fed into it and printed material is disgorged at the output end. When the press is in full operation, it has many large parts that are moving rapidly, each contributing to the sound level. By relocating controls, gauges, readouts, and sensors in an enclosure, workers can continue to operate and monitor the operation of the press in a low-level sound environment. The only time it is necessary to venture into the vicinity of the equipment is when there is some type of breakdown, at which the time the equipment has been slowed, thus causing reduction of the sound being generated. Similar enclosures for workers have become common in sawmills and in paper-making plants.

6. REMOVE THE RECEIVER FROM THE NOISE (Figure 7–1G).

This approach, usually termed *administrative noise control* takes advantage of the fact that worker rotation in and out of noisy areas will decrease the total noise exposure of each one. For example, consider a situation wherein workers are located at each end of a large machine. The in-feed end of the machine has a constant sound level of 94 dBA and the take-off end has a lower overall sound exposure level of 89 dBA. To keep workers at the in-feed end all day each day would cause them to be overexposed to noise. However, by rotating workers from one end of the machine to the other at midday, no one is exposed to more than a time-weighted 8 hour equivalent sound of 90 dBA. There are reasons that administrative controls are not feasible in all conditions in which they may seem desirable. However, administrative controls should be considered for noise exposure reduction in a noise control program.

ANECDOTAL EXAMPLES OF NOISE CONTROL ACTIVITIES

The following "war stories" are intended to illustrate both appropriate and inappropriate approaches to reducing noise by persons outside of the engineerign profession. Although it is improper for most of us to attempt engineering noise controls, the intelligent use of information from Figure 7–1 will help to make surprisingly great inroads into the reduction of some noise problems.

Example 1: A manufacturing plant that produced metallic junction boxes used in electrical wiring applications was found to have a number of locations with excessive noise exposure for employees. Much of the noise was metal-to-metal contact, a very difficult type of noise to abate. The noisiest production lines was also the fastest one. These lines started at one end with flat sheets of metal, and through a series of automated bending, punching, drilling, and shaping operations, produced the finished junction box at the output end. The plant manager indicated that the noisiest line was not operating all of the time because the other lines couldn't match its production rate. He was encouraged to slow the noisier line, which resulted in commensurate reduction in the overall noise generated by the equipment with no loss in productivity.

In that same facility, workers in the plating area experienced periodic high noise exposure when the fork lift operator arrived to dump junction boxes into the large hopper above the location where women were placing the boxes on hooks. The young man on the fork lift could have released the boxes much more quietly, but as a form of flirtation with the women in the area, he dropped the boxes from far too high into the hopper, so they fell with a loud crash. Then, to finish the job, he would jockey the lift to and fro, rapidly rattling the hopper and its contents, and making even more unnecessary noise. That activity was called to the attention of the manager, and it was brought to a halt.

These examples used "operational noise control." Often, the operation of equipment

can be altered to significantly reduce noise levels. The impact on communities by aircraft noise is being reduced by changing flight paths—operational control. In riding a motorcycle up a hill in a residential area, the annoyance can be made negligible by moving up the hill gently—operational control. Choosing to mow the lawn on Saturday afternoon rather than early Sunday morning reduces the potential noise impact on the neighborhood—operational control.

A third location in this plant was occupied by assembly workers who were removed from the noisy equipment by quite some distance, but their noise exposure condition was excessive due to the manner in which materials were delivered to their area. A conveyer belt brought the junction boxes to the assembly area and discharged the boxes from an 8 foot height onto a metal collector about 4 feet in diameter. The clang that occurred when each box dropped created an unnecessary increase in overall noise level in the area. The plant manager agreed to lower the pitch of the conveyor belt so that it released the boxes only a few inches above the collecting pan. It was also suggested that the collecting pan be lined with some material, such as thick rubber sheeting, to reduce metal-to-metal contact. By making these changes, workers in the area were reclassified to low noise exposure conditions because their time-weighted average noise exposure fell to a level below 85 dBA.

In this plant, noise in three areas was reduced without the use of "engineering controls." Two areas experienced noise reduction by operational controls and the third by slight redesign of the area. Engineering techniques play a vital part in noise control and the work of engineers can never be replaced by those of us who are not appropriately trained. Yet, as these examples show, application of intelligent understanding of noise control principles can reduce the overall noise exposures of many workers.

Example 2: A textile form had a large section devoted to making stockings.

Numerous machines were located in one room. At the ear location of each equipment operator, the sound level was a fairly constant 92 dBA. Noting this, the plant engineer determined that something should be done to reduce worker exposure to below 90 dBA. Unfortunately, this plant engineer did not have a grasp of the near- and far-field concepts, thus, he believed that much of the noise exposure each employee experienced was due to the accumulated sound of all machines, including the one being cared for by the operator. His strategy was to isolate each machine by building a small partial enclosure. To try out his idea, he obtained three sheets of 4 foot by 8 foot plywood and stood them up, attaching the edges such that a three-sided barrier was located around one of the stocking stitching machines. He heard that one of the other manufacturers in the industrial park used a product called "sound deadener," so he obtained some of that compound and coated the inside of the "enclosure" with that material. Thinking he had developed a reasonable solution, the plant manager proceeded to measure the sound level at the ear of the machine operator. To his dismay, the sound level was 95 dBA—an *increase* of 3 dB. His problems were:

1. "Sound deadener" was an enamelized compound intended to be placed on the underside of sinks to decrease the "ring" of the bowl when objects struck the sink. It was hard, creating a highly reflective surface.

2. He did not recognize that the noise reaching each machine operator was far from the near-field, thus, barrier treatment did not work. In actuality, he created a reverberant condition which elevated the overall sound level in the vicinity of the machine operator.

The plant engineer then called a noise-control engineer. This specialist pointed out the problem with the barrier. Then he observed that the machinery did not operate for a full 8 hour shift. In fact, with downtime for worker breaks and for threading the machine, it was in operation less than 6 hours per shift. The 92 dBA level at the ears of workers was acceptable when total exposure time was taken into consideration. As the two men were discussing this in the plant engineer's office, the consultant noted a stack of bearings in one corner of the room. In response to the consultant's inquiry, the plant engineer indicated that they had to order bearings for the stocking machines by the carload. The machines "ate them up." Thinking this was unusual, the consultant inquired about the type of lubricant being used. It was discovered that an inappropriate maintenance schedule and lubricant was being used, resulting in excessive bearing wear. The ultimate result of this visit by a knowledgeable consultant was that the proper lubricant was obtained, bearing life was increased dramatically, and, due to less bearing noise, the overall sound output of the machines was reduced to a level less than 90 dBA.

Example 3: A large seat-belt maker used huge stamping presses to produce metal parts. These stamped pieces were then tumbled with abrasives to smooth the sharp edges, sent to the "heat-treat" furnace, plated, and moved on to the assembly area. Each stamping press was bolted firmly to the floor, therefore, with each thump, noise was emitted to the vicinity of the worker and structure-borne vibration developed. It was possible to set a glass of water on the floor of the offices quite a distance away from the stamping area and see the effect of structure-borne vibration, which caused wave motion on the surface of the water in the glass. This vibration, in turn, caused sound emission from the floor, resulting in considerable bothersome, although not oto-hazardous, noise in the offices. It was suggested that the presses be "shock mounted" using appropriate isolating material to reduce the structure-borne noise. When this was accomplished, it did not change the near-field exposure condition at the ear of the stamping press operator, but the noise level in the office spaces was noticeably lower.

Another area of the plant contained about a dozen very large tumblers. Metal pieces were tossed around inside these devices along with abrasives. The noise level in the vicinity of the tumblers was over 100 dBA. This impacted adjacent stamping press operators, whose noise exposure condition was compounded by the combination of the stamping press operation and the radiating noise from the tumblers. Except for loading and unloading tumblers, it was not necessary for anyone to be present around these machines. The suggestion was made to isolate the tumblers by enclosing them in their own room. A room of cement block walls was constructed and capped with a ceiling. The noise level inside the enclosure automatically increased, which, in this instance, was acceptable since workers were not stationed inside. A control switch was located outside the enclosure so the tumblers could be turned on and off from that location, hence, making ingress into the room less noise hazardous than prior to the alteration.

The combination of noises in initial processes within the plant (stamping, tumbling, heat-treat) were sufficient to elevate sound levels in the vicinity of the plating area where workers placed the metal pieces on hooks so the pieces could be moved into the plating vats. Since these workers were located at designated areas, and since they were in the far-field from noise sources, a barrier was placed near the work station for the plating area workers to provide sufficient sound shadow in their area so the noise exposure was reduced to safe levels. At the time this project was initiated, a tall lead curtain was erected. Building codes in many cities may not allow that treatment any longer, but currently available substitute sound curtain materials are sufficient for that approach.

None of the remedies for the seat belt manufacturer required extensive equipment redesign or replacement. Appropriate applications of some of the fundamental principles of noise control resulted in low-cost noise reduction. One specialist was required for these procedures. The supplier of shock mounting equipment for the presses called upon engineering data and

skills to prescribe the appropriate vibration isolation devices for each stamping press.

Example 4: The personnel director for a large metal fabricating company requested impulse sound level measurement of one particularly bothersome operation in the plant. In that process, large, heavy-gauge sheets of steel were rolled between rollers and curved. Once the appropriate curvature had been achieved, the workers discharged the steel from the rollers and allowed it to drop about 4 feet onto the concrete floor. It was then retrieved by the workers using a hoist and placed on a transfer cart. The noise of a 350 lb sheet dropping to the floor with such force startled the workers across the entire enclosed space, which was about the size of a football field. The suggestion was made that a sturdy wood frame should be fashioned on which an old mattress could be placed. Then, when the steel was removed from the rollers, it would land on a soft surface with comparably less sound. In that way, a disturbing industrial noise was eliminated inexpensively, but effectively.

CONCLUSION

The principles outlined in this chapter and the examples given offer basic information on noise control techniques. The good consultant knows when the outer limits of his or her knowledge have been reached and will defer further action to more qualified persons. If engineering controls are required, then the work is best left to the qualified engineer. By understanding these principles, it is probable that the hearing conservationist can:

1. Recognize when noise control is necessary;
2. Speak more intelligently with the engineer(s) being contracted to effect noise control;
3. Undertake to plan noise control strategies at a "common sense" level.

SUGGESTED READINGS

Beranek, L. L. (1971). *Noise and vibration control.* New York: McGraw-Hill.

Cheever, C. L. (1975). Basic principles of noise control. In J. B. Olishifski & E. R. Harford (Eds.), *Industrial noise and hearing conservation* (pp. 380–409). Chicago: National Safety Council.

Harris, C. (1979). *Handbook of noise control.* (2nd ed.). New York: McGraw-Hill.

Inde, W. (1975). Application of engineering noise control measures. In J. B. Olishifski & E. R. Harford (Eds.), *Industrial noise and hearing conservation,* (pp. 445–480). Chicago: National Safety Council.

Kamperman, G. W. (1975). Problem-solving techniques. In J. B. Olishifski & E. R. Harford (Eds.), *Industrial noise and hearing conservation* (pp. 410–444). Chicago: National Safety Council.

Milligan, M. W. (1978). Introduction to basic principles of noise control. In D. M. Lipscomb & A. C. Taylor, Jr. (Eds.), *Noise control: Handbook of principles and practices* (pp. 119–130). New York: Van Nostrand Reinhold.

Nabelek, I. V. (1985). Noise measurement and engineering controls. In A. S. Feldman & C. T. Grimes (Eds.), *Hearing conservation in industry* (pp. 27–76). Baltimore: Williams & Wilkins.

Olishifski, J. B. (1975a). Vibration measurement and control. In J. B. Olishifski & E. R. Harford (Eds.), *Industrial noise and hearing conservation* (pp. 481–502). Chicago: National Safety Council.

Olishifski, J. B. (1975b). Sound level specifications. In J. B. Olishifski & E. R. Harford (Eds.), *Industrial noise and hearing conservation* (pp. 503–525). Chicago: National Safety Council.

Treplitzky, A. M., & Paolillo, A. W. (1984). Engineering aspects of noise control. In M. M. Miller & C. A. Silverman (Eds.), *Occupational hearing conservation* (pp. 28–61). Englewood Cliffs, NJ: Prentice-Hall, Inc.

Hearing Testing and Interpretation[1]

David M. Lipscomb

A distinction must be made between *identification audiometry* and *diagnostic audiometry*. These two purposes of hearing testing have their rightful places and make unique and important contributions. Industrial hearing testing employs identification audiometry exclusively. The goal there is to find those persons whose hearing has changed and to initiate requisite procedures to arrest the progression of hearing impairment—if possible. Part of the follow-up steps may well involve diagnostic audiometry, but industry is ill-equipped, inadequately staffed, and unprepared to offer full-fledged audiometric evaluations to all workers. Therefore, pure tone air-conducted

testing under earphones is the extent of identification audiometry one is likely to encounter in the industrial milieu.

Reasons for this are largely practical:

1. Cost. In industry, time is money. Therefore, to remove workers from their work stations for the length of time required to conduct any but the most basic hearing (and other health) testing will not meet with management's approval.

2. Staff. Personnel who conduct by far the largest number of industrial hearing tests are persons who are certified as "Industrial Hearing Test Technicians," having completed a two and one-half day training course sponsored by the Council for Accreditation of Occupational Hearing Conservationists (CAOHC) or the equivalent. These persons are not qualified to conduct any but pure tone air-conducted evaluations of hearing.

3. Resources. The industrial plant is not given to the development of extensive

[1]The author is indebted to the owners of Watauga Hearing Conservation, Inc. for their support and encouragement. They provided the wherewithall for developing some ideas for hearing conservation programs that had been conceived, but not brought to action. Further, a debt of thanks is offered to Mr. Russell Fankhouser for his creative development of forms comprising Figures 8-2 and 8-5.

audiometric facilities on its premises any more than medical capabilities exceed fundamental first aid treatment. For those needing more extensive hearing test workups, follow-up referral policies should be initiated.

4. Philosophy. Any activity or program that does not contribute to the infamous "bottom line" is viewed by management as being a burden, and therefore, managers feel it should be eliminated. Extensive hearing testing would fall into that category and a push to install diagnostic facilities might be countered with elimination or crippling of the hearing conservation program by management.

EQUIPMENT AND TECHNIQUES

Equipment

A more detailed discussion of facilities for industrial hearing testing is found in Chapter 17. Suffice to say here that adequately trained hearing test personnel, properly calibrated hearing test equipment, and an appropriate hearing test environment are essential to obtaining reliable hearing test data on personnel. These criteria are also appropriate for hearing conservation activities in a school system or in the military (see Chapters 13 and 14).

Equipment Selection

Often, the hearing conservationist encounters equipment already in-house and has no choice but to use those units. If equipment purchases for the purpose of hearing testing are to be considered, one of the first decisions to be made concerns the *type* of audiometer to be purchased. Generally, there are three classifications of hearing test equipment suitable for industrial applications.

Conventional Audiometers

These units come in a variety of degrees of complexity, but all have one thing in common: they are operated by a testing staff member one-to-one with the person under test. Much of the concentration in industrial hearing test technician training courses is on techniques of administering hearing tests using a conventional audiometer.

Self-recording Audiometers

These devices are often erroneously called "automatic audiometers." Originally designed by Dr. G. von Bekesy, the self-recording audiometer uses a variable attenuator which is activated by the person under test. Once the subject has been instructed on response protocol, the test begins. Test tones are presented to the ears in sequence and the subject "gives himself a test" by operating a response switch. When the switch is not activated, the test signal increases in intensity. Upon hearing the tone, the subject depresses the switch which reverses the variable attenuator and reduces the sound level until the tone is no longer heard and the subject releases the switch. A recording pen is attached to the motor-driven variable attentuator on specially printed forms. An advantage of this equipment is that numerous units can be located in the hearing test facility so that several persons can be tested simultaneously by a single technician. Drawbacks include the time required to conduct a bilateral hearing test (usually about 14 minutes), reliability of response, and accuracy problems in interpreting results.

Microprocessor-Based (Automatic) Audiometers

Rather new on the scene as a direct result of the technology explosion of the past decade, microprocessors are being built into audiometers. They are programmed to conduct hearing tests using similar protocol to that used in conventional audiometry. OSHA has recognized the existence of these devices, and in the 1983 hearing conservation guidelines indicated that one who administers hearing tests using this equipment does not need to meet the CAOHC training level or its equivalent. I take strong exception to that position. Industrial hearing testing encompasses far more than the ritual of administering the test; therefore, there is need for comprehensive training for all persons responsible for hearing testing, regardless of the type of equipment in use.

See Chapters 13, 14, and 17 for details on specific applications of hearing equipment. Also, useful reviews of hearing test equipment applications will be found in Miller and Silverman, 1984; Gasaway, 1985; and Grimes, Feldman, and Joseph, 1985.

Calibration

Audiometers are designed and manufactured to meet exacting standards with respect to all aspects of hearing test signal generation, amplification, and presentation (American National Standards Institute [ANSI] S3.6–1969). As is the case with electronic devices, some components are subject to change with use, resulting in imprecise output. Thus, it is imperative that periodic checks of the accuracy of the equipment be conducted.

Of primary concern here is the determination that audiometer "absolute output" meets standard reference values. This means that when the attenuator of the audiometer is set for 25 dB the output is 25 dB. A source of some confusion is the fact that two "references" come into play during the calibration process for hearing test equipment. One reference, sound pressure level (SPL), is based on the value of 20 μP for "0" dB (see Chapter 2). It is to this reference that sound measurement devices are calibrated. However, audiometer output is adjusted according to the normal curve of hearing, designated as Hearing Threshold Level (HTL)—sometimes cited only as HL. It is important to realize that the two types of devices, sound level meters and audiometers, are calibrated to different values. A correction is built into hearing test equipment to account for the nonlinearity of the ear's frequency response. More sound is required to evoke an auditory sensation in a normally hearing person when using low-

or high-frequency stimuli than is required for mid-frequency tones. In order to plot hearing test results on a straight line graph (audiogram) as shown in Figure 8–1, adjustments in audiometer output must be made to compensate for human hearing frequency nonlinearity. Therefore, a given setting of the audiometer attenuator does not mean that the same acoustic energy is output by the equipment checks and calibration are available to the hearing conservationist. All are intended to assure that the equipment is presenting test signals with accuracy within acceptable tolerance limits.

Listening Check

Each day, and occasionally during the day that hearing testing is underway, the hearing test technician or audiologist should listen to the test stimuli.

• Put the phones on, present audible test tones, and listen for any problem with the sound quality (fuzzy sound, distortion, static, etc.). The tone should be clean and free of contamination by other sound.
• Move the earphone wires while the tone is on to note any intermittency due to broken wires or poor connections.
• Present the tone and move the attenuator over a 50+ dB range to note any attenuator noise (scratchy sound through the earphones).
• Listen to the tone at a fairly high sensation level, press the interrupter switch several times, and listen for switching transients (pops, clicks, etc.).
• Switch the tone back and forth between ears to be sure that the earphones are connected to the proper ports. On occasion, reviewers of audiograms note that one year an employee had a severe loss in the right ear, only to find the next year that severe loss transferred to the left ear. Chances are, the earphones were switched or the plugs misplaced.

This quick check of the equipment only uncovers rather gross problems with the audiometer. To have some indication of the

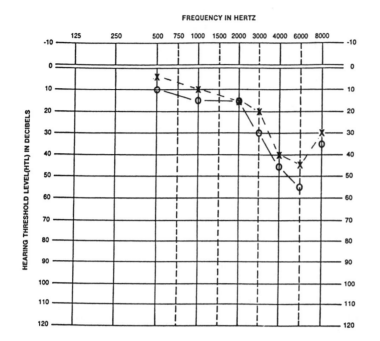
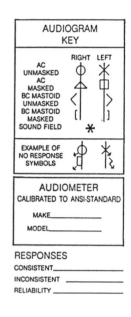

Figure 8-1. Audiogram of a person exposed to high-level noise. Note the characteristic configuration.

accuracy of the equipment, at least a biological check is necessary.

Biological Check

A person with stable hearing and who responds reliably to hearing test presentations should be identified and periodically tested to provide continuing assurance that the equipment is performing within limits. Keep on hand the records of repeat testing for this person so that the weekly or monthly biological calibration checks can be monitored. If there is a deviation from the biological baseline of the individual of more than 10 dB in any frequency for either ear, electrical calibration checks are indicated. The testing program should be put on hold at the time any deviation is noted. Any subsequent hearing test results using equipment of questionable calibration will be suspect.

Periodic Electrical Calibration

At least once each year, audiometric equipment should be subjected to objective

measures of absolute output to assess amplitude accuracy. This is accomplished using sound measuring equipment and a standard earphone coupler. University audiology clinics, consulting organizations, acoustic laboratories, and audiometer sales outlets or manufacturers are sources for this more sophisticated measurement. If equipment is out of calibration for absolute output, most of these facilities can make appropriate adjustments to restore output to proper calibration.

Exhaustive Calibration

According to the OSHA guidelines, an exhaustive calibration check must be conducted on hearing test equipment every 5 years, or sooner if indicated. In this even more sophisticated evaluation, the parameters of audiometer performance cited in the ANSI standard (S3.6–1977) are put to the test. These include: absolute output of the earphones, frequency accuracy, attenuator linearity, cross-talk between phones, and harmonic distortion of the output. Each of these measures must meet specifications set forth in the ANSI standard, or the equipment must be adjusted to perform up to specifications. Generally, manufacturers are asked to conduct this evaluation, but some university facilities, acoustic laboratories, and equipment outlets can adequately meet the demands of this most rigid of all calibration procedures.

Hearing Test Environment

Background noise must be kept to an acceptably low level during hearing testing in order to determine pure tone thresholds without masking influence from the noise. For clinical applications and diagnostic audiometry, the requisite background noise suppression is quite stringent. Those levels have been relaxed in industrial applications (Table 8–1).

Worker Noise Exposure and Hearing Health History

Noise is a significant contributor to hearing impairment of people living in most of the civilized world, but it is not the only one. Numerous other influences can create hearing problems and should be identified to the extent possible. Most hearing conservation programs utilize a case history form to help the conservationist discover any possible historical contributors to a worker's hearing condition. Usually, the worker is asked to complete the form while waiting to have a hearing test. The case history data are then kept on record and are updated as necessary. One occasional problem is encountered in that some employees are not

sufficiently literate to read and understand the form. These people must be guided through the form in interview fashion.

The Hearing Test

For the purpose of meeting OSHA regulatory guidelines, the frequencies to be presented to each ear are mandated to be 500, 1000, 2000, 3000, 4000, and 6000 Hz. In my view, it is highly recommended that 8000 Hz be added to that list. There are precious few clues to allow one to distinguish a noise-induced hearing impairment from some other etiologic factors. About the only consistent one is the well-known "notch" that bottoms out at 4k or 6k Hz with some improvement in hearing at 8k Hz (Figure 8–1). Without testing 8k Hz, the upward "slope" of the hearing test result may not be known and the distinction between possible causes is made even more difficult. History alone is not always sufficient to tie a hearing condition to noise exposure. For example, a physician in an American coal mining district found in 410 claimants for NIPTS that over 60 percent of the people manifested hearing conditions other than those attributable to noise (J. A. Spencer, personal communication, 1986). The more restricted frequencies required to meet OSHA regulations are a historical fact and we do not have any authority to amend the number of frequencies tested, but in discussing hearing test activities with employers, it is well to point out the value of testing 8k Hz. The time involvement is minimal, considering the value received.

In a text intended primarily for students in training to become audiologists, it would be a waste of space to provide details on how to obtain hearing test thresholds. OSHA offers no specifics as to how this should be done. The assumption is that once a person has an appropriate training background, and uses properly conditioned equipment in a quiet test environment, hearing test thresholds are assumed to be reliably obtained. The Council for the Accreditation of Occupational Hearing Conservationists (CAOHC) has published a

TABLE 8–1.
Permissible Background Noise Levels

	Octave-band Center Frequency				
Hertz	500	1000	2000	4000	8000
dB (SPL)	40	40	47	57	62

From *Federal Register*. (1983, March 8). *Occupational noise exposure; Hearing conservation amendment; Final rule*. 48(46). 9737–9785.

manual for training industrial hearing test technicians. The author, Dr. Maurice Miller (1985), has prepared a very useful guide for obtaining requisite information on industrial hearing conservation activities. That publication gives specific information on recommended hearing test procedures in the industrial setting. The manual may be obtained from:

CAOHC
66 Morris Avenue
P.O. Box 359
Springfield, NJ 07081
Attn: Mr. Richard Hall,
      Executive Secretary

Who Does the Testing?

The majority of hearing tests in industrial, military, and school settings are conducted by persons without expanded training or degrees in audiology. Particularly in industry, most hearing test personnel are products of brief training programs geared singularly to prepare them to assess pure tone air conduction hearing only. In an effort to assure some degree of consistency in hearing test technique and qualifications of the technicians, the CAOHC has designed a 20 hour training program for the technicians. Upon successful completion of a CAOHC-sanctioned training program, trainees may apply to the CAOHC for a certificate as a certified industrial hearing conservationist. CAOHC certification must be renewed every 5 years by the completion of a 1 day refresher course. All training courses under the auspices of the CAOHC are to be conducted by at least one person who is certified as a course director by the CAOHC. In addition, at least three representatives of associated specialties must be part of the instructional team for the course. The organizations to which those specialists belong include: the American Association of Occupational Health Nurses; the American Academy of Occupational Medicine; the American Academy of Otolaryngology— Head and Neck Surgery; the National Safety Council; the American Industrial Hygiene Association; the American Speech-Language-Hearing Association; the American Occupational Medical Association; and the Military Audiology and Speech Pathology Society.

Details of the course outline are found in the CAOHC manual (Miller, 1985). The following abbreviated content outline is intended to illustrate the general scheme of the industrial hearing test technician course, showing the relative emphasis of each topic by the time required for its presentation:

- Hearing conservation in noise (60 min)
- Anatomy, physiology, and disease of the human ear (60 min)
- Sound, psychophysics, and audition (60 min)
- Federal and state regulations relating to noise and hearing loss (60 min)
- The audiometer (90 min)
- Audiometric technique (60 min)
- The audiogram (30 min)
- Review—questions and answers (60 min)
- Supervised audiometric testing (150 min)
- Review of audiometric techniques (60 min)
- Principles of noise analysis (60 min)
- The occupational hearing conservationist in the industrial setting (60 min)
- Personal hearing protective devices (60 min)
- Record keeping (60 min)
- Review of hearing conservation program (60 min)
- Examination (60 min)

The training course is primarily an introduction to the subject with emphasis on hearing testing, and development of technique through practicum activities. Many of the topics covered in minutes constitute a full academic course in most audiology training programs. Thus, it is well not to expect too much of the trainees of this course beyond their being able to obtain adequate, but limited, hearing test results and to fit/issue hearing protective devices. Because the test personnel are limited in the scope of their knowledge, it is requisite that hearing test results be analyzed and interpreted by a person with more experience and training (e.g., a physician or audiologist).

Who Is to be Tested?

One purpose of sound level surveys (Chapter 4) is to identify those persons in the environment whose exposure appears to exceed a time-weighted average of 85 dB—the "action level." Of course, some employers or school systems may choose to provide an annual hearing test for all persons in their jurisdiction. That policy has its benefits, but the cost is increased due to higher levels of participation. For installations coming under the guidelines of OSHA, it is *requisite* that annual hearing testing be conducted for those whose exposure places them at the "action level."

A very important population to include in hearing test procedures are new employees—even young persons. One should not assume that a healthy-appearing new employee is blessed with normal hearing. High level noises in the nonoccupational environment and other etiologies can cause early hearing impairment in young people which must be detected prior to entry into the work place (Lipscomb, 1972). Previous exposure to occupational noise must be determined and recognized in new employees. Otherwise, it is possible that a given employer will ultimately be held responsible for a worker's total hearing impairment. Whenever possible, new employees should receive a hearing test prior to their first day at work. In many cases, this is a logistical problem. OSHA recognized that and offered a 6 month grace period for hearing testing or compensation claims. However, with respect to possible litigation or compensation claims, it is potentially dangerous not to test the hearing of new employees as early as possible in their employment.

The question often arises: Should people be hired if found to have pre-existing hearing impairment? This is a difficult policy decision to make. It is not possible to offer an answer that will be appropriate for all conditions in all settings. For example, from a human standpoint, it is unfortunate to disallow a person the chance for a livelihood in the name of hearing conservation. Yet, some persons can afford little or no progression in hearing impairment without suffering extreme problems with social adequacy in terms of speech communication. The person who is marginally able to use a telephone may no longer be able to converse by phone if his or her hearing impairment becomes worse by an average of 5 dB. It must be emphasized tht those persons who enter the work force with a pre-existing hearing impairment must be "flagged" for concentrated hearing conservation efforts.

Industrial hearing test surveys show that the growth of hearing impairment over a work life decelerates (Newby, 1979, p. 325). The first 5 years of a given noise exposure will cause more shift in hearing than will be noted in the subsequent 5 years of exposure. Later years will generally result in even less total hearing shift. This may suggest that one should hire persons with pre-existing hearing impairment because they will probably lose less hearing during their work experience. However, deceleration of NIPTS progression over time should not be taken as a license to expose older workers to high level noise without benefit of a comprehensive hearing conservation program.

In sum, it is appropriate to advise that, in general, hearing condition should not be made a condition of employment in terms of noise exposure alone. However, for those personnel whose job performance is negatively influenced by the inability to hear necessary sounds, alternate policies must be derived.

Preparation for the Hearing Test

It is an inadequate program that simply shifts workers into the test facility and back to the work station without appropriate precautions and preparation. A great contaminating influence on the accuracy and value of monitoring audiometry is the temporary threshold shift (TTS). To account for this, the OSHA guidelines specify that a worker should not be tested for hearing unless 16 hours have elapsed from the last time at the work station. On the surface, this seems like a fine idea because it seems to assure that all or most of any TTS will have dissolved by the time a hearing test is conducted. It does introduce certain logistic problems and makes an assumption that is tenuous at best. The 16 hour factor would dictate that workers be tested before returning to their job, meaning that only a limited number could be tested any given day, they are to be kept off their job until tested, or they are to be paid extra to come on non-work days. Further, this recommendation is based on the assumption that the worker lives in quiet in the interim between the last work time and the time of arrival at work the next day. Those who engage in high level nonoccupational noisy activities, or those who go home and return to work using noisy conveyances may actually arrive at work with more TTS than they would obtain during the work day. Therefore, one policy that has been put in force to counter the problems of TTS is to assure that each worker is outfitted with hearing protectors during work the day preceding the test. They are advised to avoid noisy activity prior to the test or to use hearing protectors if such activity cannot be avoided. Hearing protectors are also used at work the day of the test prior to the time the worker is called to the hearing test facility. This plan is certainly not foolproof, but it is considerably more promising than simply adhering to the 16 hour recommendation.

Another important phase in preparing the worker for a hearing test is to present simple but complete instructions prior to administering the test. Many workers have no idea what is expected of them when they enter the sound booth. They must be told what they will experience. They should know that one ear will be tested at a time, and that they should indicate when they hear the sound, even if the tone is very faint. Some test personnel prefer that subjects under test raise their hand with a rather large movement (right hand when the right ear is being tested, left hand when the left ear is under test). Others are happy with finger movement. Still others like to use indicator lights that signal response when the subject presses a button in the booth.

Instructions for self-recorded hearing testing are complex because the task is more complicated. In this case, the people are to be alerted that they have control over the signal and when they press (in Tennessee, we "mash') the button, the tone will become softer and go away, whereas when the button is released, the level of the tone will increase. A problem with self-recording audiometry is that response of some subjects is slow, resulting in large excursions of the recording pen and an inadequate number of threshold crossings at the test frequencies. Therefore, this problem must be addressed by requesting that the subject stay alert and press (mash) the button as soon as the sound is heard and release the button the moment it goes away.

It is curious how few people engaged in hearing testing are adept in the placement of earphones. Yet, clumsy earphone placement does harm to the subject's confidence in the tester, thus, in the hearing testing program. Test personnel should practice placing earphones on heads of various sizes with a single movement, making the adjustment of the headband at the same time. Upon removing the headphones, personnel should again be smooth. Pulling hair, letting the earphone snap onto a person's cheek, or snatching off hair pieces will damage the goodwill a hearing test program is intended to produce.

Otoscopic Examination

Most clinical audiologists use an otoscope to view the ear canal prior to hearing testing. This allows one to observe if obstructions in the canal will artifactually influence the test results. It further indicates whether the person under test has a collapsing ear canal which might cause lowered hearing response when the earphones are in place. The question arises as to whether industrial hearing test technicians or other test personnel should conduct a brief oto-scopic examination of subjects' ears prior to the test. Perhaps the best response to that question is to assess the feelings of the supervising physician. Use of the otoscope should be a matter of policy, and that policy is generally in the province of the medical department. Some physicians encourage otoscopic inspection prior to hearing testing. Others feel that the use of otoscopes is entirely the province of the physician. Once a statement of policy is obtained, then the question is answered.

Some Problems Encountered during Testing

Persons working in industry are not accustomed to being "tested." This is frightening to many because they may perceive the "test" as being some sort of threat to them or to their status as an employee. Most industrial workers are honest, but they wish to "look good." In their anxiety, some problems with subject response reliability may be encountered. For some workers, the first test is mysterious and they really do not know what to listen for, thus, their response may vary widely, especially when self-recording or microchip-based (automatic) audiometers are used. Re-instruction will normally accomplish the goal of obtaining a reliable response. On occasion, the jokester is encountered who thinks it is cute to mess up the test results. Sometimes, persons like this will have to be referred out because it would take an inor-dinant amount of time to obtain reliable hearing test results. Other times, the out-and-out malingerer will be found. As a general rule, if working with a given individual consumes more than twice the usual test time, skip the worker and retest later or refer to a hearing testing center.

At the conclusion of the test, it is more common than not for the employee to query: "How did I do?" This is not the time to present test results to the employee. Further analysis and review by other per-sonnel are necessary before a statement of hearing condition is sent or given to the person. Therefore, the suggested response is: "You responded just as I requested." This says nothing about the hearing test results, but actually responds to the inquiry directly. Chance comments such as: "Boy, you really have some hearing loss!" or "It doesn't look too good" serve no good purpose, exceeds the authority of the testing person, and has, in some instances, invoked legal action in the form of compensation claims. If the worker presses for more infor-mation, another appropriate "dodge" is to say: "The test results have to be reviewed by our audiologist or physician. You will be receiving word on the results of their evalua-tion within a month." OSHA specifies that the employee is to be notified within 20 working days if a Standard Threshold Shift was found. The point here is to avoid conducting a counseling session in the test facility. When dozens of people are to be tested each day, one cannot maintain appropriate efficiency by discussing each hearing test result at length.

INTERPRETATION OF HEARING TEST RESULTS

Magnitude of Hearing Impairment

The most obvious result obtained from hearing testing is a set of dB values that reflect the hearing of a person for a range of test frequencies. It is generally assumed that a person whose hearing is no worse than 25 dB (HTL) can be considered to have hearing "within the normal range." There are, of course, exceptions to that assumption because other factors can certainly influence the hearing function of one found to be on the low end of the normal range. Other descriptions of a hearing condition with respect to magnitude are needed when hearing falls outside the normal range. Unfortunately, there is no universal agree-ment on the parameters involved in describ-ing a hearing condition as "mild," "mild-to-moderate," "moderate," etc. In an effort to encourage some consistency among the hearing health professionals in the hearing conservation arena, I would like to propose wide use of the interpretative statements advanced by Grimes and colleagues (1985). Their hearing condition categories are included here as Table 8–2. The categories are described both in terms of the magnitude of any hearing impairment and also with respect to whether the person under test should be considered to have passed or failed the test. This initial determination contributes strongly to a consistent program of hearing test result interpretation.

OSHA Guidelines

It would seem on the surface that such simple hearing tests would make interpretation equally straightforward. However, that is not the case, because the test protocol includes single pure tones and the response required is binary in nature: ''I hear it'' or ''I don't hear it.'' The sensitivity of the test to loss of cochlear sensory cells is not as great as one would like. Therefore, industrial hearing test results must be interpreted in light of these shortcomings.

There are some guidelines offered in the OSHA publication of March, 1983 (*Federal Register*, 1983). These relate to the so-called Standard Threshold Shift (STS), defined as a change from the baseline hearing test (usually the first one in the hearing test history) of more than an average of 10 dB for the test frequencies of 2000, 3000, and 4000 Hz for either ear. This concept is quite appropriate because it utilizes those hearing test frequencies that are more sensitive to hearing changes resulting from noise exposure. These ''flag frequencies'' should be monitored closely in order to detect any progression of hearing from the baseline measures obtained early in the hearing test history of each worker. Some thought has gone into developing a single-frequency index for detecting noise-induced change, but no recommendations are currently available. At the time of this writing, a study group has been empaneled to attempt the development of such an index for both individual employee hearing assessment

and for evaluation of the adequacy of a given hearing conservation program.

Once STS has been noted, it is allowed by OSHA that the most recent test at which time the shift was detected be redefined as the new baseline. There is an inherent danger in following this recommendation literally. Without making note of the refinement of a baseline, the worker's hearing impairment might progress through several STSs with inadequate attention being paid to the problem. Thus, some steps should be followed at the time STS is noted:

1. OSHA guidelines state that the worker is to be informed within 20 days of the determination of the STS (this is taken to mean 20 working days or approximately one month).

2. Appropriate follow-up procedures should be initiated, including possible medical referral, retest, or other steps as warranted. (See Chapter 12).

3. Some notation should be made in the record of the worker, indicating the STS and the date it was detected. This can be accomplished manually, or, if computer data handling is utilized, the identification of STS can be programmed into the printout.

4. Some attempt should be made to determine the reason for a change in hearing. Often, it is due to some nonoccupational or medical reason. Over the years, the aging process reduces hearing sensitivity to higher frequency test stimuli and that may be the cause. If the STS can be pinned to such failures in the hearing conservation program as inadequate hearing protectors, lack of protector use, or exposure to sound without the benefit of hearing protective devices, a solution to the causative factor can usually be instituted, reducing the likelihood of further progression of hearing impairment due to noise exposure.

After all, the STS is actually what is being sought in hearing conservation audiometric monitoring programs. Although it is discouraging to note when one occurs, the program cannot be considered adequate if an STS is not noted and then treated

TABLE 8-2.
Categories of Hearing Conditions

Category	Description
1. Normal hearing	**Hearing thresholds are within limits established for normal hearing;** i.e. no worse than 25 dB at any frequency
2. Mild to moderate high-frequency loss	**Hearing for communication purposes is essentially unimpaired.** There is hearing loss present in the high frequencies. Category 2A denotes no loss greater than 25 dB at 500, 1000, 2000 Hz, and no worse than 50 dB in the higher frequencies. Category 2B exceeds 50 dB at 4000, 6000, and/or 8000 Hz. Very few losses in category 2A or 2B will be compensable
3. Moderate to severe high-frequency loss	**Provisional Pass. Significant hearing loss exists.** The adequacy of hearing for communication purposes is questionable and the individual is borderline for aural rehabilitation. There is no threshold worse than 25 dB at 500, 1000 and/or 2000 Hz, but it is 55 dB or more at 3000 Hz. Most losses in this category will be minimally compensable
4. Possibly medically related hearing loss	**Fail. Significant hearing loss exists of undetermined type and origin.** The individual should be referred for complete examination. Thresholds exceed 25 dB at 500, 1000 or 2000 Hz. This category has a high potential for compensation if not identified prior to employment
5. Inconsistent test	**Test responses inconsistent.** A retest is indicated since better test results are necessary for the reviewer to make reliable interpretations
6. Previously professionally evaluated hearing loss	**Significant hearing loss exists which is known to the individual.** The employee has seen a professional about the hearing status. A copy of the professional's (physician/audiologist) report should be obtained for the employee's record
7A. STS: further professional evaluation optional*	**Significant change from baseline audiogram; no referral needed.** Additional professional evaluation is not likely to provide further helpful information. If exposed to noise on the job, employee should be rechecked and reoriented about hearing protection use
7B. STS: further professional evaluation is advised*	**Significant change from baseline. Should be referred for further professional evaluation.** This change is likely to be other than noise related. If exposed to noise on the job, employee needs to be rechecked and reoriented about hearing protection use
8. No change	**No significant change from baseline audiogram.** Original designation applies

From Grimes, C. T., Feldman, A. S., & Joseph D. (1985). Audiometric testing, review and referral: Protocol and problems. In A. S. Feldman & C. T. Grimes (Eds.), *Hearing conservation in industry* (pp. 164–177). Baltimore: Williams & Wilkins. Reprinted with permission.

*Criteria for category 7A and 7B are 10 dB or more (average) loss at 2000, 3000 and 4000 Hz, and/or 25 dB or more at any frequency.

appropriately. A somewhat higher exposure baseline (90 dBA for 8 hours) was retained in the final promulgation of OSHA industrial hearing conservation guidelines because monitoring audiometry was to be introduced to workers exposed to a lower level (action level). The STS provides the most useful component in that monitoring provision.

Identification of Other Hearing Problems

Review of audiometric results is relegated to either physicians or audiologist because it is presumed that they will be able to detect subtle indications of problems that warrant medical and/or remedial action. In the extreme, the development of unilateral hearing deficit can be a harbinger of central auditory pathway complications such as acoustic neuromas. Low-frequency hearing threshold changes might signal a medically or surgically correctable conductive hearing impairment. Metabolic imbalances in the body can create hearing impairment. Metabolic imbalances in the body can create hearing deficits in one or both ears which should be identified and treated when they are first noted. Thus, other auditory anomalies can be recognized in the review of monitoring audiometry test results as well as noise-related hearing shifts. For this reason, it is not uncommon to note that otologic medical practitioners are developing industrial services for the purpose of providing early identification of treatable hearing conditions. Doubtless, the advent of comprehensive hearing conservation activities, including monitoring audiometry, has enhanced early recognition of serious and sometimes life-threatening conditions. This fortunate by-product of the monitoring audiometry program is a bonus not to be overlooked by either the consultants serving industry or by management.

At the beginning of this chapter, the point was made that monitoring hearing test programs are limited in scope because their purpose is "identification audiometry." Yet, there can be useful clues of diagnostic value:

1. The hearing test history will indicate whether time-related changes in hearing are occurring, suggesting the need for more rigid control of a person's noise exposure (both occupational and nonoccupational).

2. Comparison of hearing response characteristics between ears can offer indications of problems deserving medical management or possible misfitting of hearing protection for one ear.

3. Configuration of hearing responses can offer further indication of possible causative factors.

All of this information can be gained with only pure air-conducted testing. Thus, sharpening awareness of a range of interpretations will increase one's ability to make appropriate recommendations.

Referral Policies

In all but a relatively small number of settings, it is necessary to refer certain of the tested people for more extensive evaluation of the hearing condition. Several concerns must be addressed.

Who Pays?

At what point has the hearing testing program discharged its responsibilies with respect to financial support of hearing evaluation? This is also a matter of policy. Some hearing conservation programs are comprehensive in that follow-up procedures will be financed by the program. Others take the position that identification is the limit of the program and remediation or correction is the responsibility of the person. Still other programs take some middle ground position. This policy should be clear to hearing test personnel and to the people being tested. When referral is made for some follow-up activity, the policy statement should be included in the referral. It stands to reason that people will tend to act on the recommendation for additional testing or treatment if the cost will not be theirs to bear. However, comprehensive medical follow-up of workers in the industry is the exception rather than the rule.

Over Referral

The fable about the little boy who cried ''wolf'' comes to mind. A ''shotgun'' approach to referrals will be effective only the first year it is in place. Because most persons will have to bear the expense of follow-up procedures, they become disenchanted when they spend sizable amounts of money for medical examinations or comprehensive hearing evaluations, only to be told that there is nothing that can be done. If the recommendation is made repeatedly, it will be disregarded. Then, if a suggestion for follow-up is made for good cause, it is likely that the person will not heed the advice. The fear of missing a small number of persons who need referral may lead to a policy of over referral. However, that policy will not be effective in the long run.

Under Referral

Miller and Silverman (1984) specify two pass/fail criteria:

1. Criteria based upon the results of monitoring audiometry
2. Criteria derived from case history information

Armed with these two sources of information, one should formulate policies that will assure that the greatest proportion of persons needing some follow-up action receive the recommendation while minimizing the amount of over referral. Earlier, mention was made of a comprehensive guide for interpretation of the magnitude of hearing impairment which can also direct one's referral policies (see Table 8–2). The primary danger of under referral is missing a person for whom adequate follow-up is requisite to health or job performance. In such cases, there is a degree of liability assumed by the program. Occasionally, serious, sometimes life-threatening, conditions are identified by hearing test programs.

It is essential that appropriate referral be made in these cases.

Percent Hearing Impairment

Theoretically, calculation of percent of hearing impairment makes little or no sense. The person with a 30 percent hearing loss is far more than twice as impacted as one with 15 percent loss. Yet, percentage of impairment is a necessary fact of life, escpecially in the hearing conservation area. Compensation awards are usually based on the calculation of hearing impairment, which is then adjudicated and finalized by the use of a compensation formula. All jurisdictions do not utilize the same method for calculating the compensation award, nor do they all use the same procedure to calculate percentage of hearing impairment. Most, however, are similar to the method advanced by the American Academy of Otolaryngology in May of 1979:

- For one ear:
  - Step 1: Find the average pure tone thresholds for 500, 1000, 2000, and 3000 Hz
  - Step 2: Subtract 25 dB from that average
  - Step 3: Multiply the remainder by 1.5 to obtain the percentage of hearing impairment for that ear
- For the other ear:
  - Repeat the previous three steps
- For binaural percentage of hearing impairment:
  - Step 1: Multiply the better ear percentage by 5
  - Step 2: Add the percentage of the poorer ear
  - Step 3: Divide the total by 6 to determine the binaural percentage of hearing impairment.

Problems in Hearing Test Result Interpretation

Although a surprising amount of useful information can be gleaned from even limited monitoring audiometric information, it is well to counsel that one proceed with caution. The pure-tone hearing test using only air-conducted signals provides no more than partial information requisite to adequate diagnosis. An audiometric maxim is that *both* air- and bone-conducted pure tone test results are necessary for distinctions between sites of lesion peripheral or central to the footplate of the stapes (conductive and sensory/central impairment respectively). Magnitude of sensory affectation is usually not known without speech audiometry. Further distinction of site of lesion must depend on other tests, sometimes quite sophisticated ones.

A further limitation posed by the type and scope of testing undertaken in the hearing conservation program is that pure tones are not consistently sensitive to cochlear sensory hair cell damage. Hearing test frequencies 2000 Hz and higher seem to be quite sensitive to even small patches of sensory hair cell damage in the basal third of the cochlea. But, lower test frequencies are limited in use as indicators of actual hair cell damage unless the destruction is sufficiently extensive to influence the mechanical response of the inner ear to stimulus sounds in that frequency range (Pollack & Lipscomb, 1979). The emerging awareness of differential sensitivity of hearing test stimuli to locations of hair cell lesions came as the direct result of proliferation of histological studies of the past two decades (summarized in Lipscomb, Axelsson, Vertes, Roettger, & Carroll, 1977). This lack of sensitivity for many of the audiometric test frequencies reinforces the importance of concentrating on changes in response reflected in higher frequency test results (3000, 4000, and 6000 Hz). For those who engage in otological or audiological clinical practice, recall the patients who manifest only slight loss of hearing for frequencies above 1000 Hz and who have hearing within the normal range for the middle and low frequency test tones. It is common for these persons to present with complaints of hearing problems that are greater than one might assume with no more hearing impairment than these people seem to have. Sometimes, it is tempting to

classify them as "otoneurotic," thinking that they have slight hearing problems but they are making far too much of them. However, if these people have suffered hair cell loss which is not causing depression of hearing for the frequencies below 2000 Hz, they can be having considerable difficulty in difficult listening conditions, for example, hearing speech from a distance or detecting speech in the presence of background or competing noise. So, rather than express some type of emotional overstatement of their hearing condition, these patients may be relating more valuable clinical information than one might think. The measurement of so-called minimal hearing impairment is in its infancy, but it will become a useful adjunct for future evaluation of a person's auditory adequacy.

In summary, interpretations of hearing conservation program monitoring audiometry are handicapped because the testing is narrow in scope. However, there are several types of valuable information to be obtained:

1. Progression of hearing impairment can be noted.

2. Preliminary "diagnostic" indications can be detected by the trained eye.

3. Paying careful attention to test results for frequencies above 2000 Hz (the "flag frequencies') can give some hint of the potential communication ability impact the person might be experiencing.

RECORD KEEPING

Details of required data retention are spelled out specifically in the OSHA guidelines and are outlined in Chapter 12. The regulation does not determine how one is to go about the record keeping chores, and at times, this single element of a hearing conservation program can appear overwhelming. For example, consider that hearing test results contain information for as many as seven frequencies for each ear. Testing both ears means that the hearing test totals 14 data points. Demographic information (name, I.D. number, work area, date of test) adds at least four more data points for each worker. In a moderately sized company that employs 1,000 workers who qualify for hearing conservation work, the total number of data points obtained each year is in excess of 18,000. Add to that all of the sound measurement information, audiometer calibration records, hearing test environment acoustical quantification, records of hearing protectors issued (and reissued), citations of noncooperation by any workers, and personal medical history for each employee. For a modestly sized program like the one described here, after 10 years in operation, there will be close to 200,000 data points to organize and retain.

Early in the history of hearing conservation, the sheer weight of numbers would have mediated against adequate record keeping by any but the few firms with comprehensive computer capabilities. Now, the appearance of personal computers (PCs) with astounding power means that data storage and manipulation are within the reach of even the smallest employer. There are a growing number of hearing conservation management programs written for the PC which essentially removes the problem of paperwork from management. In fact, it should be the intent of each person who consults with industry or school systems to offer a service program that includes record keeping so that the operatives of the hearing conservation program are released from the burden of paperwork (Lipscomb & Fankhouser, 1986).

Audiometer Calibration Records

An example of audiometer calibration forms is shown as Figure 8–2. Each time some form of calibration procedure is accomplished, the information should be recorded and the form placed in an appropriate storage location on site.

Hearing Test Environment Assessment

The allowable background sound level in hearing test enclosures was shown in Tabel 8–1. These requisite levels are taken from the OSHA guidelines and should not be interpreted as being appropriate for clinical diagnostic settings. The sound level in each hearing test enclosure should be measured at least once each year, preferably at the outset of the annual hearing test program. The information should be filed for ready access by consultants or by enforcement officials.

Hearing Test Records

As indicated, great quantities of numbers on many sheets of paper can accumulate if this phase of the hearing conservation program is not streamlined. The typical audiogram form has largely been replaced by the *serial* audiogram form. Figure 8–3 combines audiometric data and hearing health history information. There is room for recording 10 annual hearing tests.

Once hearing testing has been completed for an employee each year, most programs input the data into a computer. Hearing conservation computer programs vary rather widely and reflect differing philosphies. There are two major categories. The best of both categories are sometimes provided in large and comprehensive computer programs.

Global

In this type of program, individual records are combined to provide a summary of the total hearing test program, including: the number of persons tested, the number of STSs noted, categories of hearing condition (normal, mild, moderate, etc.), the number of medical referrals recommended, etc. In such a summary, the individual is lost, but management obtains a very useful overview of the hearing conditions pertaining to their work force.

Individualized

The computer program prints out information pertaining to each individual's hearing test data and interpretations of the results (Figure 8–4A). Programs of this type

Watauga Hearing Conservation, Inc.
208½ East Watauga Avenue
Johnson City, TN 37601
615-928-1901

Audiometer Calibration Certification
For Puretone Air Conduction Output

Calibration Instrumentation: Precision Type 1

Sound Level Meter: Bruel & Kjaer 2209	S/N: 477741
Octave Band Filter Set: Bruel & Kjaer 1613	S/N: 480667
Artificial Ear: Bruel & Kjaer 4152	S/N: 490341
Microphone: 4144 audiometer calibration	S/N: 499223
Pistonphone: Bruel & Kjaer 4220	S/N: 501508

Personnel: _____   Date: _____

Audiometer: _____   Model: _____   S/N: _____

Earphone: AD ☐ AS ☐ TDH-50 ☐ TDH-49 ☐ TDH-39 ☐   Audiometer Dial Set: 70 dBHL

Sound Level Meter Readings:

Frequency in Hertz	125	250	500	1000	1500	2000	3000	4000	6000	8000
Actual* SPL in dB re: 20 uPA										
ANSI 1969 SPL in dB TDH-50/49 Earphone__	117.6	96.7	83.5	77.4	77.6	81.1	79.7	80.7	83.4	83
TDH-39 Earphone__ re: 20 uPA	115	95.5	81.5	77	76.5	79	80	79.5	85.5	83
Difference in dB										
Corrections										

*Actual - includes mic correction factors in dB @ 3K,-.5; 4K,-1; 6K,-1; 8K,+1

Linearity Check @ 1K & 2K: YES ☐ NO ☐   Meets ANSI S-3.6 1969 YES ☐ NO ☐

Figure 8-2. Audiometer calibration form. Printed with permission from Watauga Hearing Conservation, Inc.

are most useful if the total hearing test history is listed on the form along with appropriate demographic data and test result interpretation. If the program includes a modest interpretative section, so much the better. It would save many hours of writing out interpretations. Marvelous as they are, computers cannot think. Therefore, it is common for the program to allow space for some type of individualized comments by the qualified person who will be reviewing the hearing test results. That person should be identified on the printout.

Policies regarding counseling personnel about their hearing test results vary widely. Some locations arrange individual face-to-face interviews on an annual basis during which time discussions take place regarding health data obtained, job performance, salary adjustments, etc. Others prefer to notify employees by letter or memo. In either case, it is likely that a personalized letter to each employee will be beneficial to supplement any other contact the employee may have. It also provides a nice touch. Some hearing conservation computer programs provide a "letter" to each employee (Figure 8-4B).

The letter to the employee must contain language that informs but does not cause anxiety or anger. Interestingly, we have found that four messages are sufficient to communicate the range of information gained in hearing conservation hearing test

WATAUGA HEARING CONSERVATION, INC.

215 EAST WATAUGA AVENUE
JOHNSON CITY, TENNESSEE 37601
615/928-1901

AUDIOGRAM RECORD

Use ink only

FULL NAME (LAST, FIRST, MIDDLE) | CLOCK NO | BIRTH DATE | SERVICE DATE | DEPT NO | SHIFT | SOC SEC NO | COMPANY

DATE	RIGHT EAR 500	1000	2000	3000	4000	6000	8000	LEFT EAR 500	1000	2000	3000	4000	6000	8000	SERIAL NUMBER STANDARD	JOB TITLE	YEARS ON PRES JOB	NOISE LEVEL dBA	HRS PER DAY	EXPOSE LAPSE	PROTECT USED	TESTER (SIGNATURE)	RE-TEST

HEARING HISTORY

Is your hearing: ☐ Good ☐ Fair ☐ Poor

Does anyone in your family have a hearing loss? ☐ Yes ☐ No If yes, who

Have you ever had your hearing tested? ☐ Yes ☐ No If yes, when and where

Have you ever had any ear infections? ☐ Yes ☐ No If yes, explain

Have you ever been exposed to gunfire? ☐ Yes ☐ No Have you been in the military? ☐ Yes ☐ No

Have you ever had 'surgery on either ear? ☐ Yes ☐ No If yes explain

Have you ever had: Mumps ☐ Yes ☐ No, Measles ☐ Yes ☐ No, Chickenpox ☐ Yes ☐ No, Scarlet Fever ☐ Yes ☐ No

Other infectious diseases: explain

Have you ever worked at a noisy job? ☐ Yes ☐ No If yes: where ____ Did you wear hearing protection? ☐ Yes ☐ No

Dates

Have you ever had dizziness? ☐ Yes ☐ No If yes, describe

Have you ever had noises in your ears? ☐ Yes ☐ No If yes, describe

Do you hunt? ☐ Yes ☐ No Do you ride a motorcycle? ☐ Yes ☐ No

Any other noisy hobbies? ☐ Yes ☐ No If yes, describe

Do you have a second job? ☐ Yes ☐ No If yes, where

Have you ever taken Kanamycin ☐ Yes ☐ No, Quinine ☐ Yes ☐ No, Streptomycin ☐ Yes ☐ No, Garamycin ☐ Yes ☐ No

To the best of my knowledge the above information is true. (Signed)

(Use reverse side for any comments)

Figure 8-3. Serial audiogram and hearing health history form. Courtesy Watauga Hearing Conservation, Inc.

REPORT OF HEARING TEST INTERPRETATIONS
FOR
XYZ MANUFACTURING COMPANY

NAME: SAMPLE, IMA   IDENTIFICATION NUMBER: 000-00-000   BIRTHDATE: 2/29/51

DATE	RIGHT EAR							LEFT EAR						
	.5K	1K	2K	3K	4K	6K	8K	.5K	1K	2K	3K	4K	6K	8K
8/19/82	5	5	10	5	10	15	15	0	0	0	5	5	10	10
7/30/83	5	10	10	10	15	15	20	5	5	5	10	10	15	20
8/4/84	10	10	20	25	30	30	25	5	5	10	15	25	30	25
SHIFT	5	5	10	20	20	15	10	5	5	10	10	20	20	15

BRIEF SUMMARY OF THE MOST RECENT TEST:

A STANDARD THRESHOLD SHIFT HAS BEEN NOTED FOR THE RIGHT EAR.
A STANDARD THRESHOLD SHIFT HAS BEEN NOTED FOR THE LEFT EAR.
REVISED BASELINE TEST: 8/4/84

PERCENTAGE OF HEARING IMPAIRMENT:
RIGHT EAR: 0 % -- LEFT EAR: 0 % -- BINAURAL: 0

INTERPRETATION OF MOST RECENT HEARING TEST

There is a mild high frequency hearing impairment.
The use of hearing protection is strongly advised.
NOTE -- There has been a shift from the baseline test of
an average of 10 dB or more. If this is the first time a shift
of this magnitude has been noted, be sure to concentrate on
this worker. Whether the shift is due to work related noise or
other causes, steps for prevention are advisable.

ADDITIONAL COMMENTS:

DATA REVIEWED BY:

Code 2
Code 4   **A**

TO: IMA   SAMPLE

EMPLOYEE I.D. NUMBER: 000-00-000

The hearing test you received is part of our comprehensive hearing
conservation program. The purpose of this effort is to monitor your
hearing from year to year and to recognize if there is any change.
In the case of hearing disorders, as with many other physical
problems, early identification is important.

For your most recent hearing test, the results can be summarized
as follows:

The hearing test shows you have some reduction in hearing
sensitivity. Use of hearing protection is mandatory while you are
working in high noise areas. Although we have no control over your
activities off the job, we strongly urge that you avoid excessive
noise exposure then and use hearing protection if the noise cannot
be avoided.

We have noted that your hearing has changed since your
baseline hearing test. Since there are many causes resulting in
hearing decrease, we suggest that if you haven't done so recently,
you should consult an ear specialist (Otolaryngologist) to determine
if there is a medical reason for the change.

We urge that you comply with these recommendations. If you have
questions regarding this information, feel free to contact the   **B**
Personnel Office.

Figure 8-4. (A) Hearing test history example printout showing results of audiometric results and certain interpretative notations. (B) Example letter composed by the computer to be given to the worker whose hearing test history appears in Figure 8-4A.

activities (Figure 8-5, codes 1-4). The computer program I use has a system of ''codes'' to indicate which messages have been included in the letter to the employee. These codes are indicated on the hearing test history printout (see Figure 8-4A, lower left corner). Therefore, it is possible for the director of a hearing conservation program to know precisely what information has been given to the employee with each letter. To facilitate use of the coded information, an interpretation of test results letter is presented to each hearing conservationist for whom computer work is conducted (Figure 8-5). This information summarizes the pertinent features of the computer

MESSAGES TO EMPLOYEES:

Code 1:

We are very pleased to find that your hearing is within the normal range. In view of this, we urge that you use hearing protective devices when in high noise areas both on and off the job.

Code 2:

The hearing test shows you have some reduction in hearing sensitivity. Use of hearing protection is mandatory while you are working in high noise areas. Although we have no control over your activities off the job, we strongly urge that you avoid excessive noise exposure if the noise cannot be avoided.

Code 3:

Our audiological analysis further suggests that you see an ear specialist (Otolaryngologist).

Code 4: (STS)

We have noted that your hearing has changed since your baseline hearing test. Since there are many causes resulting in hearing decrease, we suggest that if you haven't done so recently, you should consult an ear specialist (Otolaryngologist) to determine if there is a medical reason for the change.

ADDITIONAL EMPLOYER AUDIOGRAM INFORMATION:

StandardThreshold Shift (STS)

Is defined as an average shift of 10 dB or more at 2K, 3K, and 4K relative to the baseline in either ear. If identified, the following measures are recommended in the hearing conservation mandate by OSHA:

1. Employee must be fitted or refitted with adequate hearing protection.
2. Required to wear the protection.
3. Employee must be notified within 21 days from time of determination of STS (Code 4).
4. If subsequent tests show the STS is not persistent, employees whose exposure is less than a TWA of 90 dBA may discontinue wearing hearing protection.

HEARING THRESHOLD LEVELS:

        -10 to 25 dB   —   Normal Range
        26 to 40 dB   —   Mild Hearing Loss
        41 to 55 dB   —   Moderate Hearing Loss
        56 to 70 dB   —   Moderate/Severe Hearing Loss
        71 to 90 dB   —   Severe Hearing Loss
        91+ dB   —   Profound Hearing Loss

A threshold recorded as 99 on the computer audiogram denotes no response by the employee at the limits of the audiometer (90 dB). Thus, a threshold shift from or to 99 could be greater than is represented.

AU represents both ears. AD represents the right ear. AS represents the left ear.

REFERRALS ARE SUGGESTED WHEN:

1. A standard threshold shift has been noted.
2. A low frequency (.5kHz &/or 1kHz) hearing impairment has been noted.
3. A unilateral hearing impairment has been noted.

Figure 8-5. Interpretation of information given in the computer printout shown as Figure 8-4A.

printout information and provides a clear understanding of some of the rather technical features included in the data. The hearing conservationist can review hearing test printouts (figure 8–4A) along with the explanations found in Figure 8–5 and gain an understanding of the information contained in printouts for each employee.

CONCLUSION

The audiometric testing program is the backbone of any hearing conservation program. It provides valuable indications of the success of the program in achieving the goal of hearing preservation among personnel. Further, it can be used to identify areas of the program that are failing to meet that stated objective.

REFERENCES

American National Standards Institute. (1969). *Standard for audiometers.* (ANSI S3.6–1969). New York: American National Standards Institute.

American National Standards Institute. (1977). *Criteria for permissible ambient noise during audiometric testing.* (ANSI S3.6–1977). New York: American National Standards Institute.

Federal Register. (1983, March 8). *Occupational noise exposure; hearing conservation amendment; Final rule.* 48(46), 9737–9785.

Gasaway, D. C. (1985). *Hearing conservation: A practical manual and guide.* Englewood Cliffs, NJ: Prentice-Hall, Inc.

Grimes, C. T., Feldman, A. S., & Joseph, D. (1985). Audiometric testing, review and referral: Protocol and problems. In A. S. Feldman, & C. T. Grimes (Eds.), *Hearing conservation in industry* (pp. 164–177). Baltimore: Williams & Wilkins.

Lipscomb, D. M. (1972). Environmental noise is growing—is it damaging to our hearing? *Clinical Pediatrics, 11*(7), (July), 374–375.

Lipscomb, D. M., Axelsson, A., Vertes, D., Roettger, R. L., & Carroll, J. (1977). The effect of high level sound on hearing sensitivity, cochlear sensorineuroepithelium and vasculature of the chinchilla. *Acta Otolaryngolica, 84*, 44–56.

Lipscomb, D. M., & Fankhouser, R. J. (1986). Computer technology in hearing conservation programs: Schools and industry. In J. L. Northern (Ed.), *The personal computer for speech, language and hearing professions* (pp. 135–155). Boston: Little, Brown and Company.

Miller, M. M. (Ed.). (1985). *Council for accreditation in occupational hearing conservation manual* (2nd ed.). Springfield, N.J.: Association Management Corporation.

Miller, M. H., & Silverman, C. A. (1984). Audiometric and audiological aspects of occupational hearing conservation programs. In M. H. Miller & C. A. Silverman (Eds.), *Occupational hearing conservation* (pp. 151–181). Englewood Cliffs, NJ: Prentice-Hall, Inc.

Newby, H.A. (1979). *Audiology.* Englewood Cliffs, NJ: Prentice-Hall, Inc.

Pollack, M. C., & Lipscomb, D. M. (1979, February/March). Implications of hair cell-pure tone discrepancies for oto-audiologic practice. *Audiology and Hearing Education,* pp. 16–23.

9

CHAPTER

Management of Hearing Conservation Data with Microcomputers

John R. Franks

The moment that an audiogram is taken for anyone, the question immediately arises, ''What do I do with this now?'' The answer depends upon the environment in which the audiogram was administered, but it usually distills to the essence of, ''File it!''

An audiogram may be viewed as a record containing information about the person to whom a hearing test was administered. The need to file the audiogram, in such a manner that it may be retrieved as some time in the future, creates the need for a data management system. This chapter addresses the means by which microcomputers may be used for data management systems.

THE MICROCOMPUTER

A microcomputer suitable for managing a hearing conservation database may be defined as a computing system that contains no less than the following components:

1. Adequate Random Access Memory (RAM) to hold the main program and a reasonable amount of data.

2. A Disk Operating System (DOS) that will support the use of at least two disk drives.

3. A Basic Input/Output System (BIOS) that will support the use of a keyboard, floppy or fixed disk drives, and at least one serial communications port. The BIOS must also support the output to a printer and the display screen.

4. A typewriter-style keyboard that allows easy entry of text in both upper and lower case, as well as entry of numbers. The keyboard should also have screen cursor control keys.

5. At least two disk drives. If dual-floppy diskette format is employed, the

---

Mention of the name of any company or product does not constitute endorsement by the National Institute for Occupational Safety and Health.

programming instructions usually reside on the default drive (primary drive), and data will be written to and read from the secondary drive. If a fixed-disk drive is employed, it will be the default drive containing both programming instructions and data and the floppy drive will be used for copying in new programming instructions, transporting and importing new data from other computers, and for backing up data files from the fixed-disk drive.

6. A display screen capable of at least 24 lines of characters in line widths of 80 characters.

7. A parallel port to send output to a printer.

8. A serial port to send output to another computer or to receive information from another computer.

The types of computers that are included are almost all microcomputers introduced since 1981 which rely on versions 2.xx and 3.xx of the MicroSoft Disk Operating System (MS-DOS)*. Additionally, the newer Apple MacIntosh** computers and any microcomputer capable of running a variation of UNIX*** operating system are included. At the end of the chapter, Appendix 9–1 lists commercially available hearing conservation data management computer programs along with the operating system environments for which they were written.

WHY A MICROCOMPUTER?

What makes a microcomputer unique from the older style mainframe or large computer is that it can be made accessible directly to the user such that the user may quickly see the results of his or her efforts. Until the proliferation of microcomputers, most computer systems were not tailored to interact with the user, since that took up

*MS-DOS is a registered trademark of Microsoft Corporation.
**Apple and McIntosh are registered trademarks of Apple Corporation.
***UNIX is a registered trademark of American Telephone and Telegraph.

valuable time the computer could spend doing other tasks.

On mainframe systems, the user put in a program, then put in the data to be manipulated by the program, and waited for the results to show up on a printer someplace. This seemed to take a lot of time. The time was really taken up because the program and the data submitted to the computer were waiting in line to be processed (in the input queue) and because the results were waiting in line to be printed (in the print queue). The actual processing of the program and data happened rapidly on a large mainframe computer.

Most microcomputers are not time-shared, but are used by one person at a time, hence the term *personal computer* or *PC*. However, when microcomputers are time-shared, they can slow down immensely. When data are put in at the keyboard, they appear to be instantly accepted. When a printout is requested, it appears instantly. The definition of instant depends upon the user's state of mind. Most people are happy if the computer can accept data faster than they can type it in. They are happier yet if the computer can print a report faster than they can read it.

The microcomputer is dedicated to only the user's task and responds only as the user asks. This sense of control and self-esteem provided to the user makes the microcomputer a very desirable tool. Since the microcomputer is framed around input and output to communicate with the user, it lends itself to management of data such as audiometric records. The cost of microcomputers has been declining at a rate of 16 percent per year since 1981, so they have become affordable data management tools for almost anyone with data to manage.

REQUIRED RECORD KEEPING

Keeping information pertinent to a hearing conservation program by using the microcomputer as an aide can make large volume data analysis a relatively simple task. Records can be kept in any fashion, from highly organized in files to random placement in shoeboxes without the computer. However, the computer can't be as flexible.

A record is any collection of data kept about any one item. A simple audiometric record might be a person's name and his or her pure tone air conduction audiogram. The complexity of the record may be increased by the addition of the date the record was created. This would allow one to chronologically sort audiograms for all persons with the same name. An increase in the complexity by the addition of a unique identification number, such as the social security number (SSN), would allow all audiograms to be chronologically sorted for each person for whom a record existed and be displayed in alphabetical order of the names of the persons, but with no name confusion.

Before going on to increase the complexity of the record, it may be helpful to see what is required of audiometric records. Paragraph m of 29 Code of Federal Regulations (CFR) 1910.95 (the Hearing Conservation Amendment) mandates four areas of record keeping:

1. Exposure measurements
2. Audiometric tests
3. Audiometric room tests
4. Audiometer calibration

Exposure Measurements

The employer is required to maintain accurate records of employee exposure measurements. The employer must maintain a record of the names and job classifications of all employees for whom exposure measurements were made and of all other employees whose exposures the measurements represent. In addition, the employer must maintain records of the date, location of measurements, number of measurements taken using sound level meters, and a description of the noise measurement equipment used during the last laboratory calibration.

If this list were thoroughly maintained,

it would, for example, be possible to go into the records and pull a list of all employees exposed to time-weighted averages (TWAs) from 95 to 100 dB. Many companies have elected to shortcut and use a two-tiered system. Those employees with TWAs of 85 dB or greater are included in the mandatory hearing conservation program, and those with TWAs of less than 85 dB are not. Although this reduces record keeping demands, it also reduces the effective use of the overall record set.

Audiometric Tests

The audiometric test record must include no less than the following:

1. The name and job classification of the employee
2. The date of the audiogram
3. The examiner's name and qualifications, such as certification number from the Council on Accreditation of Occupational Hearing Conservationists or similar credential
4. The manufacturer and model number of the audiometer used for the hearing test
5. The date of the last electroacoustic or exhaustive calibration of the audiometer used for the hearing test
6. The employee's most recent noise exposure assessment (can be expressed in TWA or percentage of dose)
7. A statement of whether the sound pressure levels in the test room were acceptably low enough to allow valid testing near 0 dB HTL
8. Pure tone air conduction hearing threshold levels for both ears for no less than the frequencies 500, 1000, 2000, 3000, 4000, and 6000 Hz

Most of this information has been traditionally kept with the employee's audiogram, with the possible exception of the most recent exposure level and test room background noise information.

Audiometric Test Rooms

The employer or the testing service retained by the employer is required to maintain accurate records of the measurements of the background sound pressure levels in the audiometric testing space (see Chapter 8).

Audiometer Calibration

The records from the annual acoustic or exhaustive calibration must be maintained and may include no less than the following:

1. Type of calibration—exhaustive or electroacoustic
2. Date of calibration
3. Numerical results of the acoustical calibration

All of this information about the employee, his or her exposure level, (the tester, the audiogram, the audiometer, and the testing environment for each and every audiogram administered) creates a large record keeping responsibility. Because one day the records may require more square footage of storage space than the employer's facility itself, it is fortunate that OSHA placed minimum record retention times on each of the four categories. Employee noise exposure measurement records need to be maintained for only 2 years, audiometric records need to be retained for the duration of the employee's tenure with the company plus 5 years, records of background sound pressure levels in the test room need to be retained for 5 years, and records of audiometer calibrations need to be maintained for 5 years.

If an employee has been with the company for 15 years, the noise exposure information, test room information, and audiometer calibration data may be destroyed. This does not render the older audiograms useless from a medical-legal viewpoint. The employee audiometric record, which will supersede the employee by 5 years, should contain noise exposure levels for each year, a statement that the background noise level was adequately low enough for testing, and the calibration date of the audiometer as well as audiometer identification information.

Other Information

There is additional information that OSHA does not require but that may be helpful in the management of a hearing conservation program. The primary practiced method for reducing employee exposure to noise is through the use of personal hearing protective devices. It might be helpful to know which devices the employee is using on a year-to-year or test-to-test basis.

Employees come to a company with a history, which might include prior exposure to potentially damaging noise. The prior exposure may have occurred during military service or on a previous job. The prior noise exposure history may correlate well with any hearing loss seen on the baseline audiogram.

Many employees have second jobs. It would be helpful to determine whether or not the second job involves exposure to noise. An employer should not be responsible for the expense of a worker's compensation claim for hearing loss when the hearing loss was caused by the noise from a second, part-time job.

Many employees have had prior ear disorders, some of which may have caused loss of hearing. Many other employees have histories of general disorders from which hearing loss is a secondary symptom. Having such a hearing-related history as part of the record could improve the management of an employee, particularly if the employee begins to show loss of hearing.

Whenever a large group of employees is sampled, there will be those with allergies and head colds. Their hearing thresholds may be elevated by their general condition. If it can be seen when evaluating an audiogram for shift that the employee has a head cold or allergies, then steps may be taken to resolve the shift as secondary to the head cold instead of necessarily resolving the shift as due to noise exposure.

There are many other factors that would

be of value in the hearing conservation records. They include date of birth, date of hire, identification number (be it SSN or company identification number), and type of audiogram (annual, pre-employment, retest following shift, or exit).

DEFINING A DATABASE

A database can be viewed as a single file or as a group of files. In each file there are *records*. The entries on the records can be considered as *data fields*. For example, all the information available for an employee could be placed in a file. Each sheet containing information about the employee can be considered to be a record. An audiogram sheet is a record. Each entry on a record, such as a hearing threshold level, can be considered to be a field.

Flat Files

"Flat" does not imply that the files are placed horizontal to the ground rather than framed to hang on the wall. The descriptor "flat" denotes that only the information in that file is used and that there is no easy means by which, or there is no need, to pull information from other files. A very common type of flat file would be a list of names, addresses, and telephone numbers. The telephone directory can be considered as a flat file with the directory being the file itself. Each entry for a telephone user is a record within the file. The person's name, address, and telephone number would constitute fields within the record.

In hearing conservation programs, flat files can be approximated if we consider all carry-forward information to be part of the file. Imagine a file for an employee. Within the file is one record for each year the employee is tested. Within the record are contained all relevant employee identification, noise exposure levels, all audiometric data from the current test as well as audiometric data from the baseline test, a statement as to the adequacy of the test environment, and required information about the audiometer and technician. The record would contain many data fields and could become very cumbersome.

An example of such a flat file is the military DD Form 2216, a sample of which appears in Figure 9–1. At the top of the form are fields for entering all of the necessary identification information about the examinee. Next is a data field for entry of the type of hearing test. Following are fields for entry of current hearing threshold levels, listing the date of test administration. Following that are fields for the entry of the reference audiogram (military equivalent to the OSHA baseline audiogram). Following are fields to enter calculations of shift and shown below are the criteria for determining Significant Threshold Shift. The next field allows entry of examiner and audiometer identification.

This flat file also supports retesting, when necessary, for a follow-up audiogram after 15 noise-free hours and after 40 additional noise-free hours, if required. At the bottom are fields for identification of the reviewer of the record. This record can stand on its own. It only requires that the threshold levels of the reference audiogram be known and be included as part of this record.

Relational Files

In order to perform comparisons or correlations between records within a file and between records in various files, one must create a database with relational files. In order to have a relational file system, it is necessary to have links common to all files. The link must be unique to each person whose records are contained in the various files. The most commonly used link in the United States is the SSN, since, by law, each person has a uniquely assigned number. Since the SSN cannot be mandated for use as a unique number outside the Social Security Administration and the Internal Revenue Service, some companies assign unique employee identification numbers which may be used as the link common to all files about that person.

HEARING CONSERVATION DATA

ZIP CODE/APO

| DOD COMPONENT | A-ARMY / N-NAVY / F-AIR FORCE | M-MARINE CORPS / 1-OTHER DOD ACTIVITY | SERVICE COMPONENT | R-REGULAR / V-RESERVE | G-NATIONAL GUARD / 1-OTHER |

| SSN | LAST NAME—FIRST NAME—MIDDLE INITIAL | SEX M-MALE F-FEMALE | DATE OF BIRTH | year | month | day |

| PAY GRADE, UNIF SVCS | GRADE CIVILIAN | SERVICE DUTY OCCUPATION CODE | MAILING ADDRESS OF ASSIGNMENT |

| LOCATION—PLACE OF WORK | MAJOR COMMAND | DUTY PHONE |

AUDIOMETRY

| PURPOSE | 1—90 DAY | 2—ANNUAL | 3—TERMINATION | 4—OTHER |

AUDIOMETRIC DATA RE: ANSI S3.6	LEFT						RIGHT					
	500	1000	2000	3000	4000	6000	500	1000	2000	3000	4000	6000
CURRENT AUDIOGRAM DATE year month day												
REFERENCE AUDIOGRAM DATE year month day												
THRESHOLD SHIFT + = Poorer − = Better												

| 1-No Significant threshold shift / 2-Yes ± 20dB or greater | STS NO • Counsel • Return to duty • Retest in 12 mo | • Validated by reviewer • Orig in health record • Send copy to registry | STS YES • Notify supervisor • Followup No 1 after minimum 15 hours noise free |

| NAME OF EXAMINER (Last, first, MI) | TRAINING CERT NO | SSN | SERVICE DUTY OCCUPATION CODE | OFC SYMBOL |

| TYPE 1-Manual 2-Self-recording (auto) 3-Microprocessor | MODEL | MANUFACTURER | SERIAL NO | LAST ELECTROACOUSTIC CALIB DATE year month day |

FOLLOWUP NO 1 Minimum 15 hours noise free

AUDIOMETRIC DATA RE: ANSI S3.6	LEFT						RIGHT					
	500	1000	2000	3000	4000	6000	500	1000	2000	3000	4000	6000
CURRENT AUDIOGRAM DATE year month day												
REFERENCE AUDIOGRAM DATE year month day												
THRESHOLD SHIFT + = Poorer − = Better												

| 1-No Significant threshold shift / 2-Yes ± 20dB or greater | STS NO • Counsel • Return to duty • Retest in 12 mo | • Validated by reviewer • Orig in health record • Send copy to registry | STS YES • Notify Supervisor • Cleared by medical reviewer before Followup No 2 |

| NAME OF EXAMINER (Last, first, MI) | TRAINING CERT NO | SSN | SERVICE DUTY OCCUPATION CODE | OFC SYMBOL |

| TYPE 1-Manual 2-Self-recording (auto) 3-Microprocessor | MODEL | MANUFACTURER | SERIAL NO | LAST ELECTROACOUSTIC CALIB DATE year month day |

FOLLOWUP NO 2 Minimum 40 hours noise free since Followup No 1

AUDIOMETRIC DATA RE: ANSI S3.6	LEFT						RIGHT					
	500	1000	2000	3000	4000	6000	500	1000	2000	3000	4000	6000
CURRENT AUDIOGRAM DATE year month day												
REFERENCE AUDIOGRAM DATE year month day												
THRESHOLD SHIFT + = Poorer − = Better												

| Significant threshold shift ± 20dB or greater 1-No 2-Yes | STS NO • Counsel • Return to duty • Retest in 12 mo | • Validated by reviewer • Orig in health record • Send copy to registry | STS YES • Refer to appro directive • Requires medical disposition | • Validated by reviewer • Orig in health record • Send copy to appro registry |

| NAME OF EXAMINER (Last, first, MI) | TRAINING CERT NO | SSN | SERVICE DUTY OCCUPATION CODE | OFC SYMBOL |

| TYPE 1-Manual 2-Self-recording (auto) 3-Microprocessor | MODEL | MANUFACTURER | SERIAL NO | LAST ELECTROACOUSTIC CALIB DATE year month day |

| REVIEWED & VALIDATED BY | SERVICE DUTY OCCUPATION CODE | AUTOVON | SSN | OFC SYMBOL |

DD Form 2216
1 SEP 79

Figure 9–1. DD Form 2216. This military form is used to record personal information as well as audiograms. If the results of the first audiogram indicate a significant threshold shift, a second audiogram is performed after 15 noise-free hours. If the shift is substantiated, a third audiogram is performed after 40 additional noise-free hours. Since all information is contained on this one sheet of paper, this is considered to be a flat file system.

Imagine a system that maintains records related to the hearing conservation program in four separate files. The first file contains identification information about all employees. The second file contains noise exposure records for all employees. The third file contains audiometric information about all employees. The fourth file contains information about the audiometers and the audiometric testing environment.

In the first file there is a record for each employee. The fields in the record contain the employee's SSN, name, job description, sex, date of birth, date of employement, job description, and relevant medical history. All of this information originates in this file and must be maintained current in this file. If there is ever any change, the record is updated. An example of the type of data screen that would be used to gather and display such information is shown in Figure 9–2.

For each employee record in this file there are fields for noise exposure level, type

Figure 9–2. Employee Data Screen. Notice that space is provided for personal identification information as well as information about employment conditions. Noise exposure levels may be entered directly or may be automatically recorded from noise data records (see Figure 9–3). Entries for personal data are those that are commonly related to hearing losses seen in the work place.

KEY: C - Character field   Letters Only
       A - Alphanumeric field   Letters or Numbers
       9 - Numeric Field   Numbers Only
       L - Logical Field   Yes(True) or No(False) Only
       Date - YY/MM/DD   YY - Year such as 87 for 1987
                                 MM - Month such as 11 for November or 06 for June
                                 DD - such as 06 for 6th day

of hearing protection used, if any, and whether or not the employee has been tagged as showing Standard Threshold Shift. This information does not originate in this file but is maintained and updated from records contained in other files.

In the second file there is a record for the employee which has fields containing his or her SSN and name. In this record, noise exposure measurements are tracked. Noted are employee's job description and work location at the time of the most recent noise level survey. There is also recorded in each record the information about how the noise level survey was done, including type of equipment and date of measurement.

Since hearing protection may be dispensed by areas of noise impact, at least in the example here, there are fields noting the style and manufacturer of hearing protection issued.

The employee records in this file would be updated every time a noise level survey is taken. They would also be updated any time an employee's hearing protection is changed. This file would also be updated any time the employee's place of work or job description is changed. Figure 9–3 displays the type of data screen that would be used to show or gather this type of information.

The third file contains records of audiograms. Each record is headed by an employee SSN and then contains for that employee an audiometric record showing hearing threshold levels. It also contains examiner information, audiometer identification

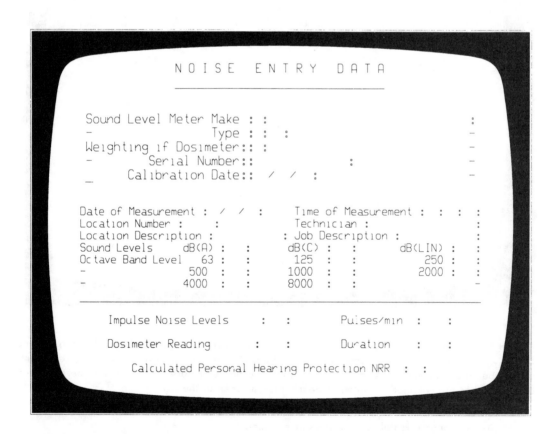

Figure 9–3. Noise Data Screen. This screen allows entry of the results of noise level surveys performed with equipment such as sound level meters, impact/impulse meters, and dosimeters. The entry for Calculated Personal Hearing Protector NRR is performed by the computer program once data has been entered.

information, reviewer information, and the status of the audiogram (baseline, shifted reference, Standard Threshold Shift flagged). Figure 9–4 shows the type of display screen that could be used to display or gather this type of information.

The fourth file contains records for each audiometer being used in the hearing conservation program. There are records of its calibration (one record per calibration). Also included are the date the unit was put into service and the date it was removed from service. The backgound noise level in the testing room where that audiometer is used is noted (assuming a fixed rather than mobile testing location). Figure 9–5 shows the type of display screen that could be used to display this type of information.

If the data contained in these four files are pulled together as necessary to develop reports and analyses, then the information should allow effective management of a hearing conservation program for virtually any size company. In the case of a hearing conservation service provider, information could be pooled across companies.

The difficulty lies in pulling this information together with reasonable ease so that it is possible to use it. Four physical files with all of these records in a company with even less than 100 employees would constitute a substantial data management headache. Here is where the management of the hearing conservation database by the microcomputer begins to become attractive.

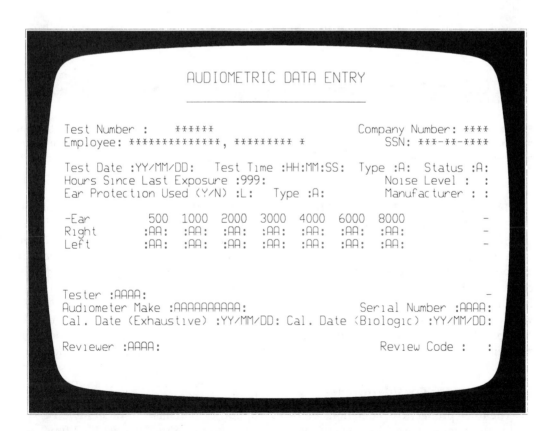

Figure 9–4. Audiometric Data Screen. This screen may be used for the entry of audiometric data, for editing audiometric data, or for reviewing audiometric data for one audiogram for one employee.

KEY: See Figure 9–2 for other codes

*Automatically entered by computer from other records

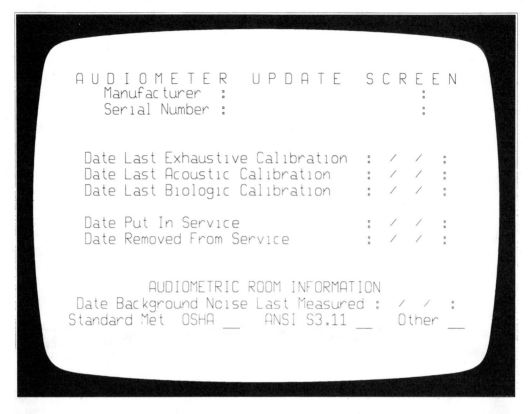

Figure 9–5. Audiometer Update Screen. This data screen allows entry of the most recent calibration data for the audiometers that are in use with the program. It is particularly important to have such a screen if the computer is controlling one or more audiometers for the direct transference of data.

MANAGING A DATABASE WITH MICROCOMPUTERS

The right computer software running on an appropriately configured microcomputer can reduce the severity of the data management headache. The wrong computer software running on an inappropriately configured microcomputer can intensify the problem.

Computer Considerations

The three primary considerations pertinent to the computer are speed, speed, and speed. Speed of the central processing unit is important. A computer with a clock rate of 4.77 megahertz (mHz), the standard IBM PC* clock rate, will run slower than a computer with a clock rate of 8, 10, or 12 mHz.

PC is the abbreviation for *personal computer*, AT is the abbreviation for *advanced technology*. Other terms such as XT, for *extended technology*—there is an alien who owns the copyright on ET—and *turbo* also imply increased speed. To make matters more interesting, on April 4, 1987, IBM introduced its new computer line, the Personal System 2—PS/2, and discontinued manufacture of the PC and XT models.

In considering a central processor type, choose one with 16 or 32 bits supporting no less than 16-bit data paths. If one has the option to purchase an older style PC-type computer or a newer-style PS/2-type computer, one should purchase the PS/2 just for

---

*IBM-PC, IMB-XT, IBM-AT, PS/2, and PC-DOS are registered trademarks of International Business Machines, Inc.

its increased speed. However, the speed of the central processing unit is not the most important factor.

The time it takes for a computer to access a file is significantly reduced for fixed-disk drives, and reduced further for PS/2-type fixed-disk drives over conventional PC-type fixed-disk drives. Four different files have been described previously. In the process of putting together a report or performing an analysis of stored data, these files will be read. The faster the computer can access and read the files, the faster the report and analysis times.

The fastest file read and write times are found with so-called RAM disks. In this case, a portion of the computer memory has been set aside to act as if it were a disk drive. As a rule of thumb, hard disks allow reading of data four times faster than do floppy disks. RAM disks allow reading of data eight times faster than do hard disks. Thus, RAM disks are 32 times faster than floppy disks when reading data. So-called aboveboard memories are available which will allow 2 to 15 megabytes of RAM disk and should be given serious consideration if the task is to handle large hearing conservation data files (such as 20 years of records for 1,500 employees). When relying on RAM disks, it is important to remember that the information is lost when the power is interrupted and so files must be backed up on magnetic media or be battery-backed for safe use.

Speed becomes important again when one asks the computer to print reports. A report can be printed no faster than the printer can respond. In the ideal situation, the printer will be ready to accept more data immediately after the computer is ready to provide it. It is not good to have the printer waiting on the computer, but it is also not good to have the computer waiting on the printer, though the latter case may be unavoidable.

In purchasing a new computer system for a hearing conservation database management project, make purchase decisions in order to get the most speed for the least money. Also, make sure that speed is a stable operating speed. That is, a speed at which all tested software will run without clock-speed induced error. If one must prioritize spending for speed, choose first a fast printer, then a fast disk drive system, and last a fast computer.

With an older computer, make add-on purchase decisions in terms of speed, for example, upgrade to a fixed-disk drive. Consider an extended memory card to obtain sufficient RAM disk to hold data, leaving the program resident on the default drive. When replacing a printer, choose a faster printer.

Software Considerations

Regardless of how a computer system is configured, the true effectiveness of any hearing conservation database management system will be determined by the computer software. The type and style of the internal architecture of the computer software has bearing on how effectively the computer can be used to manage the database.

Assembled, Compiled, and Interpreted Computer Programs

At the core of the computer is a processor that thinks of the universe in terms of variations of the number 0 and the number 1 (binary code). The bit size of the computer (8, 16, or 32) determines how big a "word" it can handle. The word for the computer is made of the combination of 0 and 1. The more digits within a word, the more information the computer can process at one time. Hence, running at the same speed (which they never have), a 32-bit computer can out-process an 8-bit computer by a factor of 4; that is, handle the same data manipulations four times faster. Since 32-bit processors usually run at higher clock speeds, they are faster yet.

Data handling is even more rapid if the programming language for the computer is composed of words made of 0 and 1 or as close as possible. A program can be written in what is known as *machine language*, combinations of words made of 0 and 1 with the exact location of each instruction and each data value also deteremined by the programmer. From 1966 to 1973 most microcomputers were programmed in just that way, thus, it could take up to one year to write a small application program.

The next level up is to write the program in what is called *assembly language*, where the program will be written as a set of instructions (such as NOP for No Operation and JMP -1 for JUMP back one memory location from the present one). These instructions must be then translated for the computer into words made of 0 and 1. To do this, an intermediate program known as an *assembler* is used. Hence, the computer will run the same program written in assembly language a bit slower than it would if the programmer had written the program in words made of 0 and 1.

At the next level are *compiled programs*. These programs are written as some higher level computer language such as Pascal, C, FORTRAN, BASIC, or dBASE. The programs are then submitted to a compiler, which translates the programs to lower level language structure. Depending upon what compiler is used and how machine independent the program must be, the compiled program will approximate assembly language. The closer the compiler-translated program approximates assembly language, the faster the program will run.

At the next level are *interpreted programs*. An *interpreter* is a memory resident program that provides the user with the ability to write commands to the computer and see them executed on the spot. The most common examples of interpreted computer languages are BASIC and dBASE. To write a program or even one instruction in BASIC or dBASE, one must own the language and have its interpreter loaded into the computer. The interpreter acts as an on-line intermediary, translating BASIC or dBASE instructions into some instruction to which the computer can respond. As with simultaneous translation from Russian to English, or from spoken English to signed English, interpretation takes time.

If the computer program runs from an interpreter such as BASIC or dBASE, it will run in the slowest possible mode. The user will be waiting for the computer to complete a task more often than the computer will be left waiting. If the program uses what is called a *Run-Time package*, it still is interpreted and will be just as slow. If the computer program runs in a *compiled mode* (and both BASIC and dBASE programs can be compiled), generally it will run noticeably faster. If the program was written in a language which only runs compiled, for which there is no interpreter version, it will run faster than the compiled version of a normally interpreted language. How much faster depends on many variables.

Since time and money can be proportionally related, programs that run faster will save money. The closer the compiled version of the program approaches assembly language, the faster the program will run and the more cost effective its operation will be.

Passwords

Since the data to be kept are part of a larger medical record for the employee, it might be wise to restrict access to the computer program. This can typically be done with passwords. Password structures can be created so that some operators can perform all functions whereas others are restricted to viewing data and obtaining reports. Password structures can also be set so that operators may have access to some files and not others. A combination of the two tiering systems is also possible.

Medical records contain confidential information about the employee. Access to those records must be restricted to those who represent the employer, the employee, or persons authorized by the employee, and some federal agencies. The password can be used to restrict entry into the database. The password, however, cannot restrict access to the printed record and so a means to control access to these must be as rigorously established and maintained as is the password system for the database.

Command versus Menu Driven

How the program interacts with the user via instruction sets and data entry screens will also effect how fast or slow the program runs. Command-driven programs are generally faster than menu-driven programs, but they also require that the user know more about how to operate the program. A compromise may be a command-driven program with a limited command set for any particular operation. This program could be designed with built-in menus prompting the operator, for cases when the operator makes an error.

Figure 9–6 is a display of a combination of the use of limited choice menu followed by a command. This and all other examples

assume a computer that operates in an MS-DOS, PC-DOS, or OS/2* environment will be used, where the 10 special function keys are employed. The special function keys are labeled F1 through F10. The F1 key is reserved for help screens that are context sensitive and will overwrite the main screen whenever the key is pressed by the operator. The F10 key is reserved as a "cancel" key, to take the operator back to a previous screen or back to the main menu of the program. As shown in this display, the operator must make choices about which functions to perform to edit records. The menu choice is made by pressing either the

---

*OS/2 is a registered trademark of Microsoft Corporation.

Figure 9–6. Limited Choice Screen. Rather than a traditional menu presentation, this screen presents a "bounce bar" of selections. When an item of the bounce bar is highlighted (shown in reverse video), the choices appear in a "pop-up" menu. The item in the pop-up menu may be highlighted for selection or the capitalized letter may be typed to indicate the choice.

first letter of the item or by moving the inverse video highlight bar over the selection with the cursor control keys and pressing the <ENTER> or <RETURN> key. This hybrid command/menu system is typical of software programs for word processing, spread sheets, and database management systems from many publishers.

Data Screens

In some cases, it will be necessary for the operator to sit at the computer console and make many data entries through the keyboard. In these cases, it will be helpful to the operator if the hard copy forms and the data screens have the same style, same location of information, and the same logical flow.

Many data fields have a limited number of legal entries. A person's SSN is limited to nine numbers and may even be forced into a ###-##-#### format. A data field can be programmed to insert the dashes and can also be programmed to accept only numeric entries from 0 to 9 for each place. A prompt could be made to appear, telling the operator of the error before an attempt to enter an improper SSN, for example, "4e1-2-23456" makes its way into the database.

The form of the test date may be set as MM/DD/YY (MONTH/DAY/YEAR) or YY/MM/DD (YEAR/MONTH/DAY), the latter being the preferred format for most computer applications. Any other entry in the data field could result in a clearing of the field and the appearance of an error notice on the screen showing correct date entry format.

Keyboard Entries

Many companies have employee audiograms that predate the use of a computerized database management program. In other cases, the baseline audiogram may have been administered by an outside facility. In yet other cases, it may not be possible to have audiograms entered directly from the audiometer because of logistical reasons.

Thus, keyboard entry of audiometric values is required.

If it is intended that values between 0 dB HTL (in increments of 5 dB) for air conduction thresholds are to be maintained, then the program can check each entry to verify its legitimacy from 0 to 90 and that it is a whole-number multiple of 5. Any other entries would be refused before they become part of the record. The program can also be set to screen for "NR" for No Response, "NT" for Not Tested, or any other legal, nonnumeric entry. Error checking at the time of data entry can save hours of searching data for errors already stored in the database. Data entry can be performed by low-skilled personnel if error checking is employed for critical data fields which can be limited to a specific set of allowed entries.

Batch Processing

The computer program should allow both batch processing and individual processing of information. Batch processing involves having the computer perform the same task for an entire grouping of files or records within files. An example of batch processing would be printing out full reports for employees last tested within a given date range. Once the operator calls the report function and provides both the lower and upper date range, the computer takes over and prints reports until the entire report set is completed. During this time, the program will need no further information from the operator.

Certain functions are very intensive in their use of processing time and file read/write time. The program should allow these functions to be performed in batch mode. Examples of candidates for batch processing are:

1. Evaluation of audiograms for assignment to hearing loss categories
2. Assignment of noise exposure levels to employees by job classification and/or work location
3. Reassignment of reference shift

audiograms for employees after new baselines have been entered

4. All reports generated for a group of employees
5. Form letters written for a group of employees

Individual Processing

An example of individual processing would be an active edit of an employee file. Consider the case of a name or job description change for one employee. The operator would enter the "Edit Employee Record" mode and change the entries in those fields. The process would affect only the fields the operator chose to change for only one employee.

System Growth

Often, hearing conservation data management programs will be started on a simple computer system. The intent is to increase the size of the computer as more storage space is required. A good computer program will allow the user to expand from dual floppy disk to fixed-disk environments without having to return to the source for an upgrade.

Networking

Many systems often start using one computer. One day the need arises for networking—allowing multiple operators to have some form of access to the computer program and files at the same time. A good computer program will allow movement from single to muli-user environments without requiring an extra-fee upgrade. Figure 9-7 shows a connectivity decision chart which can be used to determine when or when not to consider a network. Also shown is when or when not to change computers from the microcomputer to a minicomputer or a mainframe. This decision tree is made in light of the processing of abilities of all 8-, 16-, and 32-bit computers available for purchase in 1987.

Licenses versus Ownership

The computer program remains the property of those who developed it, yet is licensed to the user with certain restrictions. A good computer program will not be copy protected, thus, it will be easy for the user to make backup copies. A good computer program will provide a site license and will alow as many copies as are necessary to operate at that site. As more individual computers are added at a site, the user will be allowed by the license to make copies for use on those new computers.

Programming Bugs

Almost all computer programs of any degree contain "undocumented features," otherwise known as *bugs*. Many of these features will be discovered as the program is used by persons who have purchased it. A good computer program source will send any upgrades that provide solutions to undocumented features, and they will send these automatically with complete documentation at no additional charge.

Program Improvements and Upgrades

Over time, changes will be made in the computer program to allow implementation of new features. These upgrades should be made available at no more than a nominal charge and all users should be informed of their ability. If it is the intent of the computer program source to discontinue support of an earlier version of a program, all users should be notified and given ample opportunity to upgrade to the version which will be supported.

Data maintained in a hearing conservation computer program are medical records. They represent confidential information and they also represent a substantial cost on the part of the user to create them. Any changes made in the computer program should honor the data format of the earlier versions or should allow batch processing of old records into the new format.

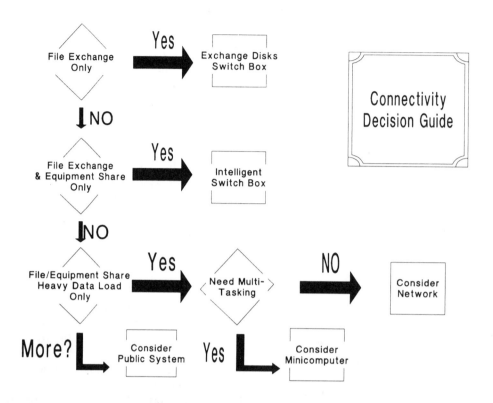

Figure 9-7. Computer Connectivity Decision Guide. Use of this chart can assist in deciding whether to use a printer switching system, a network of computers and peripherals, or a larger computer system.

Archiving Procedures

Records that are accessed for reading and that may potentially be overwritten are not safe on any medium being actively used. For protection of these records, they should be copied over to a storage medium to be used only for archive purpose. This is called *backup*. Small files of records may be copied to floppy disk for backup. Large files of records should be copied to a media such as removable fixed-disk cartridge, bubble memory, optical disk, or tape streamer. Backup should be done on a scheduled basis so that in the event of loss of data on the primary storage system, only a limited effort will be required to add new data to the backup data once they are copied back into the system.

A good computer program will support all backup procedures internally. The various disk operating systems come with utility programs to support backup. These often work well, but just as often, they work poorly. A good computer program will perform the type of backup procedures that are best suited for the type of records being copied.

Report Styles

A report is an organized and sorted listing of information explaining how hearing conservation data should be organized and sorted. A hearing conservation data management computer program should offer a variety of report formats. The most commonly used report is the full set of data on all persons tested between given dates. Consider the following example:

Annual hearing testing has been completed for all the noise-impacted employees at XZY Corporation for a calendar year. All of the audiograms have been entered. Records for all new employees have been entered and those for previous employees have been updated. A noise survey was performed and all changes have been entered. In addition, the audiograms for employees requiring retests have been entered. It is now time to get a full report so that management decisions can be made.

The time period covered will be from one year ago until today. Assume that today is November 14, 1987. All data for persons who received hearing tests during that period, regardless of type of hearing test, will be reported. This means that any employee not tested during that period will be absent from the report.

An example of a report menu appears in Figure 9–8. A full test/site report set will be selected. The next data screen will prompt for dates. The beginning date will be 86/11/14 for November 14, 1986. The ending date will be 87/11/14. To get the report for all persons ever tested at any time, the beginning date could be set to 00/00/00 and the ending date could be set to 99/99/99, which to the computer means from the beginning of time to the end of time.

FULL REPORT SET. Our example report will contain the following sections: First will appear a list of all employees tested within the date range, sorted in alphabetical order. Listed for each employee will be his or her

Figure 9–8. Sample Report Options Menu. In this case, the item number corresponding to the desired report option is entered so that the program may either generate the appropriate report (Options 1 through 9) or display a second menu for special report options (Option 10). This is a traditional menu allowing no entries other than the item number selection. Many commercial programs use this menu format.

SSN and the dates of the first and hearing test on record.

Second will appear a full report for each employee tested (shown in Figure 9–9). This will appear in the format of one employee per page. On the page will be shown all employee information, relevant history, noise exposure level, and type of hearing protection used. In addition, the results of all hearing tests ever administered will be displayed. The baseline audiogram will be so tagged. If the employee has shown Significant Threshold Shift, that audiogram will be so flagged. If the audiogram showing shift was determined to be stable, then it shall also be flagged as the current reference audiogram to which all new audiograms are compared.

Following the listing of audiograms is a statement about the nature of the hearing

```
                         TEST INSTALLATION SITE
                            605 RACE STREET
                          CINCINNATI, OH 45201
                            (513) 555-5555


NAME:  ANDREWS, JAMES A                               COMPANY:  A Large Company
SSN:   111-11-1111      EMPL. NO.:  A6709                       James Smith, Safety Dir
BIRTH DATE:  43/02/25   EMPL. DATE:  78/04/16                   3456 Grocter Road
SEX:  M                                                        Cincinnati, OH 45333
DEPARTMENT:  PLATING           NOISE EXPOSURE LEVEL:  90
JOB DESC.:                     WORK SHIFT:  1ST


MED. HISTORY:  HIGH FEVER        MIL./HOBBY HISTORY:  MILITARY     ADDITIONAL INFO:  NOISY PAST JOB
               MEASLES                               ARMY
               MUMPS
               HIGH BLOOD PRESSURE
```

	TEST DT.	TIME	RIGHT EAR (HZ) .5K 1K 2K 3K 4K 6K 8K	LEFT EAR (HZ) .5K 1K 2K 3K 4K 6K 8K	EXP	TLE	EAR PROTECT T CODE	TEST TYPE	TESTER	REVIEW RVR CDS	SHIFT
B	83/10/15		10 10 10 10 10 10 10	10 10 10 10 10 10 10	85	15	N	B	0001	0001	B
S	84/10/11		10 15 20 25 30 55 40	15 20 25 30 45 60 40	95	12	P 0	A	0001	0001	F *
S	85/11/11		40 40 45 50 55 60 60	40 40 45 50 55 60 55	105	6	P 0	A	4884	0001	D *
			AAO 44 DOL 45 OSHA 50	AAO 44 DOL 45 OSHA 50	AUDIOMETER:	ᴎᵗ		B/CAL: 85/10/11			
					SN: 30603			E/CAL: 85/06/04			

D   Overall loss of hearing in both ears. Hearing protection should be provided employee if workplace noise levels exceed 85 dB(A) for six (6) hours of workshift. Because of magnification of hearing loss by hearing protection, on-job listening should not be made important.

STS: A standard threshold shift has occurred. This is defined as a change in threshold level relative to the baseline audiogram of 10 dB or more at 2000, 3000, and 4000 Hz in either ear. It is required that the employee be notified, in writing, of this change within 21 days of its determination. Unless the change is determined by a physician to be non-work related the following steps should be taken: 1. If the employee is not wearing hearing protection (s)he should be fitted with them, trained in their use and care, and required to use them. 2. If the employee is presently using ear protection (s)he should be refitted and retrained. Hearing protection with greater attenuation should be provided if necessary. 3. The employee will be referred for additional evaluation if a medical pathology is caused or aggravated by the wearing of hearing protectors. 4. The employee is informed of the need for an otologic examination if a medical pathology of the ear (unrelated to the use of hearing protection is suspected.

Figure 9–9. Full Employee Report. This is a full audiometric record for one employee. All identification information is shown, as are relevant history, exposure levels, and audiometric records. This report charts a progressive loss of hearing, changing from normal hearing, to bilateral high frequency, to overall loss of hearing. Notice that the text notes the overall loss of hearing for the last audiogram and the fact that Standard Threshold Shift has occurred.

loss, if any, the employee is showing. If this is a "shifted" audiogram, the instructions for management of the employee are also shown. Other information related to each audiogram is also listed. Items addressed are: time since last noise exposure before the hearing test; type of hearing protection used; make, model, and serial number of audiometer; tester identification information; and reviewer finding and identification codes.

Third will appear a list of employees showing a shift. In a good year, there will be no names on this list; in other than a good year, employees on this list will be sorted in alphabetical order.

Fourth will appear a listing of employees in alphabetical order, sorted by type of hearing loss or reviewer assignment category from the most recent audiogram.

This type of report can take a long time to print. The data required to print the employee report page are located in no less than two files. They are also located in many records in the audiogram file. Thus, the computer will use time reading the files and organizing the data before printing it. Assuming that the computer runs fast enough to wait on the printer and that the printer is printing at 200 characters per second at burst rate, it will take 400 minutes to print out just the full employee report sheets for 400 employees (that's the majority of a day). Many persons wil elect to start this type of report at the end of the working day, hoping to come back in the morning to find it completed. Given that there is sufficient paper loaded, the printer does not jam, and the power does not fail, expectations will be met. If a printer buffer is used, the computer will finish formatting the report and send it to the printer more quickly, and can thus be used for other tasks while the report is being printed.

SINGLE EMPLOYEE TEST REPORT. On occasion, it will be desirable to obtain a full report on only one employee. The program should allow this at any time. This report could take as long as 5 minutes to organize and print, depending upon the number of employees in the total file and the number of audiograms on file for the employee in question.

LIST OF EMPLOYEES BY REVIEWER HEARING LOSS CATEGORY. At times, although a full report will not be needed, a limited listing will be. Such a case arises when looking at total hearing loss in a company. A listing of employees sorted by reviewer category can provide a synopsis of hearing loss. This is a subset of the full test/site report, but it takes less time. In most cases, the sort will be done on the basis of the most recent audiogram on file. In addition, it should be possible to limit employees in the listing to only those who were tested within a desired date range.

LIST OF EMPLOYEES SHOWING SIGNIFICANT THRESHOLD SHIFT. Just as a list of employees sorted by reviewer category may be helpful, a list of employees showing significant threshold shift will also be helpful. This is also a subset of the full test/site report.

LIST OF EMPLOYEES SORTED BY SPECIAL FACTORS. This can be a helpful tool in studying a particular problem. It can also be useful in looking for trends if there is a sufficient number of employees. Unfortunately, most commercially available computer programs for microcomputers do not support this feature. However, if they did, a Special Factor Selection Menu would appear as in Figure 9–10. Remember, the more stringent the specifications, the fewer the number of employees in the listing, and the longer it takes to get the report.

OTHER CONSIDERATIONS

To have data in the files for a computer program to manage, it is necessary to enter the data. The expense of entering data can quickly exceed the expense of the entire computer system and software. Therefore, the manner in which the data are entered into the computer becomes a very important issue. The most common method for

Figure 9-10. Special Factor Selection Menu. This menu is in the "bounce-bar, pop-up" menu format.

entering data into a microcomputer is by typing the data through the keyboard. This is the method by which this chapter was written, utilizing a word processing program, and may be the only method by which it is possible to enter data for the four different file types discussed previously. This is called *direct data entry*.

Importing Data

When data are transported to the database from an external source directly, it is called *importing data*. These data exist in the other source in some format and may be transformed so that they may be used by the computer program in its format.

Pre-Existing Data

Often one reviews older data for the hearing conservation program which existed before the present computer program was instituted. If these data were in hard-copy form, on sheets of paper, the only realistic way to transfer it to the existing database is through keyboard entry. If these data were in another computer application, they may be moved electronically or via electromagnetic media. What this means is that these data are transmitted from computer to computer via electronic link-up between the communication ports on the two computers, a process known as electronic importation of data. If the data from the old computer were stored on diskette or tape they may be directly read into the new computer.

Neither of these two approaches is simple. Both will require matching computer communication protocols or disk/tape formats. Both will require transformation of the file and record structures from the older structure to the newer. If the database is large, it will be worth the effort spent to avoid keyboard input, as once the communication and transformation problems are solved, the data will be imported without chance of error.

Optical scanners and associated software are becoming more commonly available and less expensive. It may be possible that hearing conservation data now in hardcopy format can be read by an optical scanner, reformatted by the scanning software, and imported into the hearing conservation database.

New Data

The microprocessor audiometer was first introduced in 1975. This audiometer uses a small microprocessor, in some cases the same one as the microcomputer, to control signal creation, gating attenuation, and response measurement for the audiometric test. It is programmed to administer a audiogram much as would a qualified tester. In addition to printing out the results on its own or a peripheral printer, it can also send test results directly to a microcomputer.

It follows that the less data are handled, the less the chance for error to appear. Thus, the hearing conservation software should be able to handle direct import of audiograms from microprocessor audiometers. This may be done on a batch or individual audiogram level.

On the batch level, a means would exist for gathering a group of audiograms and storing them in preparation for importing to the computer and its hearing conservation software. The storage could be electronic, as is the case with internal audiometer memory, and there are presently microprocessor audiometers on the market which will hold between 50 and 200 audiograms in memory. The storage could also be magnetic, such as diskette or tape. Regardless of which method is used for storage, at some time the audiograms will be imported as a group into the files and records of the hearing conservation software.

On the individual level, the microprocessor audiometer would send each audiogram as obtained to the microcomputer running the hearing conservation software. This would require that the audiometers and the microcomputer be located in close proximity and be operated concurrently. Many microprocessor audiometer manufacturers offer hearing conservation software which controls the audiometers and imports audiograms, as well as managing the hearing conservation database.

Exporting Data

One may think of the hearing conservation database maintained on the microcomputer as the last stop for the data, and in most cases, the only type of data output will be the summaries generated for the various reports. However, it may be necessary to output data from the hearing conservation database so that another computer may have access to it. Consider that the hearing conservation database is only a portion of the occupational health record for an employee, and thus, the hearing conservation data must be joined with other occupational health data. Many corporations and agencies maintain occupational health records on larger mainframe-type computers. It may be necessary to send data from the hearing conservation database to those large computers. this can be done in one of two formats, electronic transmissions or magnetic medium. In either case, these data would typically be coded in American Standard Computer Information Interchange (ASCII) format so that little conversion would be necessary at the receiving end.

Another case where export of data may be necessary would be for the company which is maintaining hearing conservation databases at many sites. In this case, an employee may move from one site to another as the result of a transfer. It would be most helpful if all of the employee's hearing conservation records could be exported from the first site to the second site. Again, this could be done electronically or via electromagnetic medium. If both sites were using the same hearing conservation database software, then these data could be sent in whatever encoding scheme the software uses. If the sites were using different software, then it would be best to have the data for export coded in ASCII.

Getting Data from the Test Site to the Computer

In many cases, the location of the hearing testing site and the location of the computer will be different, often separated by many miles. This is particularly the case faced by a service provider with a mobile facility. Getting these data from the test site to the computer becomes a problem. Consider the following two methods for solving this problem.

Moving Data in Hard Copy—Hand Entry

Whether audiometric data are gathered on cards from self-recording audiometers, on audiogram forms from manual audiometers, or on printouts from microprocessor audiometers, the final result is a piece of paper with the audiometric information. In order to get this information into the computer for management by the hearing conservation program, the data must be entered by someone working at a keyboard. Whenever data are read from a sheet of paper for keyboard entry, there is a chance for entry error. To facilitate correct entry, the computer program can be written to screen the entries and to tag illegal entries. This is particularly true for entries of SSN, audiometric data, and the like.

When data are to be entered by hand, there should be a means to list the information for verification and correction. This can be done by making an edit mode and some type of report listing available. It is also important that the program not be a factor in determination of maximum number of entries per unit time. Thus, the program should be able to accept input as fast as the operator can provide it. A program is inadequate if the operator is waiting on the computer to finish accepting input before he or she can input the next entry. Any computer

software worth its price should never slow the speed at which a competent entry clerk can input data.

Direct Entry—On Line

In many cases the microprocessor audiometer can be situated in close proximity to the microcomputer. In these cases, it is possible to have audiometric data transmitted directly from the audiometer to the hearing conservation database. Before solutions are examined, a word of warning about computer-generated noise is given.

COMPUTER NOISE. Computers generate two types of noise, acoustic noise from the cooling fan and electromagnetic noise from the circuitry. If the computer is near the testing environment, the acoustic noise from the fan can increase the background noise level in the testing environment. There is no standard for noise levels generated by computer cooling fans. Thus, some computers have fans that generate enough noise cause a problem whereas others are relatively quiet. Since there is no easy means to quiet a computer on site without running the risk of having it overheat, care should be taken in the evaluation of computers considered for use adjacent to testing environments.

The electromagnetic noise generated by computers can be picked up by the low-impedance earphones and cables running from the audiometer. This noise can appear in the form of a high-frequency hiss or as clicks in the line, appearing in synchrony with computer operations such as disk drive access. All microcomputers sold in the United States must have the approval of the Federal Communications Commission (FCC), but this standard addresses interference with devices such as radios and televisions. Computers and audiometers that carry the FCC seal should not create a noise problem if they are properly connected and properly grounded.

Dot matrix printers generate noise in the process of printing. Presently, there is no dot matrix printer available that does not cause an elevation in the background noise level of the typical testing environment. If it is intended that the printer be in use while testing is going on, then the printer must be enclosed in its own noise-reducing cabinet.

DIRECT TRANSMISSION. It is both desirable and possible to have the computer running the hearing conservation database program immediately adjacent to the audiometer or group of audiometers. If the hearing conservation program has the ability to accept data directly from the audiometer, then the logical thing to do is to connect the audiometer to the computer. If there is more than one audiometer, then some type of switch will be necessary since the computer will typically support only one communications port. The switch may be manual, in which case it will be necessary to set the switch to the correct position before letting the audiometer send a completed audiogram. This is the least expensive way. The audiometers are run independently for the purpose of audiometry. When the audiogram is complete, the switch is set to the appropriate setting and the "send" or "print" data button is pressed. The computer program display screen should show that the data have been received. Presently, all commercially available microprocessor audiometers can be used in this mode. See Figure 9–11 for a schematic diagram of such an arrangement.

Some commercially available hearing conservation programs will support control of one or more audiometers. In this case, the

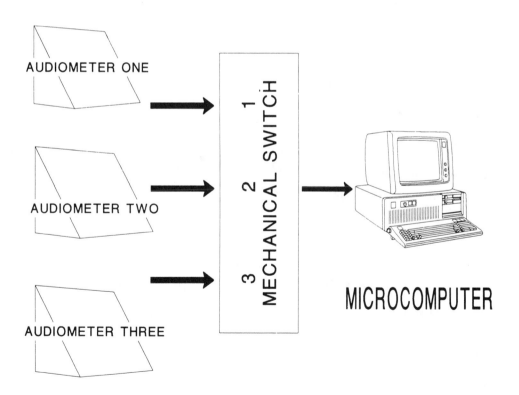

Figure 9–11. Schematic representation of a mechanical serial communications switch that connects three microprocessor audiometers to a microcomputer. The operator sets the switch to the number corresponding to the audiometer and then presses the "send" data button on the audiometer. The computer reads the incoming data passively.

software will start and stop the audiometers as well as receive audiograms from them. If one of these audiometers is used, then it only need be connected to the communication port of the computer. If more than one of these audiometers is used, then a multiplexing or interim communication switch must be used, which the computer software also controls. Presently, there are three sets of software from audiometer manufacturers that can control the audiometers and the interim communications switch, and can accept data directly. See Figure 9-12 for a schematic diagram of such an arrangement.

As one becomes more reliant upon the ability of the computer software to control audiometers as well as to accept data from them, one will become more reliant upon

one manufacturer to provide the audiometers, the interim communications switch, and the computer software. Thus, the features of the entire package should be evaluated from each manufacturer. See Appendix 9-1 at the end of this chapter for comparison details.

Indirect Entry—Off Line

It is often the case that audiometric data for a given company will be gathered concurrently in many locations. Also, a service provider with a mobile facility may gather audiograms from different locations. In both of these situations, it may not be feasible to have a computer with the central hearing conservation records located near or

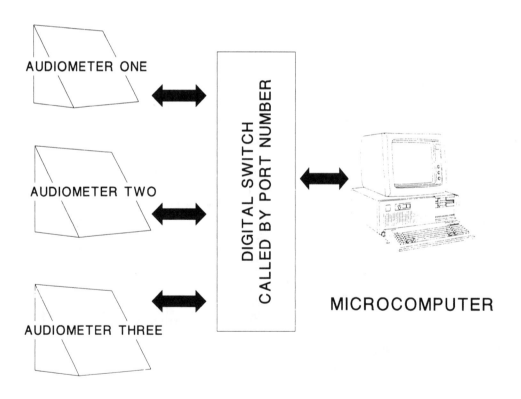

Figure 9–12. Schematic representation of a multiport computer-controlled serial switch. In this case, the computer can address any of the three audiometers on line, start and stop them, and retrieve data from them. The computer can also have the database stored on disk and the audiograms can be transferred to the database for manipulation concurrent with controlling the audiometers to perform hearing testing.

connected to the audiometers. Yet, it is desirable to have a means by which audiograms can be entered into the computer without hand entry.

In this case, an interim computer may be used to hold the audiograms. Consider the case of the *notebook computer*, so called because it is the size of a small three-ring notebook. These small, inexpensive computers can be programmed to receive data directly from microprocessor audiometers, and can also be programmed to control microprocessor audiometers. As shown in Figure 9-13, the audiometer sends its information directly to the notebook computer. In the cases where more than one audiometer is in use, a switch is installed between the notebook computer and the audiometers.

Notebook computers have battery-backed RAM. This means that they won't lose data upon power down or power failure—particularly attractive for mobile facilities where power consistency is questionable at best. Some notebook computers can hold between 500 and 700 audiograms before their memories are filled. Upon completion of testing or when filled, these data are batch transmitted to the computer holding the hearing conservation software either directly or over a telephone line using a modem. Of course, the hearing conservation program on the computer must have the ability to support the batch import of data for this to work. One may mail the notebook computer back to the main office for a batch dump of the audiograms as

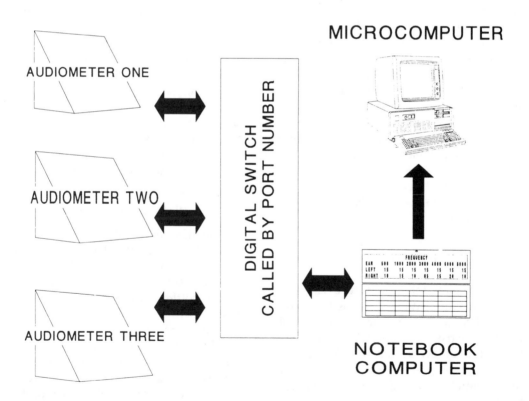

Figure 9-13. Schematic representation of a multiport computer-controlled serial switch. In this case, the notebook computer can address any of the three audiometers on line, start and stop them, and retrieve data from them. At some later time, the notebook computer can send its stored data to the larger computer used to manage the database.

needed, clear its memory, and then mail it back to the test site.

Another means for indirect entry would be a situation where a computer located in proximity to the audiometers has the ability to accept data from the audiometers, control the audiometers, and store the audiograms on diskette. The diskette containing audiograms is then sent to the location of the computer running the hearing conservation program for a batch import of the audiograms from the diskette or these data may be sent by modem.

The lap-top computer is ideally suited for this type of application because it packs the power of a desk-top computer into a small package. Lap-top computers weigh between 9 and 17 pounds, have full-sized 80-column by 24 row displays which can be read in most lighting situations, and require 9 to 15 volts to operate. The displays are of the liquid crystal display (LCD) type or plasma type. The LCD type draws low electrical current and can be run from batteries if necessary. Some of this type will survive, without program interruption or data loss, a power interruption severe enough to lockup, reset, or destroy the audiometers. The plasma display type requires higher electrical current and may not be operated from batteries. Nonetheless, this type is small enough to fit into the most cramped spaces while providing full computing power. See Figure 9–14 for a schematic diagram of such an arrangement.

THE FUTURE

Speculation into the future is risky. Guessing about future events is based on the prediction of where current trends will take us. The risk is the possibility that trends may detour or reverse. It is with full knowledge of these pitfalls that the future of microcomputing for hearing conservation is forecasted.

High-Speed Microcomputing Processing

In 1986, the speed at which microcomputers can run increased from 4.77 mHz to 16 mHz and the bit path increased from 8 to 32. This means that, given the necessary change in software, it should be possible to process data 16 times faster on a new computer than on an older style PC. Disk access times have also increased by a factor of 5, making retrieval of data from disk, or writing data to disk, a faster process.

The newer 32-bit processors have been introduced to run with present operating systems. When operating systems become available that can take full advantage of the processors, speed of computer operation will no longer be the problem it can be today. In addition, true multitasking and multistation operation will become possible on the microcomputer. In a true multitasking system, one process could control audiometers and collect data from them. That process could transfer the results from another process for database entry, analysis of the audiograms, and flagging those audiograms that showed threshold shift. The second process could then pass those audiograms requiring professional review to a process that interacted with the person conducting the review. A final process could handle the printing of reports for those audiograms that had received all necessary levels of earlier processing. Yet, the newer faster computer should cost no more than a PC-type computer costs now.

High-Speed Printing

The most commonly used printer today is the dot matrix printer. These printers have improved in speed and quality of output over the past 6 years. Yet, these printers are still the bottleneck through which all data must be sent.

The laser printer has been in use since 1984, but the cost has been high. One can presently obtain a laser printer that will compete in price with a high-quality dot matrix printer and that will allow reports to be generated 7 to 10 times faster. It is expected that the gap will narrow and that laser printers will become the printer of choice for anyone desiring high-speed, high-quality report output.

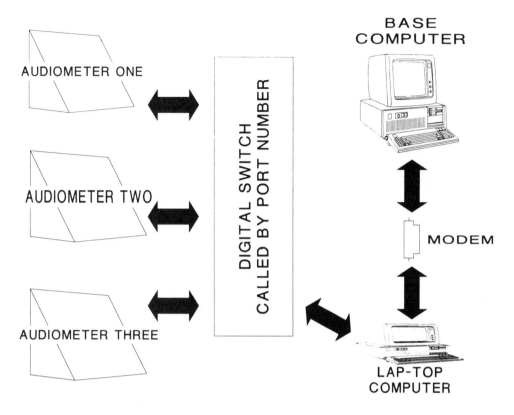

Figure 9-14. Schematic representation of a multiport computer-controlled serial switch with a lap-top computer. At some later time, the notebook computer may send its stored data to the larger computer used to manage the database. But since the lap-top is as capable as most desk-top computers, this may not be necessary.

Field Systems

Earlier in this chapter, the notebook-type microcomputer was cited for gathering field data. The advantages of this type of computer are its cost, its ability to interact with or control microprocessor audiometers, and its ability to hold data with no risk of loss.

The lap-top computer field is growing fast. For the same price as a desk-top computer, it is now possible to purchase a lap-top computer with two disk drives, a readable screen, full printer and communication port support, and the ability to run any commercial version of hearing conservation software. These lap-tops differ in features, but some will run from battery power and will survive the inconsistencies of power

availability without risk of locking up or losing data.

Whereas the average notebook computer weighs in at 6 to 7 pounds with limited memory and screen display, the average lap-top computer weighs in at 13 pounds with full memory and standard display format. Soon there will be no need to consider the notebook computer for use in the field, for the lap-top computer will be the computer of choice.

High-Speed Modems and Error Checking

Moving 750 audiograms from a remote computer to the host computer running the hearing conservation program can take up to 2 hours if there is a data-checking protocol. Phone line noise often makes the transfer impossible or risky at best. Presently, the safest speed for moving data is 300 BAUD (which is equivalent to bits/second). At 300 BAUD, moving an audiogram can take 10 seconds. At 1200 BAUD, the same transfer can take 2.5 seconds. With the advent of 4800 and higher BAUD modems and good error-checking software, it should be possible to transfer one audiogram from remote to host computer in one second, or about 750 audiograms in 12 to 13 minutes.

Better Storage Media

Presently, most computer data are stored electromagnetically on either floppy diskette or on fixed-format diskettes. This imposes limits on data storage and write/read access speeds. Around the corner are the so-called CD-ROM and CD-WORM media.

The CD-ROM is very much like the compact audio disk now growing in favor for holding music. The music track is stored in digital format and read from the disk by a laser beam. Computer data can also be stored on compact disk as read only memory, thus giving us the name of CD-ROM. This method of storage is so efficient that an entire encyclopedia with graphics and index has been stored on only $\frac{1}{3}$ of a CD-ROM. For hearing conservation, all of the data for a company as large as General Motors could be stored on one CD-ROM.

However, writing to a CD-ROM is a problem. The CD-WORM provides an alternative, its acronym standing for Compact Disk-Write Once, Read Many. With this format, it would be possible to write the hearing conservation data and updates to the compact disk and then read from it as necessary for analysis of data and the generation of reports. This will mean greater data storage possibilities at less cost. Presently, using available hearing conservation computer software, 20 years of hearing conservation data for 1,500 employees will take 15 megabytes of storage plus an additional 15 megabytes of working room, thus using up all of a 30 megabyte fixed disk.

With CD-WORM storage, only one percent of the available room would be used.

The financial institutions have developed the so-called "smart" credit card. These cards have nonvolatile RAM and are powered by available room light. In the case of the bank card, it will remember what has been charged and the credit balance. The card will not allow a purchase in excess of the credit limit. Additionally, the card can be used to hold the balance of both a checking and a savings account, and can be used as a direct debit card as well as a charge card.

The same technology could be used to hold the entire work-related medical records for the employee. Thus, the employee identification card would contain the employee's occupational medical history inclusive of all audiometric records and related histories. The card could be read by the computer administering new audiograms and all updates and recommendations could be written to the card as well as to the central database.

More Powerful Programs

Hearing conservation database programs written up to now have been designed to have the data handled within a structure determined by the program designers. With future generations of computer software, it may be possible for the user to work within a database using what is called *structured query language* (SQL). In an SQL environment, a user could present the program with a question such as, "Just how many workers over the age of 35, who have been working in 85 dBA of noise for more than 2 years while wearing hearing protection, who are male, and who have a medical history of hypertension showed threshold shift since last year?" In an SQL environment, the program would provide the answer.

Artificial intelligence has been described by skeptics as having the computer learn that "Y" means "Yes" and that "N" means "No." With present microcomputers, this description is not far off the mark. With advances in hardware design, increased operating speed, increased RAM, and rapid access mass storage, artificial intelligence will become more of a factor in microcomputer applications. In the artificial intelligence environment, it may be possible to teach the computer to perform audiometry, to analyze the resultant data, and to print necessary reports. More importantly, it may be possible to have the computer call for the attention of the hearing conservationist when it discovers a situation which is new to it so that problem cases do not go unnoticed.

CUSTOM VERSUS COMMERCIAL SOFTWARE

Most of the commercial software packages listed in Appendix 9–1 of this chapter will cost about $3000. Some of the packages support only certain audiometers, other support most of the microprocessor audiometers available as well as supporting the entry of data from other sources. It is recommended that the supplier of each program be questioned as to the ability of the program to support the features listed earlier in this chapter. No software will support all of them, but most will support many of them.

Most software is written by one person or a small group of persons. Should custom computer programs be desired, plan to spend about 1 to 2 years getting the software written in final form. Budget for 1 to 2 years of a programmer's time.

Custom software will support exactly what is needed. Commercial software, virtually without exception, will require adaptation of an operation to the features of the software. That is the trade off, balancing needs against the time and funds required from the customer.

SUMMARY

This chapter has reviewed the management of hearing conservation data with microcomputers. Many areas have been addressed with the idea of initiating thought about data handling needs. If from reading this chapter, one is better able to determine those needs, then this chapter has served its purpose. If one is better able to shop for commercial software or specify the design of custom software, then this chapter has also served its purpose.

Hearing conservation data left in a file cabinet serve the needs of no one. Hearing conservation data used to reduce the incidence of work place noise-induced hearing loss serves everyone. The appropriate management of hearing conservation data with the microcomputer is essential to achieving the latter.

SUGGESTED READINGS

A reference section in a textbook usually does not come sorted by applicable category. An exception is made here. This chapter has explored many of the considerations and ramifications of applying microcomputer database concepts to the management of hearing conservation data. The suggested readings provided here are general reference books that apply to various applications for which a microcomputer may be used, not just hearing conservation data management. These books have been written by persons who are renowned as experts and who are widely published.

In addition to books, magazines are also referenced. These magazines provide a method by which to keep current in the ever-changing field of microcomputing. They are available at major booksellers nationwide.

GENERAL COMPUTING

Norton, P. (1986). *Inside the IBM PC*. New York: Prentice-Hall, Inc.

Norton, P. (1985). *PC-DOS: Introduction to high-performance computing*. New York: Prentice-Hall, Inc.

DATABASES

Simpson, A. (1986). *Understanding dBASE III Plus.* Alameda, CA: Sybex Computer Books.

Simpson, A. (1987). *Understanding R:base System V.* Alameda, CA: Sybex Computer Books.

WORD PROCESSING

Kelly, S. (1986). *Mastering WordPerfect.* Alameda, CA: Sybex Computer Books.

Naiman, A. (1987). *Mastering WordStar on the IBM PC.* Alameda, CA: Sybex Computer Books.

Rinearson, P. (1986). *Word processing power with MicroSoft Word.* Redmond, WA: MicroSoft Press.

Sladek, J. (1987). *Using XYWrite III.* Berkeley, CA: Osborne/McGraw-Hill.

COMMUNICATIONS

Glossbrenner, A. (1985). *The complete handbook of personal computer communications.* New York: St. Martins Press.

NETWORKING

Durr, M. (1987). *Networking IBM PCs.* Indianapolis, IN: QUE Corp.

MAGAZINES

BYTE
Database Advisor
PC Magazine
PC World

APPENDIX 9–1

Advantage Health Systems, Inc.
14602 Denver Parkway West
Golden, CO 80401
(303) 277-6220
(800) 525-1352

Applied Health Physics, Inc.
2986 Industrial Boulevard
Bethel Park, PA 15102
(412) 563-2242

Computer Hearing Conservation, Inc.
P.O. Box 740013
New Orleans, LA 70174

Digital Hearing Systems Corporation
2934 Shady Lane
Ann Arbor, MI 48104
(313) 973-2658

E.L.B./Monitor, Inc.
605 Eastowne Drive
Chapel Hill, NC 27514
(800) 334-5478 (outside NC)
(800) 672-7027 (inside NC)

Environmental Health Associates, Inc.
520 Third Street
Suite 208
Oakland, CA 94607
(415) 451-1888

Environmental Research Technologies
696 Virginia Road
Concord, MA 01742
(617) 369-6758

Flow General, Inc.
7655 Old Springhouse Road
McLean, VA 22102
(703) 893-5910

Impact Hearing Conservation, Inc.
406 West 34th Street
Suite 410
Kansas City, MO 64111
(816) 531-4847

Infomed Corporation
13 Inverness Way South
Englewood, CO 80112
(303) 790-1311

Maico Hearing Instruments
7375 Bush Lake Road
Minneapolis, MN 55435
(800) 328-6366

Metpath, Inc.
One Calcolm Avenue
Teterboro, NJ 07608
(800) 222-0446

Metrosonics, Inc.
P.O. Box 23075
Rochester, NY 14692
(716) 334-7300

OSHA-Soft Corporation
48 Grandview Road
Concord, NH 03301
(603) 228-3610

OTO-DATA, Inc.
1244 Clairmont Road
Suite 100
Decatur, GA 30030
(404) 634-4729

Occupational Health Services, Inc.
400 Plaza Drive
Secaucus, NJ 07094
(201) 865-7500

Tracor Instruments, Inc.
6500 Tracor Lane
Austin, TX 78721
(512) 929-2027

Tracoustics Medical Instruments
Division of Starkey Laboratories, Inc.
411 East St. Elmo Road
Austin. TX 78745
(512) 448-4877
(800) 228-0881 (outside TX)
(800) 323-0363 (inside TX)

Hearing Protectors—Specifications, Fitting, Use, and Performance

Elliott H. Berger ■

As the number and variety of available hearing protection devices (HPDs) increased dramatically in the past 30 years, their quality also improved. The crucial variable that controls hearing protector effectiveness, however, has remained the same—the wearer, and how he or she fits and uses the device. Thus, much of this chapter is focused upon increasing the awareness of professional hearing conservationists regarding practical aspects of hearing protector selection and utilization and the details of real-world performance. Nevertheless, the more technical elements of hearing protection such as the physics of HPD performance, the results of laboratory attenuation tests, the effects of HPDs on auditory communications, and the complexities of standards and regulations will not

be overlooked. The purpose is to provide a comprehensive yet pragmatic reference source for the occupational hearing conservationist.

While reading this chapter, it is well to remember that, regardless of the cost or laboratory-measured attenuation of a particular device, there is little benefit in distributing it to a noise-exposed population without first providing adequate indoctrination concerning its fitting, use, and care. Additionally, a means of motivating the workers to properly implement their training must also be included. Therefore, education and motivation, which are reviewed in Chapter 11, are important collateral topics to the discussion at hand. (Also see Royster, L. H., & Royster, J. D., 1986.)

PHYSICS OF HEARING PROTECTOR PERFORMANCE

The acoustical performance of an HPD involves not only the physics of the protector and its interface with the ear, but the

This chapter is reproduced with minor editorial updates and modifications from *Noise and hearing conservation manual* (4th ed, pp. 319–381). E. H. Berger, W. D. Ward, J. C. Morrill, & L. H. Royster (Eds.). Akron, OH: Am. Ind. Hyg. Assoc. (1986).

anatomical/physiological limitations of the human auditory system as well.

Sound Paths to the Occluded Ear

The occluded ear receives sound energy along four primary transmission paths, as illustrated in Figure 10-1.

Air Leaks

For maximum protection, the device must make virtually an airtight seal with the canal or the side of the head. Inserts must precisely fit the contours of the ear canal and earmuff cushions must accurately conform to the areas surrounding the external ear (pinna). Air leaks can typically reduce attenuation by 5 to 15 dB over a broad frequency range (Nixon, 1979).

Hearing Protector Vibration

Due to the flexibility of the ear canal flesh, earplugs can vibrate in a piston-like manner within the ear canal. This limits their low frequency attenuation. Likewise, an earmuff cannot be attached to the head in a totally rigid manner. Its cup will vibrate against the head as a mass/spring system, with an effective stiffness governed by the flexibility of the muff cushion and the flesh surrounding the ear, as well as the air volume trapped under the cup. For earmuffs, premolded inserts, and foam inserts these limits of attenuation at 125 Hz are approximately 25 dB, 30 dB, and 40 dB, respectively.

Material Transmission

The exterior surfaces of an HPD will undergo some deformation of shape (vibration) in response to the forces applied by impinging sound waves. These vibrations are transmitted through the protector material to its inner surface where the resultant motion radiates sound, of diminished intensity, into the enclosed volume between the HPD and the wearer's eardrum. The amount of sound reduction is dependent upon the mass, stiffness, and internal damping of the hearing protector material.

Material transmission through the cup and cushion components of earmuffs is significant, normally providing a limitation to attenuation for frequencies above 1 kHz. This path is generally less important with insert HPDs, due to the fact that their exposed surface areas are much smaller. However, a special case exists for certain fibrous materials, such as cotton, which are easily permeated by air. In this latter instance, attenuation will be very low, since sound will pass through the substance of the device as though there were many tiny air leaks.

Bone and Tissue Conduction

Even if HPDs were perfectly effective in blocking the preceding three sound paths, sound energy would still reach the inner ear (see Figure 10-1A) via bone and tissue conduction (BC). Energy transmitted in this manner is said to flank or bypass the protector, imposing a limit on the real-ear attenuation that any HPD can provide. However, the level of the sound reaching the ear by such means is approximately 50 dB below the level of air-conducted sound through the open ear canal. This is illustrated in Figure 10-2. The early estimates are from Zwislocki (1957) and Nixon and von Gierke (1959); recent estimates are by Berger (1983b).

Evidence suggests that the BC limits to attenuation are constant regardless of sound level. This indicates that the relative importance of the energy transmitted via the BC pathways is independent of the noise level in which the hearing protector is worn. Unless an HPD's attenuation approaches 40 to 50 dB (depending upon frequency), which is rarely the case due to air leaks, protector vibration, and material transmission, sound transmitted via the BC paths is normally not of primary concern.

Since the regions of the head around the external ear are only a small portion of

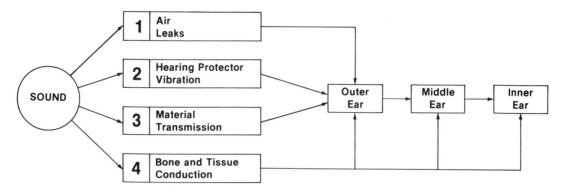

Figure 10-1A. Generalized block diagram of the four primary sound pathways to the occluded ear.

the total skull area exposed to sound, covering them with an earmuff is of little significance with respect to BC, perhaps 3 to 4 dB in the 1 to 2 kHz region (Berger & Kerivan, 1983). Thus, the relative performance of plugs compared to muffs is not, in practice, determined by the BC paths, but by factors inherent in the design of HPDs and their interface to the head. Increasing the BC limits to attenuation shown in Figure 10-2 would require completely enclosing the head in a rigid helmet with a face mask. Therefore, use of earmuffs in conjunction with a hard hat, which covers only part of the head and has many gaps through which the sound energy can penetrate, provides no more protection than the muffs alone.

Occlusion Effect

The efficiency of the BC paths in conducting energy to the inner ear is affected by the state of occlusion of the ear canal. When the ear is occluded with an HPD, the efficiency of transmission of bone-conducted energy for frequencies below 2 kHz is enhanced relative to the unoccluded ear. At those frequencies, vibratory displacements of the canal walls, concha, and/or pinna develop significantly greater pressure at the eardrum in the occluded

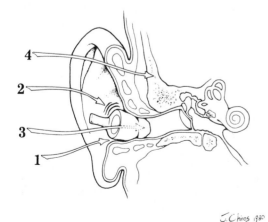

Figure 10-1B. Sound pathways for an earplug.

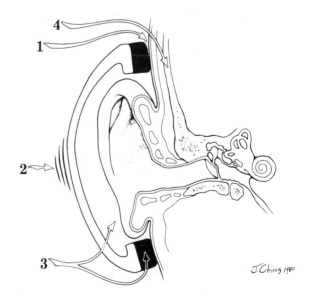

Figure 10-1C. Sound pathways for an earmuff.

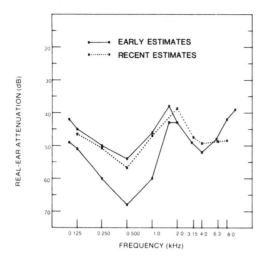

Figure 10-2. Estimates of the bone conduction limitations to hearing protector attenuation (from Berger, 1983b).

condition since the sound is unable to freely escape the ear as it can in the unoccluded case. This is called the *earplug effect* (Zwislocki, 1957), or more generally, the *occlusion effect* (Tonndorf, 1972).

An occlusion effect is observed whenever the ear is plugged or covered, either with an insert, semi-aural, or circumaural HPD. However, the magnitude of the occlusion effect is a function of how the occluding device is fitted (Berger & Kerivan, 1983). The occlusion effect is maximized when the canal is sealed at its entrance with a semi-aural or canal cap type of device. It diminishes either as the occluder is more deeply inserted (earplugs instead of semi-aurals and more deeply fitted earplugs versus less deeply fitted ones) or as the occluder begins to contain a larger internal volume (earmuffs instead of semi-aurals and larger volume earmuffs versus smaller volume earmuffs). This variation of the occlusion effect as a function of the method of occlusion is why a range of BC limits, instead of just a single value, is plotted for the frequencies below 2 kHz in Figure 10-2.

The occlusion effect can be easily demonstrated by capping one's ear canals while speaking aloud. When the canals are properly sealed or covered, one's own voice

takes on a bassy, resonant quality due to the enhancement of the BC paths by which talkers partially hear their own speech. The occlusion effect is often cited by users as one of the objectionable features of wearing HPDs. For those who are particularly sensitive to this effect, it can be minimized by inserting earplugs more deeply or using larger volume earmuffs. The occlusion effect is also a useful hearing protector fitting aid, as is discussed in the section entitled Selection, Fitting, Use, and Care of HPDs.

ESTIMATING PROTECTION

Laboratory Test Methods

More than a dozen different laboratory methods for evaluating HPD attenuation have been reported in the literature (Berger, 1986). The tests have included auditory threshold shift measures, loudness balance comparisons, masking, and lateralization procedures using human subjects. Various instrumented measurements using human subjects, artificial heads, and cadavers as acoustical test fixtures have also been employed. The most common, and definitely one of the most accurate, tests (Berger, 1986; Berger & Kerivan, 1983) is that of real-ear attenuation at threshold (REAT), which is the subject of one international and three American standards (International Organization for Standardization [ISO] 4869; American National Standards Institute [ANSI] Z24.22-1957; Acoustical Society of America [ASA] STD-1 1975; American National Standards Institute [ANSI] S12.6-1984) as well as documents in many other countries.[1] Virtually all available manufacturers' reported data have been derived via this method.

---

[1]As of this writing, a new American REAT standard, ANSI S12.6, was recently issued. It is too new for any data acquired in conformance with it to have yet been published. For that reason, and due to its strong similarity to ASA STD 1, the discussion in this chapter focuses on the 1975 standard. A detailed comparison of ASA STD 1 and ANSI S12.6 may be found in Berger (1985a).

REAT measures are based upon determination of the difference between the minimum level of sound that a subject can hear without wearing an HPD (open threshold) and the level needed when the subject is wearing an HPD (occluded threshold). The difference between these two thresholds, the threshold shift, is a measure of the REAT afforded by the device. Since the test is conducted at a relatively low sound pressure level (SPL), it cannot accurately characterize the performance of HPDs that claim to offer attenuation that increases with sound level. However, for linear HPDs (those not containing valves, orifices, diaphragms, or active electronic circuitry) the attenuation measured by an REAT evaluation will accurately represent the performance of the device regardless of sound level (Berger & Kerivan, 1983; Humes, 1983; Martin, 1979).

According to ASA STD-1, REAT measurements are to be conducted in a diffuse sound field using ⅓ octave bands of noise and a minimum of 10 subjects whose open and occluded thresholds are measured three times each. This results in 30 data points at each frequency, from which a mean attenuation ($\bar{x}$) as well as a standard deviation (SD) are computed, the latter parameter providing an indication of the variability in attenuation across subjects and replications. The mean attenuation represents protection that approximately 50 percent of the test subjects meet or exceed. When it is appropriate to estimate the protection that a greater percentage of the subjects attain, adjustments to the mean may be computed by subtracting one or more standard deviations.

Numerous data are still available that have been measured in accordance with the older test standard, ANSI Z24.22. That standard required the subject to be seated in a directional sound field, usually achieved by testing in an anechoic chamber with a loudspeaker in front of and facing the subject. The test sounds were pure tones. These conditions were less representative of typical industrial exposures than are the noise bands and nondirectional sound field

specified in the newer standard, ASA STD-1. In practice, the electroacoustic differences between the two standards have little effect on the measured attenuation of earplugs, but earmuff attenuation tends to be lower, especially in the 1 and 4 kHz region, using the newer standard (Bolka, 1972; Martin, 1977). Additionally, SDs tend to be lower for both types of HPDs when tests are conducted using noise bands in a diffuse sound field.

REAT data are normally limited to the frequency range of 125 Hz to 8 kHz. Some laboratories routinely test to frequencies as low as 63 or 75 Hz (Camp, 1979; Martin, 1977), but few data are available at frequencies above 8 kHz. In a recent study (Berger, 1983d), the attenuation of five earplugs, one semi-aural device, four earmuffs, and one earplug-plus-earmuff combination were examined at frequency extremes. The results suggested that the attenuation values of HPDs at 80 Hz, and at 12.5 and 16 kHz, could be approximated by assuming they were equivalent to the performance at the 125 Hz and 8 kHz ⅓ octave bands respectively.

Long Method Calculation of HPD Noise Reduction

If one can be assured that the user of an HPD is wearing it in the same way as did the laboratory test subjects (see *Real-World Performance of HPDs* in this chapter), then the most accurate method of applying laboratory test data to estimate a user's protected exposure is the long method calculation illustrated in Table 10–1 (Method 1 in National Institute for Occupational Safety and Health [NIOSH], 1975). At each frequency, the HPD's $\bar{x}$ minus an SD correction (2 SDs are used in Table 10–1), is subtracted from the measured A-weighted octave band SPL. The protected levels are then logarithmically summed to determine the A-weighted sound level under the protector. This computation requires that the user conduct or have available an octave band analysis (see Chapter 3) and perform the appropriate calculations for each

TABLE 10-1.
Long Method Calculation of HPD Noise Reduction

Octave Band Center Frequency (Hz)	125	250	500	1000	2000	4000	8000	dBA*
1. Measured sound pressure levels	85.0	87.0	90.0	90.0	85.0	82.0	80.0	
2. A-weighting correction	−16.1	−8.6	−3.2	0.0	+1.2	+1.0	−1.1	
3. A-weighted sound levels [step 1 − step 2]	68.9	78.4	86.8	90.0	86.2	83.0	78.9	93.5
4. Typical premolded earplug attenuation	27.4	26.6	27.5	27.0	32.0	46.0[†]	44.2[††]	
5. Standard deviation × 2	7.8	8.4	9.4	6.8	8.8	7.3[†]	12.8[††]	
6. Estimated protected A-weighted sound levels [step 3 − step 4 + step 5]	49.3	60.2	68.7	69.8	63.0	44.3	47.5	73.0

The estimated protection for 98% of the users in the noise environment, assuming they wear the device in the same manner as did the test subjects and assuming they are accurately represented by the test subjects, is

93.5 − 73.0 = 20.5 dBA.

[†]Arithmetic average of 3150 and 4000 Hz data.
[††]Arithmetic average of 6300 and 8000 Hz data.

individual noise spectrum, that is, the amount of protection afforded cannot be calculated independently of the noise spectrum in which the HPD will be worn.

It is important at this point to clearly understand the meaning of the SD correction. It is *not* a method of correcting the laboratory data to estimate performance in the real world; rather, it adjusts the mean laboratory data to reflect the attenuation achieved by 84 percent (for a 1-SD correction) or 98 percent (for a 2-SD correction) of the laboratory subjects. The correction also applies to actual users to the extent that it can be assumed they are accurately represented by the test subjects and are wearing the device in the same manner as did the laboratory subjects. However, such assumptions, especially the latter one, are not normally justified.

Single Number Calculation of HPD Noise Reduction

The primary advantage of a single number rating is that it can be precalculated by the manufacturer and supplied to the user with the device. This permits estimation of wearer noise exposures using only simple mathematics and a single noise measurement.

One of the more accurate and also one of the more common single number ratings is the Noise Reduction Rating (NRR), which was adapted by the Environmental Protection Agency ([EPA], 1979) from earlier National Institute of Occupational Safety and Health ([NIOSH], 1975) work. It is calculated in a manner similar to the long method, except that a pink noise spectrum (equal energy in each octave band from 125 Hz to 8 kHz) is used instead of the particular noise spectrum, and a negative 3-dB spectral uncertainty correction is included in the computational procedures. The 3-dB correction is a safety factor to protect against overestimates of HPD noise reduction that might arise from differences between the shape of the assumed (pink) and particular noise spectra. In computing the NRR, the x values are reduced by subtracting 2 SDs so the resultant NRR should theoretically estimate the degree of protection at the 98th percentile. The meaning of this 2-SD correction is the same as discussed for the long method calculation. The NRR calculated in this manner will be denoted by the abbreviation NRR_{98}.

A representative NRR calculation is presented in Table 10–2. This table is organized in a manner similar to Table 10–1 so that the computational procedures may be compared. The NRR, which is finally computed in step 9, is found by subtracting the protected (interior) A-weighted sound levels from the unprotected (exterior) C-weighted sound levels, less a 3-dB spectral uncertainty correction. The NRR, using the same earplug data as are found in Table 10–1, is 20.8 dB. The NRR is used to estimate wearer noise exposures by subtracting it from the C-weighted sound level as shown in equation (1).

Estimated Exposure (dBA) = Workplace Noise Level (dBC) – NRR (1)

To compare the long method calculation to the NRR, the C-weighted sound level for the spectrum shown in Table 10–1 must be computed. It is 95.2 dBC. When the NRR is subtracted from this C-weighted sound level as indicated in equation (1), the predicted exposure is 74.4 dBA versus the 73.0 dBA estimated using the long method. The more conservative estimate of protection that is provided by the NRR is expected since it includes the 3-dB spectral correction that is not found in the long method. This correction is intended to assure that, in most instances, errors arising from spectral variability will lead to under- instead of overestimates of protection.

The practice of using the NRR with C-weighted sound levels may seem peculiar since one would expect that estimation of an employee's protected A-weighted exposure should be accomplished by simply subtracting a single number rating from the A-weighted workplace noise level. However, a number of authors have examined this

TABLE 10–2.
Method of Computation of the NRR

Octave Band Center Frequency (Hz)	125	250	500	1000	2000	4000	8000	dB(X)[A]
1. Assumed sound pressure levels	100.0	100.0	100.0	100.0	100.0	100.0	100.0	
2. C-weighting correction	−0.2	0.0	0.0	0.0	−0.2	−0.8	−3.0	
3. C-weighted sound levels [step 1 − step 2]	99.8	100.0	100.0	100.0	99.8	99.2	97.0	108.0 [dBC]
4. A-weighting correction	−16.1	−8.6	−3.2	0.0	+1.2	+1.0	−1.1	
5. A-weighted sound levels [step 1 − step 4]	83.9	91.4	96.8	100.0	101.2	101.0	98.9	
6. Typical premolded earplug attenuation	27.4	26.6	27.5	27.0	32.0	46.0[B]	44.2[C]	
7. Standard deviation × 2	7.8	8.4	9.4	6.8	8.8	7.3[B]	12.8[C]	
8. Estimated protected A-weighted sound levels [step 5 − step 6 + step 7]	64.3	73.2	78.7	79.8	78.0	62.3	67.5	84.2 [dBA]
9. NRR = step 3 − step 8 − 3[D] NRR = 108.0 − 84.2 − 3 = 20.8 dB								

The NRR represents the attenuation that will be obtained by 98% of the users in typical industrial noise environments, assuming they wear the device in the same manner as did the test subjects, and assuming they are accurately represented by the test subjects.

[A]Logarithmic sum of 7 octave band levels in the row. This is a C-weighted sound level for step 3 and an A-weighted sound level for step 8. See Chapter 2 for method of computing a logarithmic sum.

[B]Arithmetic average of 3150 and 4000 Hz data.

[C]Arithmetic average of 6300 and 8000 Hz data.

[D]The 3-dB spectral uncertainty factor is to protect against overestimates of the HPD's noise reduction that could arise from potential differences between the assumed spectrum and that of the user's actual exposure.

question, and the weight of empirical evidence leads to the conclusion that "the best single-number method, representing a practical compromise between accuracy and convenience, is one which generates a rating number to be subtracted from the C-weighted SPL of the noise" (Sutton & Robinson, 1981). This relationship exists because the attenuation of most HPDs decreases as frequency decreases, falling off at approximately the same rate as the increase in difference between the C- and A-weighting curves. Alternatively, the relationship may be explained by observing that the A-weighted noise reduction of a hearing protector decreases as the proportion of low-frequency energy in the noise spectrum increases, and this may be accounted for by taking the difference between the C- and A-weighted sound levels of the noise spectrum.

If one chooses to subtract the NRR from A-weighted sound levels, considerable accuracy is lost, and an additional 7 dB safety factor must be included in the computations as shown in equation (2) (NIOSH, 1975; Occupational Safety and Health Administration [OSHA], 1983).

Estimated Exposure (dBA) =
Workplace Noise Level (dBA) − (NRR–7) (2)

The 7-dB safety factor in equation (2) accounts for the largest differences typically expected between the C- and A-weighted sound levels of industrial noises. Since it is a "worst-case" correction, in most instances it will overestimate actual C-A differences. As an alternative, one can correct the A-weighted time-weighted average (TWA) (obtained from dosimeters which are usually only capable of A-weighted measurements) by using a sound level meter to develop a C-A value for typical processes, areas, or job descriptions (Berger, 1984; Royster & Royster, 1985). This C-A value is then added to the A-weighted TWA to calculate an estimated C-weighted TWA or workplace noise level, from which the NRR can be subtracted using the procedure of equation (1). To the extent that an accurate C-A value

can be estimated, this method will provide enhanced accuracy over use of equation (2) for those situations in which C-weighted sound levels are unavailable.

Berger (1983a) has shown that the primary source of error in applying the NRR to estimate user noise exposures is not in the computation of the NRR from the basic laboratory data, but in the fact that laboratory data are not representative of the values attained by actual users. Later in this chapter (see *Real-World Performance of HPDs*) the divergence between laboratory and field data will be discussed, and suggestions for de-rating laboratory data will be presented.

Inter- and Intra-laboratory Variability

Another consideration regarding laboratory attenuation data is the reproducibility of the data between different test facilities. Berger, Kerivan, and Mintz (1982) reported the results from a round-robin test program initiated by the U.S. EPA. Four HPDs representing a wide range of commercially available products were tested. Seven laboratories participated directly; data obtained separately from an eighth laboratory also were included in the evaluation. Data from eighth laboratory were included since that facility was responsible for more than 80 percent of manufacturers' reported data on file with the EPA at the time of the study.

The results showed significant variation in both x̄ and SD, leading to substantial differences in reported NRRs among the different laboratories (see Figure 10–3). One source of the variability appeared to be the uncertainty of obtaining the proper fit to avoid acoustic leaks. Others included subject selection and training, as well as data reduction techniques.

The results demonstrated that the task of rank ordering the performance of a group of HPDs was not possible using presently available data unless the user could be assured that all of the data were from one laboratory. Therefore, all of the subsequent laboratory data that are presented in this

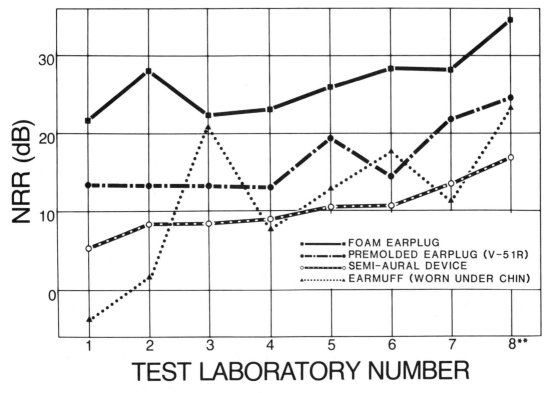

Figure 10-3. NRRs from eight U.S. laboratories. Data from Berger, Kerivan, and Mintz (1982).

**Laboratory responsible for greater than 80% of the manufacturers' label verification reports that were on file with the U.S. EPA at the time of the round robin study.

chapter are from only one facility, the E-A-R Division Acoustical Laboratory.

Even if one analyzes data from just one laboratory when comparing HPD attenuation values and/or NRRs, accuracy and repeatabilty are of concern. Over a period of years or even months, a facility's subject selection, fitting, and sizing techniques may vary, the characteristics of subsequent samples of test devices may differ from those tested initially, and other uncontrolled parameters may change and give rise to divergent results. Berger (1983c, 1986) has provided representative data illustrating many of these effects, and based upon his results and those of others, he concluded that *changes in the NRR of less than approximately 3 dB should not be regarded as having any practical importance.*

SELECTION, FITTING, USE, AND CARE OF HPDs

Hearing protection devices may be broadly categorized into earmuffs, which fit over and around the ears (circumaural) to provide an acoustic seal against the head; earplugs, which are placed into the ear canal to form a seal; and semi-aural devices (also called semi-insert, concha seated, and canal caps), which are held against the canal entrance with a headband to provide an acoustic seal at that point. Other categories that may be considered are helmets, which contain circumaural cups and are normally used only by the military or when head protection combined with communication must be provided along with hearing protection, and devices that rest upon the external ear (supra-aural), such as are commonly used for presenting stimuli for hearing testing. Due to the considerable variety among earplugs, they can be further differentiated into three types: premolded, formable, and custom molded.

As with all human attributes, there is a wide degree of variability in the anatomical characteristics of the human head and ear. Population norms for head widths have been reported as 121, 136, and 148 mm for the 5th, 50th, and 95th percentiles, respectively. Similarly, the 5th to 95th percentile head heights cover a range of 116 to 143 mm with a median value of 130 mm. Ear canals also vary widely in size, shape, position, and angle of inclination among individuals and even between the ears of the same individual. Cross sections may vary from 3 to 14 mm, with an average reported diameter of approximately 8 mm and an average length of 25 mm. Typically, ear canal cross sections are elliptical in nature, but the extremes of round and slit shaped canals are also observed. This wide range in anthropometric data must be considered by the designers of HPDs and also must be accounted for when protectors are selected for particular individuals. Obviously, no one device will be the correct choice for all concerned.

Earmuffs

Earmuffs normally consist of rigid molded plastic earcups that seal around the ear using foam or fluid-filled cushions. They are held in place with metal or plastic headbands, or by a spring-loaded assembly attached to a hard hat. The cups are lined with acoustical material, typically an open cell polyurethane foam, to absorb high frequency (>2 kHz) energy within the cup. The headbands may function in a single position only, or may be of the more versatile "universal" style suitable for use over the head, behind the head, or under the chin. Most manufacturers offer plastic or fabric crown straps for use in the alternate wearing positions to provide for a more

snug, secure, and comfortable fit. Representative earmuffs are shown in Figure 10–4.

Earmuffs are relatively easy to dispense since they are one-sized devices designed to fit *nearly* all adult users. Regardless, earmuffs must be evaluated for fit when initially issued, since not every user can be fitted by all models. Head or ear sizes may fall outside the range that the muff band or cup openings can accommodate. Furthermore, unusual anatomical features such as facial structures with prominent cheek bones (zygomatic arches) or severe depressions below the pinna and behind the jaw (posterior to the temporo-mandibular joint) are particularly hard to fit.

Earmuffs are good for intermittent exposures due to the ease with which they can be donned and removed, and they may be suitable when earplugs are contraindicated. For long-term wearing, it is often reported that earmuffs feel tight, hot, bulky, and heavy, although in cold environments their warming effect is appreciated. It is easy for supervisors to monitor that earmuffs are in use. However, earmuff attenuation can be easily compromised by a number of factors, so it is inadvisable to assume that use is synonymous with "effective protection."

Instructions for earmuff usage must be verbally reviewed when the devices are issued, since details of correct placement are often ignored or overlooked in practice. For example, many cups have a preferred orientation, either left/right or top/bottom, and this must be pointed out to the employee. Long hair, sideburns, and caps can also reduce attenuation. Employees should be advised to position the earmuffs with as much of their hair removed from the cushion-to-head interface as possible, and to be certain that caps or other clothing are not placed underneath the earmuff cushions.

Earmuff performance will be degraded by anything that compromises the cushion-to-circumaural-flesh seal. This includes other pieces of personal protective equipment such as eyewear, masks, face shields, and helmets. Typically, eyeglasses

Figure 10-4. Representative earmuffs (left to right): large volume cups with single-position metal headband, small volume cups attached to hard hat, small volume cups with "universal" plastic headband and optional crown strap.

will degrade earmuff attenuation by 3 to 7 dB, with the losses most noticeable at the low and high frequencies, although the effect varies widely among earmuffs (Nixon & Knoblach, 1974). The losses will be minimized with thinner, closer-fitting temple pieces.

Foam pads are available that fit over eyeglass temples in order to relieve the pressure of the impinging cushion and also to attempt to circumvent loss of attenuation that temples may create, but they do little to reduce leaks caused by overlength temples breaking the seal behind the ear. Unfortunately, the data indicate (Berger, unpublished data, 1982) that at least one commercially available pad device actually creates more of a loss in attenuation than simply wearing the eyeglasses alone. However, it can relieve the discomfort of temples pressing against the head, and therefore should be considered for use, since comfort is crucial in motivating employees to wear their HPDs.

Acoustical leaks may also arise due to employees modifying their earmuffs by drilling holes in the cups to promote ventilation and assist the drainage of perspiration. In some instances, this practice has been so extreme as to include the personalizing of earmuff cups by drilling initials in the cup wall, using many closely spaced holes (Riko & Alberti, 1982). Obviously, this practice must be discouraged.

When the use of protective headgear is required, hard hats with attached earmuffs provide a convenient alternative to the use of headband-attached earmuffs worn behind the head or under the chin. However, hard hat–attached muffs are more difficult to properly orient and fit since the attachment arms can never provide as adaptable an adjustment as do the headband-attached versions, nor can they fit as wide a range of head sizes. For such devices, not only must the attachment arms be properly extended and located, but the helmet's webbing must be adjusted to properly locate the hat on the head. Compromises may need to be made between comfort and safety in adjusting either the hard hat or the earmuff cups.

A number of design parameters affect the attenuation of earmuffs, including cup volume and mass, headband force, area of opening in the cushion, and the materials from which the device is constructed. The most visible of these factors to the buyer will be the cup volume, which can vary from a minimum of slightly less than 100 cm³ to the largest cups sold today with a volume of approximately 330 cm³

Volume is one of the primary factors affecting earmuff attenuation, as shown in Figure 10–5, where the attenuation of foam-cushion earmuffs (worn over the head) is illustrated. Notice that from 125 Hz to 1 kHz, earmuffs provide attenuation that increases approximately 9 dB/octave, with

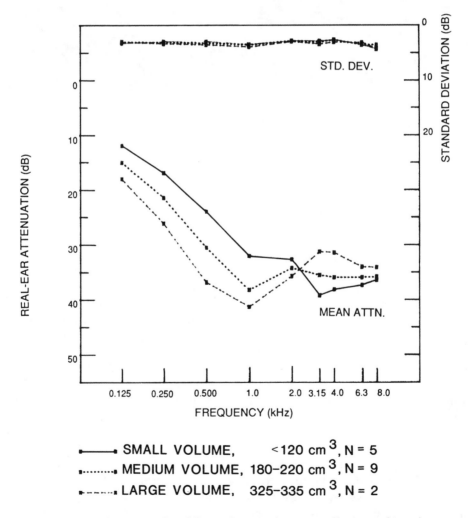

Figure 10–5. Real-ear attenuation of 16 different foam cushion earmuffs grouped into three categories.

the larger volume muffs providing better attenuation than the smaller ones, since in this region the extra volume and mass of their cups are the controlling physical parameters. Above 2 kHz however, larger volume earmuffs tend to be inferior, since in this region their increased shell surface area makes them more susceptible to developing vibrational modes within the cup walls and therefore, less capable of blocking higher frequency acoustical energy. At 2 kHz, the attenuation of all of the devices approaches 39 dB, the limit imposed by the BC pathways (cf. Figure 10–2).

Since the circumaural regions of the head are rarely flat or free from obstructions, sufficient force must be exerted so the cushion fully contours and seals to the side of the head. Manufacturers must compromise between high forces, which yield good attenuation and poor comfort, and low forces, which produce the opposite results. Band force, which can deteriorate with use and age, is also often modified by wearers if they feel a device is too tight. For many earmuffs, it is easy to spring the band without breaking it, thus, permanently reducing the force. Band force can be roughly checked (see Figure 10–6) by comparing the resting positions of suspect

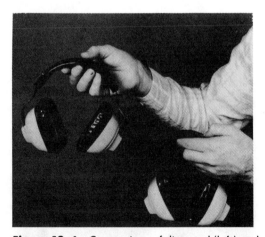

Figure 10–6. Comparison of distorted (left) and undistorted (right) headbands of two identical new earmuffs adjusted for the same headband extension. The headband force of the earmuff on the left has been reduced by 20 percent.

earmuff cups to those of new samples to make sure that the cup-to-cup separation is approximately the same.

The dimensions of the earmuff cup opening must also be considered when selecting a protector. Due to the transformation of sound pressure from the outside to the inside of the earmuff cup, the smaller the cup opening (other parameters being held constant), the greater will be the attenuation of the protector. Additionally, smaller diameter cup openings allow cushions to make circumaural contact nearer to the base of the pinna, which in turn tends to increase attenuation, since facial contours and jaw and neck motion are minimized in those regions (Guild, 1966). However, when attenuation is increased by decreasing the cup opening, it is often at the expense of comfort and ease of use.

As the cup opening is decreased in size, the difficulty of fitting it over and around the pinna increases. For example, the length and width of male pinnae at the 75th and 95th percentiles are 67 × 38 mm and 71 × 41 mm, respectively, whereas some of the higher attenuating earmuffs sold today have openings as small as 58 × 39 mm. For such devices, wearers with even average size ears must be careful to ''tuck'' their ears into the cups. Even for cups with more typical openings (65 × 41 mm), employees must still be cautioned to place their pinnae fully within the earmuff cups, since some have been observed who rest the cushion on portions of the external ear, which not only reduces comfort, but also creates a significant acoustical leak (Riko & Alberti, 1982).

The selection of whether to use foam or liquid-filled cushions is somewhat academic in today's market, since both offer similar performance, the liquid-filled cushions providing slightly better protection at the low frequencies, and the foam slightly more at the high frequencies. Tests indicate that, for at least one brand of earmuff, both types contour equally well around eyeglass temples (Berger, unpublished data, 1982). Liquid-cushion earmuffs weigh 5 to 40 percent more than their foam counterparts, generally sell at a 10 to 20 percent premium, and may be punctured or split, allowing their contents to drain out.

The cover or bladder on both foam and liquid cushions is made of a plastic material that may harden and deform with time, due to contact with body oils, perspiration, cosmetic preparations, and environmental contaminants. For this reason, cushions should be examined at least twice yearly and replaced if necessary. Absorbent cushion covers are available to enhance comfort, but they may reduce attenuation since they are porous and can create acoustical leaks.

When selecting earmuffs for particular applications, the relatively small differences in attenuation between most popular brands and types (see small and medium volume earmuffs in Figure 10-5) suggest that, except for extreme noise exposures, where the highest protection possible must be afforded, selection should be based upon other factors. Often smaller, less expensive earmuffs may provide better comfort, and therefore will be more readily accepted. Symmetrical designs that do not require a particular orientation will decrease the likelihood of misuse. In practice, more attenuation can often be gained by assuring that properly maintained earmuffs are worn correctly and consistently than can be assured by buying heavier, and perhaps more expensive, ''high performance'' models.

Earplugs

Earplugs tend to be more comfortable than earmuffs for situations in which protection must be used for extended periods, especially in warm and humid environments. They can be worn easily and effectively with other safety equipment and eyeglasses, and are convenient to wear when the head must be maneuvered in close quarters. However, they are less visible than muffs, and therefore, their use can be somewhat more difficult to monitor.

Earplugs come in a variety of sizes, shapes, and materials, but regardless of the particular model, care must be taken in inserting and sometimes preparing them for use. They generally require more skill and attention during application than do earmuffs. In fact, even under the best circumstances, a small percentage of users may never learn to wear them correctly, either due to their particular canal shapes, lack of manual dexterity, finger size, or missing digits.

As with earmuffs, when earplugs are initially dispensed, even the formable ''one-size-fits-all'' devices, the fitter must individually examine each person to be sure that a proper seal can be obtained. This can sometimes be determined by simple observation, but it almost always requires diligent participation of the fitter and the person being fitted.

Earplug insertion will be facilitated if the wearer reaches around behind the head with the hand opposite the ear that is being fitted and pulls outward and upward on the pinna while inserting the plug (see Figure 10-7). This procedure helps to straighten and enlarge the canal opening. Since ear canals usually angle upward and/or toward the front of the head, employees should be instructed to push the plugs in that direction during insertion, although due to the wide variation in human anatomy, other directions may also be appropriate. There is little likelihood of hurting the eardrum

Figure 10-7. Preferred method of pulling the pinna outward and upward while simultaneously fitting an earplug.

during insertion since the sensitivity of the adult ear canal to pressure or pain increases significantly as the eardrum is approached. The discomfort experienced when touching these deeper portions of the canal will alert the user to stop pushing the plug before a problem can occur.

Plugs that create an airtight seal, such as premolded inserts, can be painful and potentially damaging to the eardrum if they are rapidly removed. These plugs should be withdrawn with a slow, twisting motion to gently break the seal as they are extracted from the ear. With most foam and fibrous plugs, which do not create a pneumatic seal (and hence, cause less of a blocked-up feeling), there is little possibility of generating a sudden, large pressure change upon rapid removal, and thus, virtually no likelihood of damaging or rupturing the eardrum.

Finally, users must be alerted that earplugs may work loose with time and require reseating. Studies have demonstrated this effect for certain premolded and fiberglass earplugs, but have shown foam and custom molded earplug wearers did not experience this problem. (Abel & Rokas, 1986; Berger, 1981; Kasden & D'Aniello, 1976; Krutt & Mazor, 1980).

Premolded Earplugs

These devices are manufactured from flexible materials such as vinyls, cured silicones, and other elastomeric formulations. Generally, silicone formulations offer the best durability and resistance to shrinkage and hardening. Most models are available with attached cords to help prevent loss and to simplify storage by permitting hanging around the neck when not in use. Typical models are depicted in Figure 10–8.

One of the oldest and most common of the premolded devices, the V-51R earplug, was developed during World War II. It is a one-flanged, PVC insert, available in five sizes, and sold under various brand names. If this device is to be used for an entire population, all five sizes must be stocked, in which case about 95 percent of adult

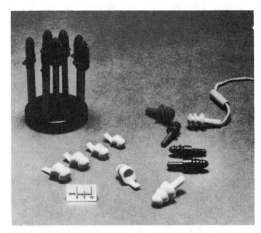

Figure 10–8. Representative premolded earplugs (clockwise from upper left): 0-flanged 8-sized hollow plug (sizing tool shown), 3-flanged 3-sized solid plug with insertion tool and cord, 2-flanged 2-sized hollow plug, 3-flanged 1-sized solid plug with swept-back flanges, and 1-flanged 5-sized plug (V-51R).

males can be adequately fitted (Blackstock & von Gierke, 1956). It is difficult to suggest what distribution of sizes to order since ear canal dimensions vary as a function of the racial and sexual characteristics of the population, with black females having the smallest ear canals and white males, the largest. Black males and white females fall in between (Royster & Holder, 1982). For example, in a predominantly black female population, nearly 40 percent would be expected to use an extra-small V-51R, whereas this number would be nearer to 5 percent for a white male group.

The V-51R is particularly prone to shrinkage, cracking, and hardening from exposure to body oils, perspiration, and cerumen. Safe use periods for plugs of this type have been estimated to be 3 months (Royster & Holder, 1982), but even more frequent replacement periods may be required. Another common failure for this plug is splitting along the mold line. This is often exacerbated by the tendency of users to store the tip of one plug inside the posterior opening of the other.

Premolded plugs are also available with fewer (zero) as well as more (up to five) flanges. Generally, the greater the number of flanges, the fewer sizes are required to fit the population. Flangeless premolded plugs normally are more difficult to size correctly and are more prone to work loose during use. Some two- and three-flanged varieties are molded with an entrapped air pocket to increase softness and improve comfort. One recently developed three-flanged variety incorporates a swept-back flange design to increase conformability and contact area with the ear canal in order to provide improved comfort and high attenuation.

As with all other HPDs, premolded earplugs are subject to user modifications to improve comfort without regard for the effects on attenuation, which will of course almost always be degraded. Common alterations include removal of flanges, punching holes through the body of the plug, and puncturing entrapped air pockets so that the plug deflates upon insertion (Gasaway, 1984).

Correctly sizing and fitting premolded inserts will almost always require that a compromise be made between attenuation and comfort. The appropriate compromise can often be achieved, but only with care and skill (see Figures 10–9A, B, C, D, and 10–10A, B). Although it is often suggested that if an ear canal falls between two sizes, the larger size plug should be selected, this may be poor advice when one considers the importance of comfort. Even though the larger size may provide better attenuation, if it is not worn or not used correctly due to discomfort, the resultant attenuation advantage will be lost. Conversely, in the *initial* fitting it is best to try a plug that is too large rather than too small (Royster & Royster, 1985). Then, if a second (smaller) size must be selected and tested, it will appear to be more comfortable to the wearer. Should the fitter proceed in the reverse order, by initially underestimating the size of plug required, if a second (larger) size must be selected, it is more likely that it will seem less pleasant, and the employee may reject it regardless of the appropriateness of its size.

When initially inserting premolded earplugs, the fitter should be able to easily detect gross errors in sizing. Ear gauges are available from some manufacturers of premolded earplugs to aid in this process. Plugs that are much too small will tend to fall into the canal, their depth of insertion being limited only by the fitter's finger and not the plug itself. Overly large plugs will either not enter the canal at all or will not penetrate far enough to allow contact of their outermost flanges with the concha (the hollow shell-like area at the canal opening). A plug that appears to make contact with the interior wall of the canal without appreciably stretching the tissues, and that is well seated, is a good size to begin wearing (Guild, 1966). If, after a couple of weeks of use, the employee still experiences problems or discomfort, then another size or type of HPD should be issued.

A properly inserted premolded insert will generally create a plugged or blocked-up feeling due to the requisite airtight seal. Additionally, due to the occlusion effect, users should experience a resonant or bassy quality to their voices, as though they were talking in a drum. This will be more pronounced for males than for females due to the lower pitch of male voices. This perception is useful as a fitting test, since if the wearer speaks aloud with only one ear correctly fitted, the voice should be more strongly heard or felt in the occluded ear (Ohlin, 1975). If this does not occur, the plug should be reseated or resized. When the second ear has been fitted correctly, the wearer should perceive his or her voice as though it were emanating from the center of the head.

Records of the plug size issued to each employee should be maintained at the fitting station. They should be consulted when plugs are reissued to replace those that are lost or worn out. Experience suggests that in approximately 2 to 10 percent of the population, different size premolded earplugs will be required for the left and right ears, although military data indicate that for the V-51R type earplug, this number may be as high as 20 percent (Department of the Air Force, 1982). As a general rule, the

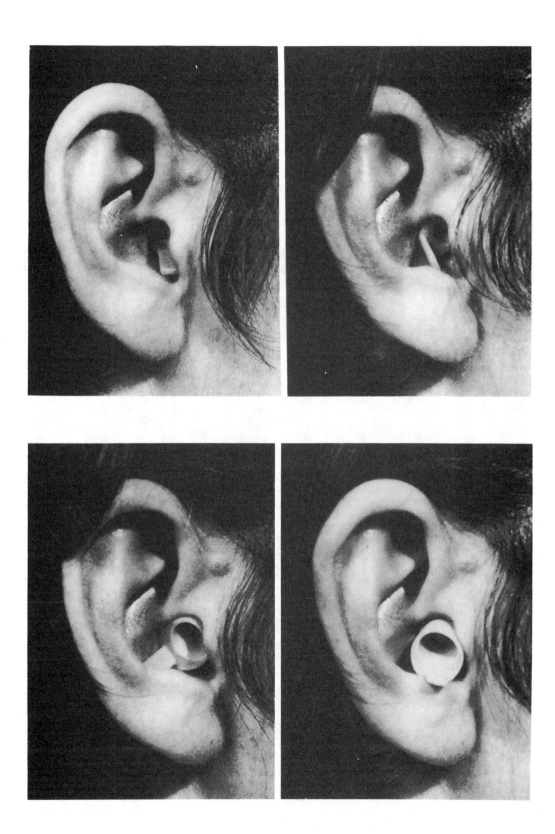

Figure 10-9. Examples of different insertions of a V-51R earplug.

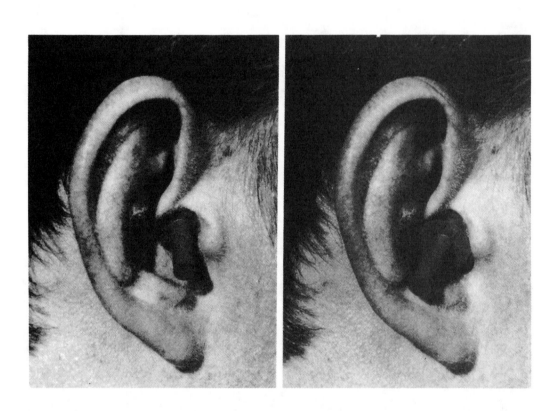

Figure 10-10. Examples of different insertions of a three-sized three-flanged earplug.

more sizes in which a particular premolded plug is manufactured, the greater will be the likelihood that different ones are required for each ear. The use of unmatched sizes for the two ears can pose a problem for those devices that are color coded to indicate size, since some employees may be hesitant to wear two different-colored plugs.

Values of attenuation for a variety of pre-molded earplugs are plotted in Figure 10–11.

The values are approximately constant up to 1 kHz clustering closely around 25 dB, and increasing to approximately 40 dB at the higher frequencies. Below 2 kHz, the lowest measured attenuation was found using a one-sized flangeless plug and the highest using three-flanged devices. The relatively small range in performance between devices suggests that selection would be most meaningfully based upon comfort, and the ease

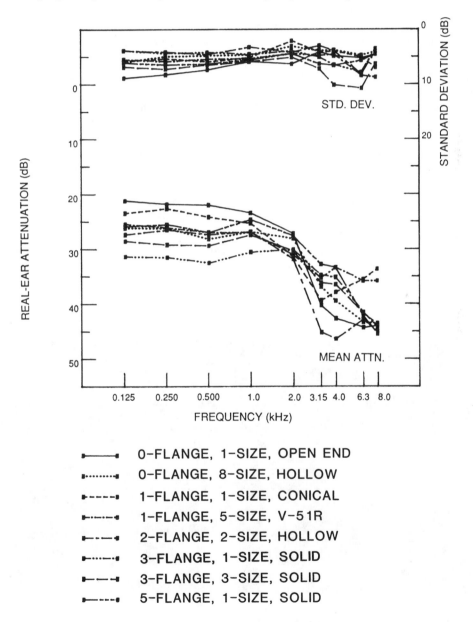

•——•	0–FLANGE, 1–SIZE, OPEN END
•········•	0–FLANGE, 8–SIZE, HOLLOW
•– – –•	1–FLANGE, 1–SIZE, CONICAL
•–·–·–•	1–FLANGE, 5–SIZE, V–51R
•––·––•	2–FLANGE, 2–SIZE, HOLLOW
•–····–•	3–FLANGE, 1–SIZE, SOLID
•——•	3–FLANGE, 3–SIZE, SOLID
•–– – –•	5–FLANGE, 1–SIZE, SOLID

Figure 10–11. Real-ear attenuation of eight different models of premolded earplugs.

with which the various devices can be sized and fitted. Comparison to Figure 10–5 indicates that the average laboratory performance can be better than earmuffs at 125 and 250 Hz and above 2 kHz, but is poorer at the intermediate frequencies.

Formable Earplugs

Earplugs of this variety may be manufactured from cotton and wax, spun fiberglass (often called fiberglass down, mineral wool, or Swedish wool), silicone putty (exposed, or encased in a bladder), and slow-recovery foams. Life expectancies vary from single-use products such as some fiberglass down earplugs, to relatively permanent plugs such as encased putties. Foam earplugs provide intermediate life expectancies varying from a few pairs per day to a pair per week or more. Except for bladder-encased putties and some brands of foam plugs, formable plugs are generally not available with attached cords.

The primary advantage of formable earplugs is comfort, some of the products in this category being among the most comfortable and user-accepted devices sold today. Additionally, formable plugs are generally sold in only one size that usually fits most, but not all, ear canals. This simplifies dispensing, record keeping, and inventory problems, but when using such products, special attention must be given to wearers with extra-small and extra-large ear canals to make sure that the plugs are not too tight or too loose. Since these plugs usually require manipulation by the user prior to insertion, during which time the hands should be relatively clean, they may not be the best choice for environments in which HPDs have to be removed or reinserted many times during a work shift by employees whose hands are contaminated with caustic or irritating substances or sharp or abrasive matter. User modifications and abuse typically consist of reducing the amount of formable materials used for each plug, or in the case of foam earplugs, cutting their length or diameter. Representative formable earplugs are shown in Figure 10–12.

Although cotton alone is a very poor hearing protector due to its low density and high porosity, when it is combined with wax, good protection can be obtained. The wax tends to make the devices somewhat

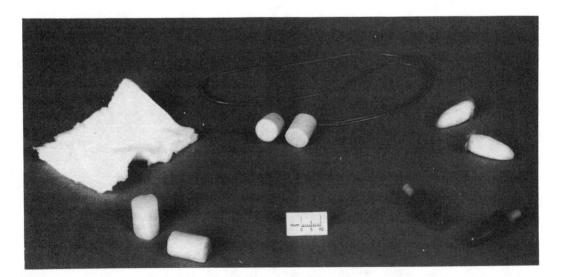

Figure 10–12. Representative formable earplugs (clockwise from upper left): fiberglass down, slow-recovery foam with attached cord, fiberglass down with taut polyethylene sheath, encased silicone putty, cotton wax.

messy to prepare and use, especially under conditions of high temperature. These plugs are not commonly found in industrial hearing conservation programs, but they have attained popularity in the consumer market. Insertion is accomplished by kneading the material until it softens, forming it into a ball, or preferably a cone, and then pressing it into the entrance of the ear canal. Since these materials lack elasticity, they may lose their seal as a result of jaw and neck motion and require frequent reseating (Ohlin, 1975).

Fiberglass down was first available in the late 1950s as a folded strip of batting, which required tearing and rolling into a cone for insertion. More recently, it has evolved into versions that are partially encased in either loose or taut polyethylene sheaths, and may also include a lightweight foam core and end cap. These modifications have improved the likelihood of properly fitting these types of plugs and have also reduced the possibility of small sections or pieces of the fiberglass breaking off from the main body of the plug and remaining in the ear canal. Sheathed fiberglass plugs are inserted by placing them into the ear canal with a slight rocking and twisting motion while using the opposite hand to pull the outer ear (see Figure 10-7).

Slow-recovery foam earplugs were first introduced in the early 1970s. They usually have a cylindrical shape and are normally about 14 mm in diameter and from 18 to 22 mm in length. Insertion is accomplished by rolling the plug into a thin, tightly compressed cylinder, which (while still fully compressed) is then placed into the ear canal and gently held in position for a few seconds until expansion begins. As with other earplugs, pulling the pinna eases insertion. The plug will be properly inserted when it is actually situated in the ear canal, as opposed to simply capping its entrance (see Figure 10-13A, B). This can be verified for most wearers if approximately one half of the plug is in the shape of the ear canal immediately after removal (see Figure 10-14A, B).

Unlike other formable or premolded earplugs, foam earplugs should not be readjusted while in the ear. If the initial fit is unacceptable, they should be removed, recompressed, and reinserted. And since the depth of insertion of foam earplugs can be comfortably varied, a maximal occlusion effect does not signify best fit. In fact, the deeper the insertion and the better the fit and attenuation, the less noticeable (and annoying) will be the occlusion effect (see section on *Occlusion Effect*).

The attenuation values for a number of formable earplugs, which are plotted in Figure 10-15, demonstrate that a fairly wide range of laboratory-rated protection is available, especially below 2 kHz. Minimum values of attenuation are provided by one type of sheathed fiberglass down, whereas the highest values are developed by the foam earplugs[2], devices that can offer the best low frequency attenuation of any type of HPD.

Custom Molded Earplugs

Custom earmolds are most often manufactured from two-part curable silicone putties, although some are available in vinyl. The silicones are either cured by the fitter using a catalyst at the time the impression is taken, or are returned to the supplier for manufacturing. Some custom earmolds are finished by inserting small handles into their exterior surface to facilitate handling whereas others are available with attached cords. Representative devices are pictured in Figure 10-16.

Most earmolds fill a portion of the ear canal as well as the concha and pinna. The canal portion of the mold is what makes the acoustical seal to block noise whereas the concha/pinna portion principally functions to hold the HPD in position. Incorrect insertions are easily detected and the molds have

---

[2]Although recently (circa 1981) foam earplugs have become increasingly available from more than one manufacturer, the laboratory and real-world data in this chapter pertain only to the E-A-R and Decidamp brands. To the author's knowledge, there are no currently available real-world performance data for other brands of foam earplugs.

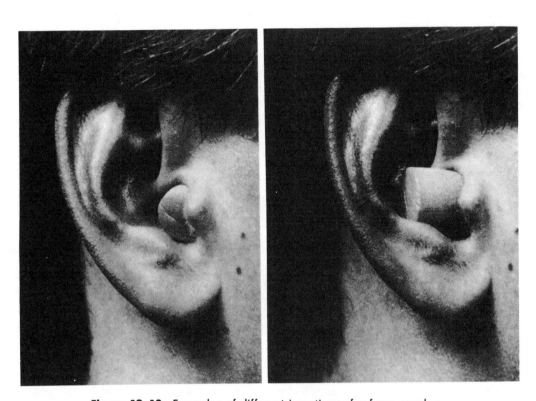

Figure 10-13. Examples of different insertions of a foam earplug.

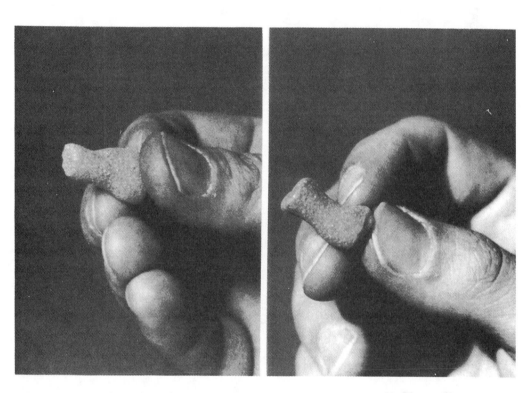

Figure 10-14. How to "read" a foam earplug after removal. Plug must have expanded in ear canal for about 1 minute prior to removal. Primarily useful for medium and smaller ear canals.

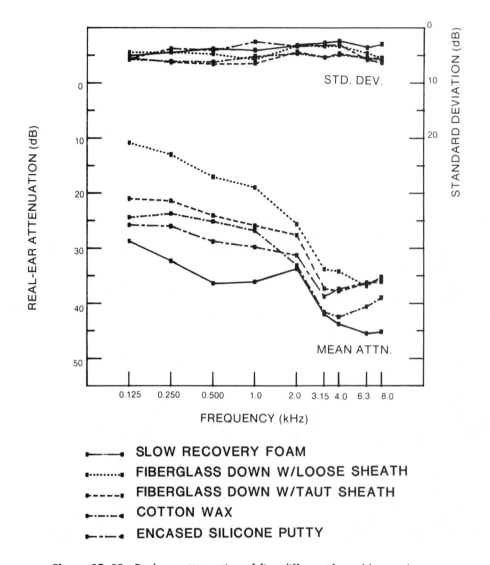

Figure 10-15. Real-ear attenuation of five different formable earplugs.

little chance of working loose with time. Some molds are manufactured with only a canal impression present, on the supposition that leaving a greater area of the pinna uncovered makes them cooler and more comfortable. Unfortunately, this may also impair the retention of these molds in the canal, allowing them to lose their seal during use.

Considerable skill and time are required to take individual impressions for each employee. Contrary to popular beliefs, the fact that custom earmolds are user-specific and intended to fit only the canal for which

they are manufactured does not assure that they will provide better protection than other well-fitted earplugs. In fact, often the opposite is the case (Guild, 1966) as can be noted by comparing laboratory attenuation for the various types of insert HPDs as illustrated in Figures 10-11, 10-15, and 10-17. This observation may be partially explained by reference to Figure 10-16, which depicts four different earmolds, all of which are impressions of one subject's right ear canal, and all of which were manufactured by "experienced" fitters. Note the significant differences in the canal portion of the

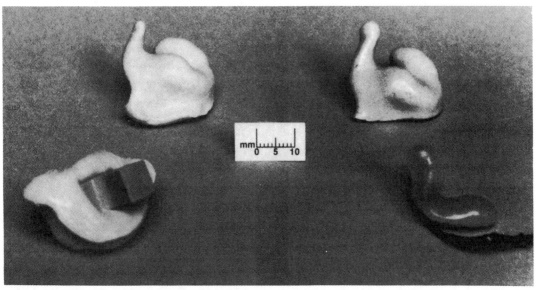

Figure 10-16. Representative custom earplugs: lower left illustrates handle, lower right is mold without concha portion. All four earmolds are impressions of one subject's right ear canal. Note the significant variability in the size and shape of the canal portion of the molds.

impressions of the three molds for which that feature is visible.

A worthwhile aspect of individually molded earplugs is that they are manufactured personally, for only one employee, so that they are "customized." This customization can be effectively utilized as an incentive for motivating employees to wear their HPDs. Another positive aspect of custom molds (those with both a canal and concha impression) is that, due to the way in which they fit into the ear, they are less subject to misinsertion under field conditions.

Custom earmolds can be very comfortable, but experience has shown that as the molds more fully and tightly fill the canal and therefore, more effectively attenuate sound, comfort deteriorates. Thus, in practice, there is a limit to how snugly a custom mold can be fitted.

In spite of the longevity claims made by some manufacturers for their "permanent" earmolds, these devices, like other earplugs, are susceptible to shrinkage, hardening, and cracking with time, and must be periodically reexamined to assure they are still soft, flexible, and wearable. They are also subject to

user modifications such as whittling or total removal of the canal portion of the mold. Additionally, since they may be easily lost or misplaced as can all earplug-type HPDs, their use may create administrative problems since time must be allowed to remanufacture them on an as-needed basis.

Custom earmold impressions are made using a viscous material with a consistency varying from that of thick syrup to soft putty. It is mixed with a curing agent and is then either formed into a cone and pressed into the canal, or placed in a large syringe and injected into the ear canal. The silicone must then cure in the ear approximately 15 minutes. When cured, it is itself the HPD. In some cases, it may be dipped into a coating liquid before use. In other cases, impressions are returned to the supplier, who then uses them to make subsequent negative and positive molds in order to create the final custom earplug.

Some manufacturers recommend placing a cotton block or eardam inside the canal prior to making the impression to assure that no silicone is forced too deeply into the canal or reaches the eardrum. Additionally,

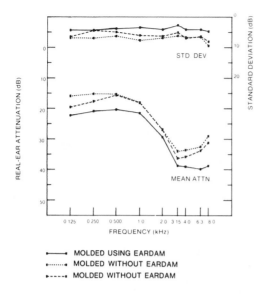

Figure 10–17. Real-ear attenuation of three different models of silicone custom earmolds.

the eardam helps to ensure that a better fitting impression will be taken. The reason is that, as the impression material is pressed into the canal, the dam forces it radially outward against the canal walls to ensure a tighter fit and more effective seal. Without the dam, the viscous mold material is permitted to simply flow further into the canal without ever being forced into contact with the canal walls (see Figure 10–16, upper left impression).

Attenuation data for three custom earmolds are shown in Figure 10–17. The variation between different brands of properly manufactured molds is minimal, with the average attenuation falling below that of some of the premolded earplugs. The best protection and lowest variability are afforded by impressions taken using cotton eardams. Selection of a particular product should be based primarily upon the users' judgment of comfort and the fitters' evaluation of the ease of manufacture of the earmolds. This latter consideration is important since the final performance of the molds is closely linked to the fitters' skill in making the initial impression.

Semi-Aural Devices

Semi-aural devices, which consist of pods or flexible tips attached to a lightweight headband, provide a compromise between earmuffs and earplugs. They can be worn in close quarters, easily removed and replaced, and conveniently carried when not in use. One size fits the majority of users. Their fit is not compromised by safety glasses or hard hats. These devices are usually available with dual position (under the chin and behind the head) or universal headbands made from either metal or plastic. The tips can be made from vinyl, silicone, or composites such as foam encased in a silicone bladder, and may cap, or in some cases, enter the canal. The tips normally have a bullet, mushroom, or conical shape. Representative devices are shown in Figure 10–18.

Semi-aural devices are principally intended for intermittent use conditions where they must be removed and replaced repeatedly. Examples include: ground crews servicing commerical aircraft, periodic equipment inspection by personnel normally located in sound-treated booths, and supervisor walk-throughs. During longer use periods, the force of the caps pressing against the canal entrance may be uncomfortable, but for those who do prefer this type of device for extended use, the better ones can offer adequate protection. Since semi-aural devices generally cap the canal at or near its entrance, they tend to create the most noticeable occlusion effect and consequently distort the wearers' perception of their own speech more than other types of HPDs. This may be objectionable to some users. Undesirable field modifications generally involve springing the band, with a consequent reduction in protection.

Semi-aural devices generally are simply pushed into place at the entrance to the ear canal, although the particulars vary for each product. For example, one type requires rolling and stretching of the pods prior to insertion, another suggests that pulling the pinna may be helpful, and others have

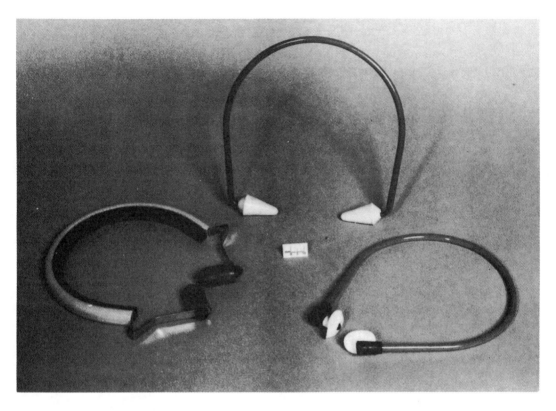

Figure 10–18. Representative semi-aural devices (clockwise from top): conical (foam in silicone bladder), mushroom, hollow bullet.

specific orientations (left/right, top/bottom) that must be attended to if proper protection is to be attained. Certain semi-aural devices, particularly those that do not enter the canal, are prone to losing their seal periodically, especially if the band is bumped and caused to skew on the head.

Attenuation values for five semi-aural devices are presented in Figure 10–19. It can be seen that a wide range in performance is available at all frequencies, with better protection being provided by those devices that not only cap, but also partially enter, the ear canal.

Special Types and Combinations of HPDs

Double Hearing Protection

For very high level noise exposures, especially when 8-hour TWAs are greater than 105 dBA, the attenuation of a single

HPD may be inadequate. For such exposures, double hearing protection, that is, earmuffs plus earplugs may be warranted. It is well recognized that double hearing protection does not simply yield overall attenuation equal to the sum of the individual attenuation of each device. This is primarily due to the BC flanking paths and the acoustical-mechanical interaction between two such closely spaced devices.

An extensive empirical study of the incremental performance to be gained by double protection was reported by Berger (1983b), who examined five earplug conditions (V-51R earplugs, fiberglass down, and three insertion depths of foam earplugs) and three types of earmuffs, with the various devices worn both singly and in combination. He found in almost all cases that a combination of "plug plus muff" outperformed either device alone. At individual

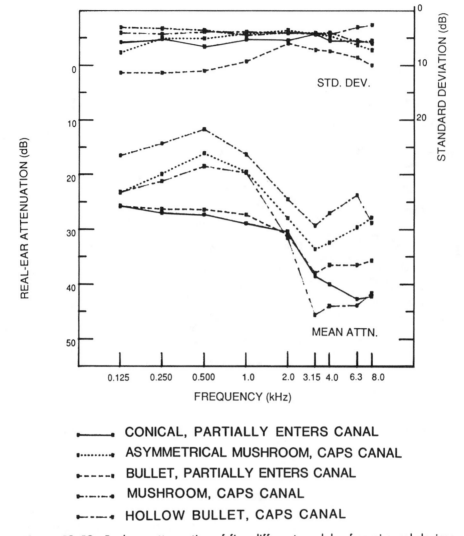

Figure 10–19. Real-ear attenuation of five different models of semi-aural devices.

frequencies the incremental gain in performance varied from approximately 0 to 15 dB over the better of the single devices, except at 2 kHz, where no combination exhibited a gain of greater than 3 dB. The gain in the NRR for the double protection combinations ranged from 7 to 17 dB when compared to the plugs alone, 3 to 14 dB when compared to the muffs alone, and 3 to 10 dB when compared to the better of the two individual devices.

The data suggested that although the attenuation provided by an insert HPD could be improved by wearing an earmuff over it, the choice of earmuff was relatively unimportant. However, when the situation was reversed and an earplug was added to the earmuff, regardless of the earmuff worn, the choice of insert was critical at the frequencies below 2 kHz, as illustrated in Figure 10–20. At and above 2 kHz all plug-plus-muff combinations that were studied provided attenuation that was approximately equal to that of the human skull, that is, the combined attenuation was limited only by the flanking BC paths to the inner ear (cf. Figure 10–2).

Amplitude Sensitive Devices

Amplitude sensitive devices, or nonlinear HPDs as they are sometimes

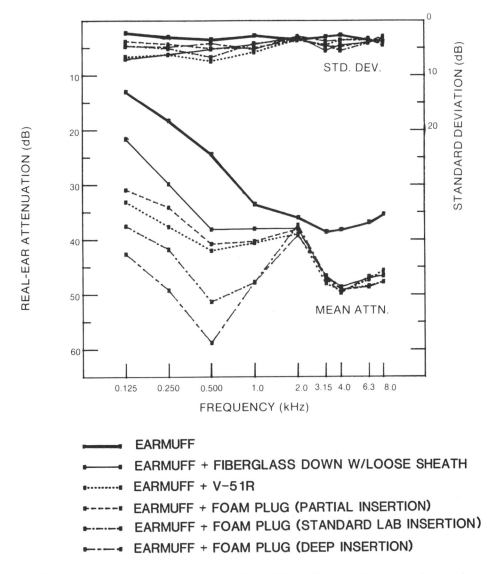

Figure 10-20. Real-ear attenuation of a small volume (100 cm³) earmuff alone, and in combination with each of five different earplugs (from Berger, 1983b).

called, are designed to provide little or no attenuation at low sound levels, with protection increasing as sound level increases. Communication is unimpaired during quiet periods (see *Effects of HPDs on Auditory Communications* in this chapter), but when high noise levels are present, additional protection is available. This can be accomplished to varying degrees with either passive devices that utilize orifices, valves, or diaphragms, or active devices containing electronic circuitry. Orifices and valves have been incorporated into earplugs and at least

one manufacturer has offered a diaphragmatic circumaural HPD. Active electronics are currently available in earmuffs, and efforts are underway to include them in an earplug type device as well.

Both theoretical and empirical research indicates that at sound levels below 110 to 120 dB, orifice and valve type passive insert devices simply behave as a vented earmold with almost no attenuation below 1 kHz and attenuation increasing to a much as 30 dB at higher frequencies (Berger, unpublished data, 1982; Forrest, 1969; Martin, 1979). At higher sound levels, steady-state or impulsive sound waves generate turbulent airflow in the orifice, which impedes the passage of sound and provides for an increase of attenuation of about 1 dB for each 2 to 4 dB increase in sound level. These devices are primarily suited for gunfire exposures, where peak SPLs range from approximately 135 to 175 dB. They are more effective in less-reverberant conditions such as outdoor ranges (Coles & Rice, 1966; Mosko & Fletcher, 1971). These devices are of little value for many occupational and recreational exposures where noise levels are rarely the appropriate type or of sufficient level for the nonlinear characteristics to become functional (Harris, personal communication, 1979; Martin 1976).

Active HPDs may be of the peak-limited sound transmission or noise-cancelling types. The sound transmission type contains a microphone external to the circumaural cups that is used to monitor the ambient noise. When the sound levels are low, it broadcasts information into the earmuff via electronic circuits and earphones. At high sound levels, the circuits do not pass the signal, permitting the physical attenuation of the cup to protect the ear. The noise-cancelling approach places both the microphone and earphone inside the HPD so that electronics can sense the sound and generate an antiphase signal through the earphone to cancel it. This latter method is generally successful only at the lower frequencies.

Active HPDs can improve speech communications in certain noisy environments, especially during intermittent quiet intervals, but at the current state of the art the cost is upward of $100 per device. This makes them too expensive for general industrial use except for specialized circumstances and environmental conditions.

Communication Headsets

In order to transmit signals to the ear, earphones may be built into both supra-aural and circumaural devices, or can be connected to canal inserts. Usually, only

circumaural devices will provide sufficient attenuation to be suitable for use in noisy environments. They are available with either wireless (FM or infrared) or wired systems, designed for one- and two-way communications and/or music transmission. The better devices provide specialized electronic circuits to limit the delivered SPLs so that the earphones themselves cannot present signals that could be hazardous to the wearer. Prices for communication headsets range from $100 to $1,000 per unit.

In extremely high noise levels, when circumaural communication headsets may not provide sufficient attenuation of ambient noise to permit clear communications, speech intelligibility can be improved by wearing earplugs under the cups. The earplug will reduce both the environmental noise and the desired signal equally, but as long as the headset has sufficient distortionless gain so that its output can be increased in order to overcome the insertion loss of the earplug, the signal-to-noise ratio in the listener's ear canal can be significantly improved.

Recreational Earphones

Often, employees request to use stereo earphones for protection against noise while they enjoy the music. The inappropriateness of relying on these devices for hearing protection is illustrated in Figure 10–21 in which the attenuation of a circumaural radio headset and a more popular set of lightweight foam supra-aural stereo earphones are compared to an industrial earmuff. The foam earphones offer almost no protection. Even the circumaural device provides no more than approximately 20 dB of attenuation at high frequencies, and actually amplifies sounds at others. This protection is inferior to that of a well-designed, properly fitted HPD (also see Skrainar, Royster, Berger, & Pearson, 1985).

Recreational earphones alone can generate equivalent sound-field noise levels up to approximately 100 dBA. Since they offer so little attenuation, a greater concern is that employees might turn up the music to mask

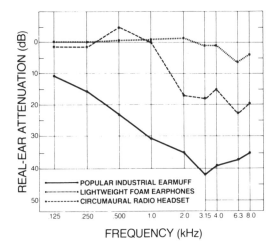

Figure 10–21. Attenuation of recreational earphones versus a popular industrial earmuff.

(i.e., "drown out") the factory noise, with a consequent reduction in their ability to communicate or hear warning sounds and a significant increase in their effective noise exposures. Thus, the use of recreational earphones should be prohibited when sound levels equal or exceed 90 dBA. At sound levels between 85 and 90 dBA, their use is problematical. The music they provide may alleviate boredom and increase productivity (Fox, 1971), but they offer little or no protection and can actually increase noise exposures, as previously cited.

In one industrial study (Royster, Royster, Berger, & Skrainar, 1984) with a noise environment characterized by a variation of sound level between 80 and 90 dBA (with approximately equal time periods at each level), and a TWA of 87 dBA, employees on the average increased their equivalent exposure by only 2 dB as a result of using recreational earphones. However, about 20% of the workers were observed to play their radios at levels of 90 dBA or higher. The authors recommended that employees continue to be allowed use of such devices, with the stipulation that a significant educational effort be directed at the proper use of personal radios, that employees exhibiting permanent threshold shifts of 20 dB or more be prohibited from further use, and that the overall audiometric database be annually analyzed for relative changes between the hearing of users and nonusers of personal radios.

Hearing-Aid Earmolds

When employees who wear hearing aids work in noise, they may request to wear their aids, turned off, in lieu of standard industrial HPDs. This may be due to comfort (since they are accustomed to their custom hearing-aid earmold), or convenience (since their hearing aids are available for use when needed), or reduced attenuation (which may help them hear better under certain conditions), or because they may wish to occasionally use their aids in the noise. The latter is uncommon since it is generally observed that present-day hearing aids are of little value in noisy environments (Gasaway, 1985).

The question is: Can an earmold that is part of a hearing-aid system provide adequate hearing protection? If so, the wearer could quickly and easily turn on and use the aid when needed, and yet turn off the aid and continue wearing it to obtain noise reduction as required.

Berger (1987) examined this question and reported that for the typical vented earmold, and even unvented earmolds depending upon how they were fabricated, attenuation is insufficient for all but the most marginal occupational exposures. However, for a tightly fitted unvented earmold or when foam earplugs are used as hearing-aid earmolds, protection equivalent to standard commercially available earplugs is achievable. He suggested that, if possible, it is best to validate the level of protection by asking the audiologist who fitted the hearing aid to estimate its attenuation using sound field audiometry, that is, measuring the difference between the individual's unaided, unoccluded thresholds and the occluded thresholds with the aid turned off.

Regardless of the amount of attenuation that is provided by the hearing-aid earmold, the aid itself, which usually supplies from 20 to 50 dB of maximum gain, can potentially cause additional noise-induced hearing loss when used in the presence of sustained high-level noise. Although no definitive answers are available, a prudent recommendation is that employees should never operate their aids without the addition of an earmuff when the sound levels exceed 80 dBA. Whenever hearing aids are worn in noise, careful employee orientation is necessary, and more frequent audiometric monitoring (twice annually) is advised until the stability of the individual's hearing threshold levels can be verified. For additional information see Berger (1987).

Whatever decision is made concerning the suitability of the earmold for use as a hearing protector, the hearing-impaired individual should be protected. Exceptions may include individuals with hearing loss so severe that the noise is inaudible, or persons with a conductive loss that exceeds in magnitude the attenuation that a hearing protector can provide.[3]

The Initial Ear Examination and HPD Hygiene

Prior to issuing HPDs, the fitter should visually examine the external ear to identify any medical or anatomical conditions that might interfere with or be aggravated by the use of the protector in question (also see Foltner, 1986). If such conditions are present, HPDs should not be worn until medical consultation and/or corrective treatment can be obtained, or the suspected condition has been shown not to constitute a problem. Areas of concern include extreme tenderness, redness or inflammation (either in or around the ears), sores, discharge, congenital or surgical ear malformations, and additionally, in the case of earplugs, canal obstructions and/or impacted or excessive cerumen. The latter condition, however, is difficult to judge since few data are available

on the effects of earplugs on the formation, buildup, and possible impaction of wax.

As with all clothing and equipment that comes in repeated and intimate contact with the body and the work environment, the cleanliness of HPDs must be considered. HPDs should be cleaned regularly in accordance with manufacturers' instructions, and extra care is warranted for environments in which employees handle potentially irritating substances. Normally, warm water and soap are recommended as cleansing agents. Solvents and disinfectants should generally be avoided.

Earplugs should be washed in their entirety and allowed to dry thoroughly before reuse or storage in their carrying containers. Earmuff cushions should be periodically wiped or washed clean. Their foam liners can also be removed for washing but must be replaced before reuse since they do affect attenuation. Earplugs and earmuff cushions should be discarded when they cannot be adequately cleaned or when they no longer retain their original appearance or resiliency.

Stressing hygiene beyond practical limits, however, can compromise the credibility of the HPD issuer/fitter. It is often difficult enough to get employees to replace or repair worn-out HPDs, let alone clean them routinely. And in spite of this, information from authorities in the field of audiology and hearing conservation (Gasaway, 1985, D. Ohlin, personal communication, 1981), as well as the available epidemiological data (Berger, 1985b) suggest that the likelihood of HPDs increasing the prevalence of outer ear infections is minimal.

If an ear irritation or infection is reported, the exact extent and etiology of the problem should be investigated firsthand by medically trained personnel to determine whether the causative agent is an HPD or another predisposing factor. Such factors include excessive cleaning of the ear, recreational water sports, habitual scratching and digging at the ears with fingernails or other objects, environmental contaminants, and systemic conditions such as stress, anemia,

---

[3]OSHA published regulations do not permit such exceptions, but according to J. Barry (Technical Support IH in OSHA Region #3, 1986) special cases can be discussed with area directors.

vitamin deficiencies, endocrine disorders, and various forms of dermatitis (Caruso & Meyerhoff, 1980). When HPDs are implicated, a common cause has been found to be earplugs or even earmuffs that are contaminated with caustic or irritating substances, or embedded sharp or abrasive matter. In one reported case of earplug contamination (Royster & Royster, 1985), more careful hygiene practices, combined with the use of corded plugs to allow removal without touching the protector, eliminated the problem.

If occurrences of external ear problems develop, it is important to determine whether they are limited to a particular department or operation, to one or more brands or types of HPDs, to a change in the HPDs being utilized, to a particular time of year, or if they are perhaps due to some other policies or procedures that may have been modified within the work environment. This will allow a reasoned approach and help to avoid an overreaction which could compromise the HCP, without necessarily resolving the problem at hand.

REAL WORLD PERFORMANCE OF HPDs

It is invariably found that the real-world performance of HPDs is significantly less than estimated by laboratory measurement methods. This effect is often exaggerated by the particular choice of laboratory data (cf. laboratory #8 in Figure 10–3), and by observance of the "experimenter fit" protocol of ASA STD 1-1975. The experimenter fit procedures are intended to develop "optimum performance" data, but this can be misleading. Optimum performance for a laboratory test wherein a trained and motivated subject sits immobile for 5 minutes, utilizing a test protocol in which the word "comfort" is never mentioned, and with equipment and expertise available to assure proper fitting of the HPD, is very different from that which can be attained in the real world. In the latter instance, when HPDs are worn for extended periods on a daily basis

by active workers who may consider them to be an inconvenience or a major burden, optimum performance takes on a very different meaning.

More explicitly, the factors that are often overlooked in hearing conservation programs and that can compromise HPD performance in the real world may be summarized as follows:

1. Comfort—This is ignored in laboratory tests but is crucial in the real world.
2. Utilization—Due to poor comfort, poor motivation, poor training, or other user problems, earplugs may be incorrectly inserted and earmuffs may be improperly adjusted.
3. Fit—Fitting and sizing of earplugs must be carefully accomplished for *each* ear, otherwise performance will be degraded.
4. Compatibility—Since not all HPDs are equally suited for all ear canal and head shapes, the proper device must be matched to each user.
5. Readjustment—Since HPDs can work loose or be jarred out of position, employees must be advised of the need for readjustment.
6. Deterioration—No HPDs are permanent or maintenance free. So-called permanent HPDs must be inspected at least twice yearly, and replaced or repaired as necessary.
7. Abuse—Employees often modify HPDs to improve comfort at the expense of protection. This must be avoided.
8. Removal—When devices become uncomfortable they are often removed to give the ears a "break." This can dramatically reduce the effective protection (Else, 1973) as illustrated in Figure 10–22. For example, if the HPD has a nominal NRR of 25, then its effective, or time-corrected, NRR would be only 20 dB if it is removed from the ear for just 15 minutes during each 8 hour noise exposure.

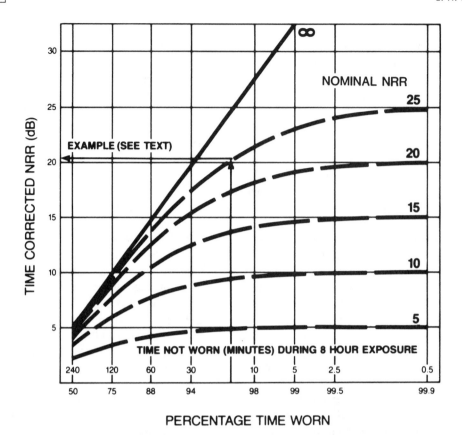

Figure 10-22. Time-corrected NRR as a function of wearing time (using OSHA 5 dB trading relationship).

Summary of Existing Field Data

This section summarizes the results of 10 field studies including data from over 50 different industrial plants with a total of 1,551 subjects. Various protocols were implemented, but all used actual employees who were participants in ongoing hearing conservation programs. The employees were either tested at their plant sites (both on the job and in special test rooms) or at remote hearing testing clinics. The studies have been summarized and critically reviewed by Berger (1983a), who amassed sufficient data to characterize the real-world performance of earmuff, premolded, fiberglass, foam, and custom molded HPDs as shown in Figure 10-23.

The data in Figure 10-23, which may be compared with the data in Figures 10-5, 10-11, 10-15, and 10-17, illustrate significant differences between laboratory and

real-world performance of HPDs. Comparative results are summarized in Figure 10-24, where the manufacturers' labeled NRR_{98}s are superimposed on NRRs computed from field data using only a 1-SD correction (NRR_{84}). Thus, the attenuation that at least 98 percent of the laboratory subjects achieved is compared to that achieved by at least 84 percent of the real-world users. The rationale for comparing laboratory-measured NRR_{98}s to field-measured NRR_{84}s has been described by Berger (1983a).

Consideration of the protection afforded to 84 percent of the real-world users indicates that earmuffs can offer around 10 to 12 dB of protection, whereas earplugs, with the exception of foam plugs, offer less than 10 dB. The average difference between the labeled NRR_{98} and the real-world NRR_{84} across all devices was 13 dB. If the labeled NRR_{98} had been compared to a real-world NRR_{98}, the average difference would have

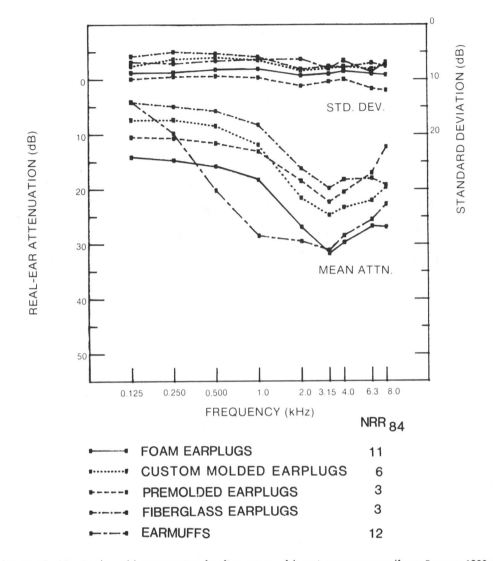

Figure 10–23. Real-world attenuation for five types of hearing protectors (from Berger, 1983a).

been greater than 20 dB.

Suggestions for De-Rating Laboratory Data

The average NRR on devices sold in North America today is approximately 24 dB. This number clearly overstates the protection that most buyers can expect their employees to achieve. It misleads the buyer into believing that almost any HPD will reduce almost every noise exposure to a safe level. (After all, how many 8 hour exposures does one find that are greater than 85 dBA

+ 24 = 109 dBC?) In turn, this belief fosters programs in which HPDs are handed out indiscriminately with little or no attempt made to train or motivate the employees.

In order for users to have a more realistic perspective concerning the probable efficacy of HPDs, NRRs must be de-rated. An approximate real-world correction that has been suggested is to reduce the NRR by 10 dB before subtracting it from the measured C-weighted sound level. This correction is smaller than the average labeled versus real-world differences which are depicted in Figure 10–24. It is the minimum correction

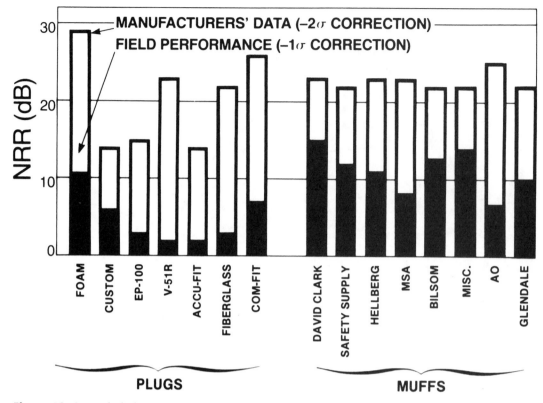

Figure 10-24. Labeled NRRs versus real-world performance for earplugs and earmuffs (from Berger, 1983a).

that is necessary to reduce laboratory-generated NRRs to potentially achievable real-world values. In many instances, especially in existing hearing conservation programs, larger corrections would be warranted. (See the section on *OSHA's Calculational Procedure*.)

Whatever real-world adjustment is utilized for the NRR, the same one should be applied with a long method computation. For example, if 10 dB is subtracted from the NRR for use in estimating employee noise exposures, 10 dB should also be subtracted from the HPD's octave-band attenuation values prior to entering them in step 4 of Table 10-1.

A Comment Concerning the Long Method

An alternative approach to estimating protected exposure levels is to use the long method. Certainly, if one can assure the

similarity of labeled and real-world attenuation for the particular user or group of users in question, and if one has the octave band analysis of the noise environment available, then the long method is preferred. In most cases, however, it is unlikely that either or both of these "ifs" will be satisfied. Most users complain of the need to measure C-weighted sound levels for use with the NRR, let alone considering conducting an octave band analysis.

The primary utility of examining the mean attenuation of a hearing protector at the individual octave bands is to provide the ability to make a gross match between the device and the environment. For example, both the laboratory and real-world data show that if significant low frequency energy is present (125 to 250 Hz), then an earmuff is a poor choice and a foam or perhaps a premolded earplug would be better. Conversely, if significant midband energy is present (primarily around 1 kHz), then an earmuff is preferred. The desire to perform this type of protector/noise spectrum matching must be tempered by the realization that assigning particular HPDs to particular jobs in a plant (based on spectrum shapes) is often impractical. It is difficult enough to assure that HPDs are worn, and worn correctly, without also trying to keep devices from being shifted among work areas with different spectral characteristics and having to assure that certain devices are worn only by employees exposed to particular spectra.

OSHA's Calculational Procedure

The Hearing Conservation Amendment (see Appendix A) specifies that employers shall evaluate hearing protector attenuation according to the methods in Appendix B of the amendment. Although a note is included in Appendix B of the amendment that states, "the employer must remember that calculated attenuation values reflect realistic values only to the extent that the protectors are properly fitted and worn," and although it is clear that calculated attenuation values based upon manufacturers' labeled data are unrealistic, OSHA chose to permit use of such data for compliance with the regulation.

The amendment specifies that employers are to utilize the labeled data to calculate protection using either NIOSH Methods #1, #2, or #3 (NIOSH, 1975), or the NRR with either a C- or A-weighted noise measurement. NIOSH Method #1 is equivalent to the long method (described previously), Method #2 is equivalent to use of the NRR with a C-weighted noise measurement (Berger, 1980), and Method #3 is equivalent to the use of the NRR with an A-weighted noise measurement.

The use of the NRR with an A-weighted noise measurement entails the subtraction of an additional 7-dB safety factor from the NRR (see *Single Number Calculation of HPD Noise Reduction* in this chapter). This is not to be confused with the 10-dB real-world correction factor that was discussed previously, since the 7-dB factor derives from the computational procedure itself and is due to the reduced accuracy of using the NRR with A- instead of C-weighted sound levels. The accuracy lost by this approach dictates that when it is used, it will add considerably more imprecision to an already rough estimate. (For additional details, see Berger, 1984.)

More recently, OSHA (1987) has, in an update to their Industrial Hygiene Technical Manual, recognized that laboratory data must be de-rated. They now recommend reducing published NRRs by 50 percent, but only for the purpose of evaluating the *relative efficacy* of HPDs and engineering noise controls. The policy was developed in conjunction with OSHA's relaxation of the engineering noise control portions of the noise regulation and their increased emphasis on the use of hearing protection as the only required protective measure in most situations.

According to a strict interpretation of the 50 percent de-rating policy, it is not applicable for determining compliance with the hearing protector attenuation requirements (Appendix B) of the Hearing Conservation Amendment. However, the distinction between using the derating to evaluate *relative efficacy* versus using it to determine *HPD adequacy*, eludes many hearing conservationists, including most compliance officers. Thus, it is quite possible that during an inspection, OSHA might invoke such a de-rating.

Legal considerations aside, a 50 percent de-rating is certainly defensible on purely technical grounds. Although it is more stringent (by up to about 5 dB) for many protectors than is the 10-dB de-rating proposed earlier in this chapter, either of the approaches can be justified based upon examination of available real-world data, and one or the other of them must be implemented in order to utilize manufacturers' laboratory data to develop meaningful estimates of the actual field performance of HPDs.[4]

---

[4]When using both the 50 percent de-rating *and* A-weighted sound levels, OSHA will subtract 7 dB from the NRR and then de-rate by 50 percent.

Evaluation of
the Audiometric Database

The final arbiter of HPD efficacy will be the shifting or lack thereof in the hearing levels of noise-exposed employees. This may be assessed using any of a number of techniques that have been proposed for the evaluation of audiometric databases (Royster, J. D., & Royster, L. H., 1986). Regardless of the accuracy of any predictive scheme that attempts to match HPDs to employee noise exposures, if the hearing levels of individuals or groups of employees are deteriorating, then additional measures must be instituted. Conversely, if the hearing levels are not changing any more rapidly than those of an appropriately selected nonindustrial noise-exposed reference population, then this is an indication of a job well done.

EFFECTS OF HPDs ON
AUDITORY COMMUNICATIONS

An important consideration that arises when one recommends the use of HPDs is what effects, if any, they will have on the wearers' ability to verbally communicate, listen to operating machinery, and respond to warning sounds. On the one hand, many hearing protector manufacturers claim in their literature that their devices will block out harmful high frequency noise and yet let speech through. On the other hand, employees often complain that they can't talk to their fellow workers or hear their machinery operate when they are using HPDs. Both of these statements contain elements of truth, as will become apparent after an examination of the available data.

Understanding Speech

The level to which a particular sound will be attenuated by an HPD is dependent only upon its frequency and initial intensity. HPDs that do not contain active electronic circuitry cannot differentiate desired signals such as speech from useless information such as noise; both will be attenuated equally. At any one frequency, the resultant signal-to-noise ratio will be unaffected by use of the HPD. But hearing protector attenuation does vary with frequency, generally increasing as frequency increases, so that the frequency balance of the attenuated signal-plus-noise spectrum will differ from the unattenuated condition. Since the predominant speech energy is located at or below 2 kHz, if the noise energy is primarily above that frequency, most HPDs will reduce the noise level more than they will diminsh the overall level of the speech, thus, apparently ''letting the speech through and cutting out the noise.'' However, this is normally not the case.

The principal reason that HPDs improve the ability of normal hearing listeners to discriminate speech in high noise level environments is that, by reducing the overall level of the signal-plus-noise, the HPD permits the cochlea to respond without distortion, a characteristic that can only be assured at sound levels well below 90 dBA (Lawrence & Yantis, 1956). The effect is similar to wearing sunglasses on a very bright day. Since the total illumination of the scene is reduced, the eye is allowed to function more effectively and in a more relaxed manner. Metaphorically speaking, HPDs reduce the ''acoustical glare'' of high level sounds.

Speech discrimination is a measure of one's ability to understand speech. It is greatly affected by such factors as a person's hearing sensitivity, signal-to-noise ratio, absolute signal levels, visual cues (lip and hand motion), and the context of the message set. Speech discrimination is measured by verbally presenting to subjects one of a number of prepared standardized word lists and determining the percentage of correct responses they achieve. The effects of HPDs on speech discrimination can be evaluated by establishing a set of test conditions and measuring speech discrimination with and without HPDs on the subjects. The results of such tests on normal hearing subjects may be summarized as follows:

1. HPDs have little or no effect on the speech discrimination of normal hearing listeners in moderate background noise, approximately 80 dBA, but will decrease speech discrimination as the noise is reduced below that level (Howell & Martin, 1975; Kryter, 1946; Rink, 1979).

2. At high noise levels, greater than approximately 85 dBA, HPDs actually improve speech discrimination, as is demonstrated in Figure 10–25. For the pairs of curves shown, whenever the dashed line is above the solid line, the earplugs provide improved speech discrimination.

When the listener is hearing impaired, the situation is considerably more complex, and the answer is not as well defined.

However, it is clear that HPDs will decrease speech discrimination for hearing-impaired listeners in low to moderate noise situations, with the effects being minimized as both the noise levels and signal-to-noise ratios are increased (Abel, Alberti, Haythornthwaite, & Riko, 1982; Chung & Gannon, 1979; Coles & Rice, 1965; Froehlich, 1981; Rink, 1979). The difficulty for hearing-impaired listeners arises from the fact that HPDs may reduce the level of the speech signals below the person's threshold of audibility, especially for important higher frequency consonant sounds. No published studies have been able to unequivocally define the level of hearing loss that would be required before hearing protection will degrade instead of improve speech discrimination in noisy environments; however, a rough estimate based upon the work of Lindeman (1976)

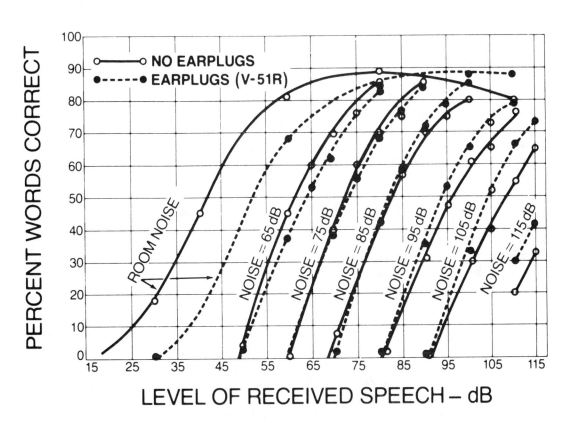

Figure 10–25. Relationship between speech discrimination and speech level with noise level as a parameter and hearing protection as a variable. Each point represents an average of the percentage of correct responses for 8 subjects to a list of 200 words read over a speaker system in a reverberant room (from Kryter, 1946).

would be a hearing threshold level of greater than about 40 dB when averaged across the frequencies of 2, 3, and 4 kHz.

The preceding generalizations can be modified in practice by additional important factors. For example, in real-world environments, communications may be either limited in scope and/or accompanied by visual cues, allowing missed words to be "filled in" and intelligibility maintained. Rink (1979) illustrated this fact for subjects listening in a 90-dBA background noise. Both his normal and hearing-impaired listeners were able to maintain speech discrimination scores regardless of the use of hearing protection, as long as visual cues were presented along with the auditory stimuli. And Acton (1970) has demonstrated that employees become accustomed to listening in noise, thus performing better with respect to speech discrimination than do laboratory subjects with equivalent hearing levels. Conversely, Howell and Martin (1975) and Hormann, Lazarus-Mainka, Schubeius, and Lazarus (1984) have shown that when the person speaking wears HPDs, speech quality is degraded, adversely affecting communications.

Howell and Martin's observation is at least partially explained by examining the effects of an HPD upon a talker's perception of his or her own voice. The HPD significantly attenuates airborne energy, but has little effect on the BC portion, except in the lower frequencies where perceived voice levels are actually amplified as a result of the occlusion effect (see *Occlusion Effect* in this chapter). This alters the frequency balance of the information monitored by the talker and causes the loudness of his or her own voice to increase or remain the same, while existing environmental noise levels are sigificantly reduced. It appears as though the talker's own voice is louder compared to the noise than actually is the case, and speech levels tend to be lowered accordingly, typically by 2 to 4 dB (Howell & Martin, 1975; Kryter, 1946). As a result, employees must be taught that while wearing HPDs in noisy environments, good communications can be assured only by talking

at what seems to be a louder than necessary level (Guild, 1966).[5]

Responding to Warning and Indicator Sounds

The effects of HPDs on the ability of normal hearing and hearing-impaired users to detect warning sounds is very similar to the findings with regard to speech intelligibility. However, when employees are engrossed in alternative tasks and not specifically attending to sounds that may warn them of danger or indicate a machine malfunction, then not only is the question of detection and discrimination of importance, but one must also inquire whether or not the sound will get their attention. Inattention may result in elevation of effective thresholds for particular test stimuli by 6 to 9 dB, or even more for certain individuals (Wilkins & Martin, 1978).

One field study was conducted that assessed the effectiveness of intentional (warning horn) and incidental (clinking of metal components spilling from their container) warning sounds under actual factory conditions, using subjects who were employees at the test site (Wilkins, 1980). It was found that there was no significant effect of wearing hearing protection on the ability of either normal hearing or hearing-impaired subjects to detect the horn or the clinking sound when they were specifically awaiting its occurrence. However, when the subjects were distracted by performing their normal job duties, the clink sound was less

---

[5]The observed reduction in speech levels as a result of wearing hearing protection in noise does not apply when wearing hearing protection in quiet. For this latter case, speech effort is not determined so much by the perceived speech-to-ambient noise ratio, but simply by the speakers' perception of their own absolute voice level. Although the BC component is still amplified by the occlusion effect (as was the case when speaking in noise), the attenuation of the airborne speech component is generally of greater significance, and thus the speakers' overall assessment of their speech levels is that they are either acceptable or too low. Thus, the tendency is to maintain speech levels or even to speak too loudly when wearing hearing protection in quiet ambient conditions.

well perceived as a warning by those with substantial hearing loss, and HPDs adversely affected its perception independent of the hearing sensitivity of the wearers. By contrast, even when distracted, there were no significant variations in response rates for the intentional warning sound (the horn) due to either the hearing ability of the listeners or their ear condition (protected or unprotected).

Since warning sounds may be adjusted in pitch and loudness to achieve optimum perceptibility, results similar to those reported should be achievable in most conditions. Warning sounds will be most effective when their primary acoustic output is located below 2 kHz, since hearing-impaired employees will exhibit their largest losses above that frequency and many HPDs will deliver less attenuation below that frequency. Additional evidence regarding any possible hazards associated with the use of hearing protection while working in noisy environments is provided by two field studies (Cohen, 1976; Schmidt, Royster, & Pearson, 1980). Those authors demonstrated that implementation of a hearing conservation program that required utilization of HPDs reduced rather than increased the number of industrial mishaps.

Localization and Depth Perception

Another effect that HPDs can have is to confuse one's ability to locate the direction of origin of sounds (Atherly & Noble, 1970; Noble & Russell, 1972). The indications are that earmuffs can interfere with localization accuracy to a greater extent than do inserts that leave the outer ear exposed. However, one recent study that measured the minimum detectable azimuthal changes of a frontally located sound source (typically <10°) found the opposite to be the case for one of the higher attenuation earplugs they studied (Lin, 1981). Furthermore, experiments indicate that subjects cannot learn to compensate for the adverse effects of earmuffs (Russell, 1977). There is also some suggestion, but no direct evidence, that HPDs may impair the ability of wearers to judge the distance to a sound source (Wilkins & Martin, 1978).

General Remarks

The preceding data indicate that HPDs can be effectively utilized for the preservation of hearing in high noise level environments with minimal negative effects on auditory communications and demonstrated advantages in certain conditions.

For hearing-impaired persons, the utilization of HPDs in lower noise levels should be carefully considered. When substantial hearing impairments are present, especially in the case of hearing-aid users, decisions regarding employment in noisy occupations and/or the use of hearing protection are not clear cut (Berger, 1987; also see *Hearing-Aid Earmolds* in this chapter). Even with individual counseling, comprehensive audiological workups, and expert consultation, ideal solutions are elusive. It sometimes becomes a decision between preserving an employee's remaining hearing and creating additional communications disadvantages while at work. Often warning and indicator sounds can be augmented or replaced by visual or tactile signals to assist the hearing impaired.

Intermittent noise poses a significant problem, since HPDs will cause degradation of communications during the intervening quiet intervals. Consequently, earmuffs or semi-aural devices are the preferred type of HPDs for such conditions, since they are easily removed and replaced as the intermittency of the noise may warrant. Unfortunately, passive amplitude-sensitive earplugs cannot normally be recommended, since when the noise is on, it will not generally be of sufficient level to activate their level sensitive characteristics. Thus, the HPD may provide inadequate protection, especially at the lower and middle frequencies. Additionally, although amplitude-sensitive earplugs can improve speech discrimination during the quiet conditions, when noise is present they may actually provide worse speech discrimination than do standard linear protectors (Coles & Rice, 1966).

HEARING PROTECTOR STANDARDS AND REGULATIONS

At this time, there are no federal or state agencies or U.S. standards-writing organizations that approve or disapprove of particular HPDs, although there is an existing EPA regulation requiring the labeling of all HPD packaging.

Standards and Related Documents

Of specific interest is ANSI S3.19–1974 (ASA STD 1–1975) and its recent revision ANSI S12.6–1984. Users often mistakenly perceive these documents to contain criteria for judging the acceptability of HPDs. In fact, these standards only describe a particular experimental method for determining hearing protector attenuation (see *Laboratory Test Methods* in this chapter). Testing an HPD by these methods in no way confers any approval or attributes any particular degree of quality to the device. It simply characterizes the laboratory attenuation of the protector, however good or bad that may be.

An agency that many program administrators mistakenly believe stipulates acceptable HPDs is the Mine Safety and Health Administration (MSHA). MSHA does publish a "Hearing Protector R & D Factor List," but the only criterion for having a device placed on the list is that it be tested according to ASA STD 1, and that the data be available in sales literature or on specification sheets (L. C. Marraccini, personal communication, 1983).

Occasionally, users will inquire about Food and Drug Administration (FDA) approval of HPDs, especially when they are manufacturing or using other products with which the FDA is involved. At this time, hearing protectors for industrial use are not considered by the FDA to be medical devices, and therefore are not subject to FDA approval (D. M. Link, correspondence with F. E. Willcher, Jr., Executive Director of the Industrial Safety Equipment Association, April, 1980). Neither does the FDA approve

the materials from which HPDs are manufactured. They only approve finished products, which of course in the case of those approved products, implies that the constituent materials are acceptable for the intended application(s).

The only other agency with a hearing protection related regulation is OSHA, whose Hearing Conservation Amendment (OSHA, 1983) requires that HPDs reduce an employee's TWA to 90 dBA or less, and in the case of employees demonstrating standard threshold shifts, to 85 dBA or less. No particular types or brands of HPDs are specifically recommended or proscribed. The method for assessing the amount of reduction to be expected is contained in Appendix B of the amendment (see section on OSHA's Calculational Procedure and Appendix A).

A proposed standard that has been discussed for over 15 years and has passed from committee to committee is American National Standard Z137.1. This document was intended to evaluate the physical characteristics of HPDs by specifying mechanical tests to which they should be subjected, such as temperature cycling, vibration, headband extension, drop testing, etc. Subsequent to these tests, the device's attenuation would be measured in conformance with ASA STD 1. Currently, work on this standard has ceased, although it is likely that interest may again be revived. One of the major problems that has plagued this document over the years is exactly what physical tests to specify and what, if any, pass/fail criteria can be objectively applied to the results.

EPA Labeling Regulation

Section 8 of the Noise Control Act of 1972 empowered the EPA to label all noise-producing and noise-reducing devices. The first and only standard that was promulgated for noise-reducing devices was the hearing protector labeling regulation (EPA, 1979). As with the preceding regulations that have been discussed, it did not specify criteria by which HPDs were to be deemed acceptable or unacceptable. The regulation did, however, specify that the attenuation of any device or material capable of being worn on the head or in the ear canal, that is sold wholly or in part on the basis of its ability to reduce the level of sound entering the ear, was to be evaluated according to ASA STD 1. An NRR was to be computed and placed on a label whose size and configuration was specified. The regulation also included requirements for submission of label verification data to the EPA and contained enforcement and compliance audit testing provisions.

In the early 1980s, budget cuts at the EPA led to elimination of the noise enforcement division. In recognition of their inability to enforce the regulation, the EPA then revoked the product verification testing and the attendant reporting and record keeping requirements for the standard (EPA, 1983). However, the remainder of the regulation remains law unless, and until, Congress deletes Section 8 from the Noise Control Act. This leaves the future of the regulation in limbo. Pending congressional action, it is possible that a voluntary labeling regulation embodying an NRR or NRR-like number and strict quality control requirements may be developed by the Safety Equipment Institute of the Industrial Safety Equipment Association.

CONCLUDING REMARKS

Recommendations

The factors most often considered by purchasers of HPDs are attenuation, comfort, human engineering, cost, durability, styling, and availability. It is not possible to rank order these items in a manner suitable for all applications, but in my opinion, attenuation and comfort should be considered foremost and given equal importance. If the laboratory attenuation for a particular HPD was found to be very good, but the comfort of the device was poor, then

actual in-use protection might be reduced or even nonexistent. Conversely, a lower attenuation HPD that is comfortable and can be worn regularly and consistently will provide greater effective protection to the using population.

Attenuation may be estimated by using manufacturers' published data, although a better predictor of in-field performance will be the available measured real-world attenuation data, and the best indicator of actual on-the-job effectiveness can be provided by analysis of the annually updated audiometric database. The natural inclination to precisely match attenuation data to a particular noise exposure spectrum must be tempered by an awareness of the large differences between rated and real-world performance and of the large inter-employee variability in HPD effectiveness. Although it is an interesting technical exercise to utilize the information in this chapter to attempt to develop ''precision'' estimates of the real-world protection of HPDs, in most instances, a hearing conservationist's time can be more fruitfully spent by gaining first-hand knowledge of the operational characteristics of HPDs and communicating this information to the workforce. Comfort, and also human engineering, should be evaluated personally by prospective buyers and their employees. Program administrators will gain a much improved appreciation of the devices under consideration if they wear them 8 hours a day for a few days, in order to conduct their own subjective evaluation.

The program administrator should be able to narrow down the types and brands of HPDs that will be offered within any one program. If this is accomplished with care and perspicacity, following the guidelines provided in this chapter, then the users will be offered a selection of the best available devices. At the same time, it is important to limit the selection, because the problems of preparing uniform high caliber training for a large variety of products can easily become unwieldy. Furthermore, a limited group of devices simplifies inventory and spare parts control and may allow buyers to negotiate higher volume, lower-cost contracts with their suppliers.

The selection of devices to be offered in a hearing conservation program should include a minimum of three devices, representing at least two types. Generally, this will consist of an earmuff and a couple of models of earplugs, but a semi-aural device or additional brands of each protector type may be warranted. The employee should be involved in the final choice of the HPD that he or she will wear. HPDs are a personal piece of protective equipment that may not adapt equally well to all head and ear canal shapes, and individual preference may vary widely. Involving employees in the selection process will increase the likelihood of maintaining their participation in the entire program. If, after a couple of weeks of daily use, an employee is still experiencing difficulties or discomfort, the protector should be resized and/or refitted, or another hearing protector issued.

And finally, it is worth stressing a widely quoted but often unappreciated axiom—The best hearing protector is the one that is worn and worn correctly. And that protector will be the one that is matched to the environment, the noise, and the person in need of protection.

Final Considerations

The contents of this chapter provide a comprehensive review and analysis of the measurement, performance, selection, and fitting of HPDs. Of equal importance is the information that is provided to hearing conservationists to assist them in developing educational materials and programs to train and motivate their workforce. Hearing protectors are not a panacea and cannot be dispensed indiscriminately, but they can and do work when utilized within the context of a well-defined and properly implemented hearing conservation program.

Evaluation of a large number of existing programs suggests that many protectors can provide at least 10 dB of noise reduction for 84 percent of the workforce. According to the best government estimates (OSHA,

1981), 92 percent of industrial noise exposures represent 8 hour equivalent levels of less than or equal to 95 dBA (and 97 percent ≤ 100 dBA). Therefore, 10 dB of attenuation is often all that is needed to reduce these exposures to acceptable levels. Furthermore, a number of studies have examined both short- and long-term changes in employees' hearing threshold levels and have shown that certain HPDs can and do provide adequate protection from high level industrial noise exposures (Hager, Hoyle, & Hermann, 1982; Royster, 1979; Royster & Royster, 1982). Thus, HPDs remain one of the most important and potentially effective tools available today for the hearing conservationist to utilize in the ongoing struggle to protect workers from the hazards of occupational noise.

REFERENCES

Abel, S. M., Alberti, P. W., Haythornthwaite, C. & Riko, K. (1982). Speech intelligibility in noise: Effects of fluency and hearing protector type. *J. Acoust. Soc. Am.*, 71(3), 708–715.

Abel, S. M., & Rokas, D. (1986). The effect of wearing time on hearing protector attenuation. *J. Otolaryngol.*, 15(5), 293–297.

Acoustical Society of America. (1975). *Method for the measurement of real-ear protection of hearing protectors and physical attenuation of earmuffs.* Standard ASA STD 1-1975 (ANSI S3.19-1974). New York, NY.

Acton, W. I. (1970). Speech intelligibility in a background noise and noise-induced hearing loss. *Ergonomics*, 13(5), 546–554.

American National Standard. (1981). *Personal hearing protective devices for use in noise environments.* Draft standard Z137.198X.

American National Standards Institute. (1957). *Method for the measurement of real-ear attenuation of ear protectors at threshold.* Standard Z24.22-1957 (R1971). New York, NY.

American National Standards Institute. (1984). *Method for the measurement of the real-ear attenuation of hearing protectors.* Standard S12.6-1984, New York, NY.

Atherley, G. R. C., & Noble, W. G. (1970). Effect of ear-defenders (ear-muffs) on the localization of sound. *Br. J. Ind. Med.*, 27, 260–265.

Berger, E. H. (1980). Suggestions for calculating

hearing protector performance. *Sound and Vibration*, 14(1), 6–7.

Berger, E. H. (1981). Details of real world hearing protector performance as measured in the laboratory. In L. H. Royster, F. D. Hart, & N. D. Stewart (Eds.), *Proceedings of Noise-Con 81* (147–152). New York: Noise Control Foundation.

Berger, E. H. (1983a). Using the NRR to estimate the real world performance of hearing protectors. *Sound and Vibration*, 17(1), 12–18.

Berger, E. H. (1983b). Laboratory attenuation of earmuffs and earplugs both singly and in combination. *Am. Ind. Hyg. Assoc J.*, 44(5), 321–329.

Berger, E. H. (1983c). Considerations regarding the laboratory measurement of hearing protector attenuation. In R. Lawrence (Ed.), *Proceedings of Inter-Noise 83* (pp. 379–382). Edinburgh, U.K.: Institute of Acoustics.

Berger, E. H. (1983d). Attenuation of hearing protectors at the frequency extremes. *11th Int. Cong. on Acoustics*, (Vol. 3, pp. 289–292). Paris, France.

Berger, E. H. (1984). EARLog #12—The hearing conservation amendment (Part II). *Am. Ind. Hyg. Assoc. J.*, 45(1), B22–B23.

Berger, E. H. (1985a). EARLog #16—A new hearing protector attenuation standard—ANSI S12.6. *Am. Ind. Hyg. Assoc. J.*, 46(5), B26–B27.

Berger, E. H. (1985b). EARLog #17—Ear infection and the use of hearing protection. *J. Occup. Med.*, 27(9), 620–623.

Berger, E. H. (1986). Review and tutorial—Methods of measuring the attenuation of hearing protection devices. *J. Acoust. Soc. Am.*, 79(6), 1655–1687.

Berger, E. H. (1987). Can hearing aids provide hearing protection? *Hearing Instr.*, 38(3), 12–14.

Berger, E. H., & Kerivan, J. E. (1983). Influence of physiological noise and the occlusion effect on the measurement of real-ear attenuation at threshold. *J. Acoust. Soc. Am.*, 74(1), 81–94.

Berger, E. H., Kerivan, J. E., & Mintz, F. (1982). Inter-laboratory variability in the measurement of hearing protector attenuation. *Sound and Vibration*, 16(1), 14–19.

Blackstock, D. T., & von Gierke, H. E. (1956). *Development of an extra small and extra large size for the V-51R earplug*. Wright-Patterson AFB, OH: Wright Air Development Center.

Bolka, D. F. (1972). *Methods of evaluating the noise*

and pure tone attenuation of hearing protectors. Doctoral dissertation, Penn State University.

Camp, R. T. (1979). Hearing protectors. *The Otolaryngological Clinics of North Am.*, 12(3), 569–584.

Caruso, V. G., & Meyerhoff, W. L. (1980). Trauma and infections of the external ear. In M. M. Paparella & D. A. Shumrick (Eds.), *Otolaryngology*. Philadelphia, PA: W. B. Saunders Co..

Chung, D. Y., & Gannon, R. P. (1979). The effect of ear protectors on word discrimination in subjects with normal hearing and subjects with noise-induced hearing loss. *J. Am. Aud. Soc.*, 5(1), 11–16.

Cohen, A. (1976). The influence of a company hearing conservation program on extra-auditory problems in workers. *J. Saf. Res.*, 8(4), 148–162.

Coles, R. R. A., & Rice, C. G. (1965). Letter to the editor: Earplugs and impaired hearing. *J. Sound Vib.*, 3(3), 521–523.

Coles, R. R. A., & Rice, C. G. (1966). Speech communications effects and TTS reduction provided by V51R and Selectone-K earplugs under conditions of high intensity and impulsive noise. *J. Sound Vib.*, 4(2), 156–171.

Department of the Air Force. (1982). *Hazardous noise exposure*. (AF Regulation 161-35). Washington, DC.

Else, D. (1973). A note on the protection afforded by hearing protectors—implications of the energy principle. *Ann. Occup. Hyg.*, 16, 81–83.

Environmental Protection Agency. (1979). *Noise labeling requirements for hearing protectors*. *Federal Register*, 42(190), (40CFR Part 211), 56139–56147.

Environmental Protection Agency. (1983). Noise emission standards for. . ., and noise labeling requirements for hearing protectors; final rule; revocation for product verification testing, reporting, and recordkeeping requirements. *Federal Register*, 47(249), (40 CFR Parts 204, 205, and 211, 57709–57717).

Foltner, K. A. (1986). Visual evaluation of the external ear and eardrum. In E. H. Berger, W. D. Ward, J. C. Morrill, & L. H. Royster (Eds.), *Noise and hearing conversation manual* (4th ed., pp. 217–232). Akron, OH: Am. Ind. Hyg. Assoc.

Forrest, M. R. (1969). *Laboratory development of an amplitude-sensitive ear plug* (Report HeS 133, Med. Res. Council). London, England: Royal Naval Personnel Research Committee.

Fox, J. G. (1971). Background music and

industrial efficiency—A review. *Appl. Ergon.*, 2(2), 70–73.

Froehlich, G. R. (1981). *The effects of ear protectors and hearing loss on sentence intelligibility in aircraft noise* (paper 16). AGARD Conf. 311. Soesterberg, Netherlands.

Gasaway, D. C. (1984). "Sabotage" can wreck hearing conservation programs. *Natl. Saf. News, 129*(5), 56–63.

Gasaway, D. C. (1985). *Hearing conservation—A practical manual and guide* (p. 173). Englewood Cliffs, NJ: Prentice-Hall, Inc.

Guild, E. (1966). Personal protection. In *Industrial noise manual* (2nd ed., pp. 84–109). Am. Ind. Hyg. Assoc..

Hager, W. L., Hoyle, E. R., & Hermann, E. R. (1982). Efficacy of enforcement in an industrial hearing conservation program. *Am. Ind. Hyg Assoc. J., 43*(6), 455–465.

Hormann, H., Lazarus-Mainka, G., Schubeius, M., & Lazarus, H. (1984). The effect of noise and the wearing of ear protectors on verbal communication. *Noise Control Eng. J., 23*(2), 69–77.

Howell, K., & Martin, A. M. (1975). An investigation of the effects of hearing protectors on vocal communication in noise. *J. Sound Vib., 41*(2), 181–196.

Humes, L. E. (1983). A psychophysical evaluation of the dependence of hearing protector attenuation on noise level. *J. Acoust. Soc. Am., 73*(1), 297–311.

International Organization for Standardization. (1981). *Acoustics—measurement of sound attenuation of hearing protectors—subjective method.* ISO 4869, Switzerland.

Kasden, S. D., & D'Aniello, A. (1976). Changes in attenuation of hearing protectors during use. In J. K. Mowry (Ed.), *Proceeding of Noisexpo* (pp. 28–29).

Krutt, J., & Mazor, M. (1980–1981, winter). Attenuation changes during the use of mineral down and polymer foam insert-type hearing protectors. *Audiology & Hearing Educ., 13–14.*

Kryter, K. D. (1946). Effects of ear protective devices on the intelligibility of speech in noise. *J. Acoust. Soc. Am., 18*(2), 413–417.

Lawrence, M., & Yantis, P. A. (1956). Onset and growth of aural harmonics in the overloaded ear. *J. Acoust. Soc. Am., 28,* 852–858.

Lin, L. (1981). *Auditory localization under conditions of high ambient noise levels with and without the use of hearing protectors.* Masters thesis, North Carolina State University, Raleigh, NC.

Lindeman, H. E. (1976). Speech intelligibility and the use of hearing protectors. *Audiology, 15,* 348–356.

Martin, A. M. (1976). Industrial hearing conservation 1: Personal hearing protection. *Noise Control Vib. and Insul., 7*(2), 42–50.

Martin, A. M. (1977). The acoustic attenuation characteristics of 26 hearing protectors evaluated following the British standard procedure. *Ann. Occup. Hyg., 20,* 229–246.

Martin, A. M. (1979). Dependence of acoustic attenuation of hearing protectors on incident sound level. *Br. J. Ind. Med., 36,* 1–14.

Mosko, J. D., & Fletcher, J. L. (1971). Evaluation of the Gundefender earplug: Temporary threshold shift and speech intelligibility. *J. Acoust. Soc. Am., 49*(6 Part 1), 1732–1733.

National Institute for Occupational Safety and Health. (1975). *List of personal hearing protectors and attenuation data.* (U.S. Dept. of HEW Report No. 76–120). Cincinnati, OH.

Nixon, C. W. (1979.) Hearing protector devices: Ear protectors. In C. M. Harris (Ed.), *Handbook of noise control* (2nd Ed., pp. 12-1–12-13). New York: McGraw-Hill.

Nixon, C. W., & Knoblach, W. C. (1974). *Hearing protection of earmuffs worn over eyeglasses* (Report No. AMRL–TR–74–61). Wright-Patterson AFB, OH: Aerospace Medical Research Laboratory.

Nixon, C. W., & von Gierke, H. E. (1959). Experiments on the bone-conduction threshold in a free sound field. *J. Acoust. Soc. Am., 31*(8), 1121–1125.

Noble, W. G., & Russell, G. (1972). Theoretical and practical implications of the effects of hearing protection devices on localization ability. *Acta Otolaryngol. 74,* 29–36.

Occupational Safety and Health Administration. (1981). Occupational noise exposure; hearing conservation amendment. *Federal Register, 46*(11), 4078–4181.

Occupational Safety and Health Administration. (1983). Occupational noise exposure; hearing conservation amendment. *Federal Register, 48*(46), 9738–9783.

Occupational Safety and Health Administration. (1987). *Industrial Hygiene Technical Manual,* Change 2, March 1, pages VI–13—VI–19.

Ohlin, D. (1975). *Personal hearing protective devices fitting, care, and use.* (Report No. AD–A021 408). Aberdeen Proving Ground, MD: U.S. Army Environmental Hygiene Agency.

Riko, K., & Alberti, P. W. (1982). How ear

protectors fail: A practical guide. In P. W. Alberti (Ed.), *Personal hearing protection in industry* (pp. 323–338). New York: Raven Press.

Rink, T. L. (1979). Hearing protection and speech discrimination in hearing-impaired persons. *Sound and Vibration, 13*(1), 22–25.

Royster, J. D., & Royster, L. H. (1986). Audiometric data base analysis. In E. H. Berger, W. D. Ward, J. C. Morrill, & L. H. Royster (Eds.), *Noise and hearing conservation manual* (4th ed., pp. 293–317). Akron, OH: Am. Ind. Hyg. Assoc.

Royster, L. H. (1979). Effectiveness of three different types of ear protectors in preventing TTS. *J. Acoust. Soc. Am., 66* (Supp. 1), S62.

Royster, L. H., & Holder, S. R. (1982). Personal hearing protection: Problems associated with the hearing protection phase of the hearing conservation program. In P. W. Alberti (Ed.), *Personal hearing protection in industry* (pp. 447–470). New York: Raven Press.

Royster, L. H., & Royster, J. D. (1982). Methods of evaluating hearing conservation program audiometric data bases. In P. W. Alberti (Ed.), *Personal hearing protection in industry* (pp. 511–540). New York: Raven Press.

Royster, L. H., & Royster, J. D. (1985). Hearing protection devices. In A. S. Feldman & C. T. Grimes (Eds.), *Hearing conservation in industry.* Baltimore, MD: Williams and Wilkins.

Royster, L. H., & Royster, J. D. (1986). Education and motivation. In E. H. Berger, W. D. Ward, J. C. Morrill, & L. H. Royster (Eds.), *Noise and hearing conservation manual* (4th ed., pp. 383–416). Akron, OH: Am. Ind. Hyg. Assoc.

Royster, L. H., Royster, J. D., Berger, E. H., &

Skrainar, S. F. (1984). Personal audio headsets and table radios in industrial noise environments. *ASHA, 26*(10), 77.

Russell, G. (1977). Limits to behavioral compensation for auditory localization in earmuff listening conditions. *J. Acoust. Soc. Am., 61*(1), 219–220.

Schmidt, J. W., Royster, L. H., & Pearson, R. G. (1980). Impact of an industrial hearing conservation program on occupational injuries for males and females. *J. Acoust. Soc. Am., 67* (Suppl. 1), S59.

Skrainar, S. F., Royster, L. H., Berger, E. H., & Pearson, R. G. (1985). Do personal radio headsets provide hearing protection? *Sound and Vibration, 19*(5), 16–19.

Sutton, G. J., & Robinson, D. W. (1981). An appraisal of methods for estimating effectiveness of hearing protectors. *J. Sound Vib., 77*(1), 79–81.

Tonndorf, J. (1972). Bone conduction. In J. V. Tobias (Ed.), *Foundations of modern auditory theory* (Vol. 2, pp. 195–238). New York: Academic Press.

Wilkins, P. A. (1980). *A field study to assess the effects of wearing hearing protectors on the perception of warning sounds in an industrial environment.* (Contract Report No. 80/18). Southampton, England: Inst. of Sound and Vibration Research.

Wilkins, P. A., & Martin, A. W. (1978). *The effect of hearing protectors on the perception of warning and indicator sounds—a general review.* (Tech. Report No. 98). Southampton, England: Inst. of Sound and Vibration Research.

Zwislocki, J. (1957). In search of the bone-conduction threshold in a free sound field. *J. Acoust. Soc. Am., 29*(7), 795–804.

The Employee Education Program

David M. Lipscomb

Although it is possible to operate a hearing conservation program without employees understanding why the program is in place, there are benefits to the hearing conservation effort if workers are aware of some basic facts. This text cannot overlook the fact that employee education is an essential element in a hearing conservation program. OSHA also recognized the importance of such a program and required in its guidelines that the workers who come under the "action level" or who are exposed to 90 dBA TWA sound should have an *annual* education program. The final version of the guidelines (*Federal Register*, 1983, p. 9778) summarizes topics to be included in the education program. A discussion of OSHA provisions in this area is contained in Chapter 12.

MANAGEMENT COOPERATION

No hearing conservation program will succeed without strong endorsement and support by management at all levels.

Educational programs should be sponsored *and attended* by management. This area is so important that a separate chapter has been devoted to the topic (Chapter 18).

WORKER COOPERATION

Especially in the early stages of a hearing conservation program, resistance from the workforce will be encountered. "Why?" asks the worker. Here is someone who is wearing a hard hat, steel-toed shoes, gloves, safety glasses, and a leather apron. The hearing protective devices (HPDs) being issued are, to that person, just another in a long line of devices forced upon him or her. Without some understanding of the reasons for a hearing conservation program, this resistance will continue. Other forms of safety apparatus are generally accepted without question, or at least, with minimal resistance. Workers know that a chip or spark flying into an eye will cause instantaneous discomfort and possible loss of sight; they realize that hot items cannot be handled without gloves; if they drop a heavy object on the foot, toes cry out immediately for attention; a blow to the head can ruin one's entire day if head protection is not in place. In each of the preceding examples, the key was: *instantaneous knowledge of injury.* Except for very rare instances of inordinately high sound levels (greater than 120 dB), sound exposure causes relatively little pain or discomfort. Sound overexposure does not cause one to realize instantaneously that hearing has been permanently injured. Noise-induced hearing injury develops insidiously, hence, it is not feared in the same way as are painful, sometimes bloody, injuries.

The primary target for noise-induced hearing damage is the delicate organ of Corti located in the cochlea. Excessive exposure to common industrial noise does not cause "pain" because there are no pain receptors in the inner ear. Any sensation of pain or discomfort arises from stress on the tympanic membrane. This site of discomfort was identified by Ades, Graybiel, Morrill, Tolhurst, and Niven, (1958). They subjected persons who had no tympanic membranes to high-intensity sounds. In these selected subjects, sound levels approaching 170 dB (SPL) did not elicit a painful sensation. Pain serves a valuable purpose as a harbinger of damage, or of the potential for damage; unfortunately, the inner ear is packed so tightly with other sensory mechanisms that room was not available to provide for this warning device. So, it is incumbent upon the hearing conservationist to resort to employee education and motivation programs for the development of awareness of other noise hazard warning signs.

Another frequently encountered feature of the resistance to avoiding noise exposure is based on the "macho" image some people like to display. Two decades ago, most airports were not equipped with passenger walkways. Thus, the plane pulled up to the gate in the vicinity of awaiting passengers and persons meeting passengers. At that time, the sound level at the ears of those on the ground reached levels well in excess of 100 dBA. Yet, very few people would cover their ears. Children, in their honesty, would squeal or make faces and place their hands over their ears. Most adults would grit their teeth noticeably and "bear it." In the foregoing example, and in many occupational areas, it is considered to be a "sissy" thing to cover one's ears. "Gotta be tough" is the image projected.

That attitude is becoming less prevalent, perhaps due to the large number of articles in the popular literature attesting to the hazards of noise in the environment. Some workers are seeing the effects of a lifetime of noise exposure on parents and other relatives or friends. Younger employees generally accept HPDs more willingly than the previous generation did. Yet, we still have a long way to go before HPDs and other forms of hearing conservation action by workers will reach the level of acceptance necessary for the desired level of success in these programs.

FORMAT OF EDUCATION PROGRAMS

An area of almost total agreement, in this field where agreement is scarce, is that employee education is necessary—yes, vital. The next point of consideration is to offer an educational program—but in what format? Each approach to employee education has strengths and weaknesses. A sample of these formats follow.

Lecture Series

The hearing conservationist or an invited speaker delivers an annual lecture to employees gathered in a meeting room. Content of the lecture varies from year to year, and includes the comprehensive hearing conservation program, descriptions of ear anatomy with reference to loci of injury by noise, the hearing test, HPD use and fitting, extending hearing conservation to the nonoccupational hours, reasons behind sound measurement, regulatory reasons for hearing conservation, principles of noise control with emphasis on how employees can assist, social effects of noise-induced hearing impairment, and similar topics. Most of these presentations take one hour, with a question and answer period at the end of the lecture.

The advantage of such programs is that information can be disseminated quite efficiently. One person can "educate" a number of others. If the lectures are well prepared and are presented by a knowledgeable person, this format can be effective.

Disadvantages also exist. The problem of employee alertness is ever-present. Sitting passively and listening to a lecture is not a favorite activity for most people, and maintaining attention is not easy. A logistical problem is that such activities require that segments of the plant be closed down or reduced in personnel during the lecture time or that the employees be paid extra for attending the lecture in their off hours.

Movies, Videos, and Slide Presentations

Manufacturers of products used in hearing conservation programs, particularly HPDs, have been actively producing educational films and video tapes in recent years. One manufacturer distributes an annotated listing of these films and videos—the 1986 version is 43 pages in length (Berger, 1986). Therefore, resources are not lacking for this presentation format. Some films use well-known movie or TV actors to convey the message of hearing conservation. Some actors are hearing impaired and speak from their own personal experience; other treatments are dramatizations of various situations involving hearing conservation information and interests. Some films are in the documentary style, whereas others seek to hammer home the need for hearing conservation through semi-shock techniques. A few of the products are not well done, a few are exceptionally high-quality productions, and the remainder are adequate. Targeting to the understanding, intelligence, and interests of a wide range of personnel is difficult. Therefore, some of the productions

insult the intelligence of some people when covering the more basic topics. A number of the films are overly technical and will not be comprehended by most of the workforce.

Generally, the more successful attempts at producing an educational film or video tape are those that entertain and simultaneously make a few points. They become less effective with length. Comprehensive treatments of the subject are not as successful as those that speak on a circumscribed topic or on a limited range of subjects. It is useful to combine the film presentation with a discussion session to bring home those points made by the "canned" presentation.

The major advantage of this format is that the work is done for the hearing conservationist. All that is necessary is previewing the film and preparing the projector and screen or video tape playback equipment. It is often an advantage to have professionally prepared presentation material if the alternative is to present lectures by a person who is not an experienced public speaker.

Disadvantages of this approach to worker education are essentially the same as those enumerated for the lecture format.

Programmed Learning Booklets

In this era of "continuing education," a mechanism of teaching that has become widely used is *programmed learning*. The format of this approach is to make some statements and follow them with one or more questions about the content of the immediately preceding materials. If the question is answered correctly, one is encouraged to proceed to the next section. Otherwise, the person reviews material again so the correct answer can be obtained. There are many variations on that theme. A sample page from a programmed learning booklet is shown in Figure 11-1. An appealing feature of this format is that it encourages the active participation of workers. They read the information, and they answer the questions. Upon returning

---

OUR HEARING CONSERVATION PROGRAM—AND YOU

There are many things done in the plant that you may know about, but you don't have to be bothered with them. Hearing conservation is different. By now, you have gotten almost through this booklet. You have probably noticed several things. One of the important lessons is that our hearing conservation program will work only with your serious attention to your part in the program.

Q: You don't need to do anything in the hearing conservation program. (choose one):

   True_____ **False**_____

Let's review the parts of the hearing conservation program so you will be sure when your part comes in:

1. Sound measures will be taken in the plant. You will be involved when your work area is being measured. Go on about your work like usual. You may be asked to run certain equipment or make changes in your routine to check something out.

Q: Can you be involved during sound measurement?

   Yes_____ **No**_____

2. Efforts to reduce noise levels will begin when and where it is possible. You may have some ideas to share. Be sure to give your thoughts to the people working on the controls.

Q: Can you be part of the noise control work?

   Yes_____ **No**_____

3. You may be called in to take a hearing test. It will be necessary for you to pay attention to the instructions for the test and to listen very carefully.

Q: Are you an important part of the hearing test?

   Yes_____ **No**_____

4. You may be required to wear hearing protection or it may be made available to you at your option. We suggest that you wear protection whenever you are in noise loud enough to cause you to raise your voice when talking to others. This means both on- and off-the-job. Be sure to wear the protectors correctly.

Q: Are you important to the hearing protector part of the hearing conservation program? (choose one):

   Yes_____ **No**_____

Each of the last four questions should be answered "yes." You will be *very* involved in the program if it is to work like we plan it. Good luck and *hang on to those ears—they are precious.*

Figure 11–1. Sample page from a programmed learning booklet used for worker education. (Anonymous, 1985, p. 10). Reprinted with permission of Hear Now Publications.

the booklet to the hearing conservationist, the answers can be reviewed for accuracy and any confusion can be cleared up at that time by the conservationist.

There are several advantages to a programmed learning booklet distribution approach. It is not necessary to stop production or to arrange for employee meetings in the off hours. Booklets can be distributed individually and they can be reviewed by workers without interruption of production schedules. Participation in the education exercise can be documented by both the hearing conservationist and the employee by completing the form at the end of the booklet (Figure 11–2).

Although encouraging, this format is not without its limitations. The use of programmed learning assumes sufficient literacy to read and understand the material, and the ability to answer questions in writing. That is an assumption that cannot

---

REGISTRATION OF PARTICIPATION

Worker's name: _____

has read *HEARING CONSERVATION—A Programmed Learning Review for Industrial Workers* and has satisfactorily answered the questions in the text.

Worker signature: _____

Program officer: _____

Date answers were reviewed: _____

Figure 11-2. Documentation form at the end of the programmed learning booklet. (Anonymous, 1985, p. 11). Reprinted with permission of Hear Now Publications.

be made, even in this modern age. Because there are individual contacts in reviewing the material and checking the answers, demands on the time of the hearing conservationist are greater than in some other formats. Further, booklets must be made available on a number of topics. It will not be sufficient to distribute the same material repeatedly.

Computer-Assisted Instruction (CAI)

The proliferation of personal computers at affordable prices has spurred a great movement toward using computers as educational tools. Interactive programming can be utilized so it is possible for the trainee to ''talk'' to the computer. Such programming languages as FORTH, LOGO, and PILOT are easy to learn and offer interactive facilities. This means that a program can be developed (written) which will respond appropriately to answers input by the person in training (one ''interacts'' with the computer). Thus, this format is quite similar to the programmed learning format and has similar advantages and disadvantages. An additional consideration can be both advantageous and disadvantageous—the computer

itself. Some workers may be intrigued with the computer and find work on it to be fascinating. Others are inhibited by any type of equipment and will be unable to relate to it or to the program being run. As computers continue to be extended into everyday life, however, it appears that CAI must be considered as one of the more promising formats for worker education. To date, no interactive training programs for employee education for hearing conservation are in existence. But it is only a matter of time until such programs will appear along with clever graphics to both enlighten and entertain.

Distribution of Printed Materials

In the past a favorite method for employee education was to simply give each employee in the hearing conservation program some pamphlets or brochures. There are numerous good information publications available for distribution. The obvious advantage of such a program is that it is not time consuming for the hearing conservationist and it does not take workers off of the production line. There is no guarantee, however, that workers will read the materials.

Or, if they do read them, there may not be adequate understanding of the information. Distribution of reading materials assumes the person's ability to read.

Individual Conferences

Some companies conduct annual interviews with each employee. During this time, results of quality control data, health records, absenteeism, productivity, and other pertinent topics are discussed. Although it is time consuming, some attempts to inform and motivate workers regarding hearing conservation may be included in the discussion. It is doubtful whether this would be an appropriate or effective format, due to the fact that the worker's mind is much more on those matters relating to job performance. Often, in times of economic downturns, there is serious concern regarding whether layoffs are in the near future. To attempt any "educational" efforts in that climate would be frustrating.

Education by Example—Negative

A large furniture manufacturer was initiating a hearing conservation program. During the "walk through" with the safety engineer and plant manager, an employee was noted operating a fork lift without a hard hat in place. The plant manager asked why the man was unprotected. The safety engineer replied that it was impossible to get the man to wear the hard hat even though he was moving material that could fall free and enter the operator's semi-enclosed cab. "Fire him!" was the immediate reply of the plant manager. "But, he's worked here nearly 20 years," the safety manager responded. "Makes no difference, let's make an example of him," was the retort by the plant manger. Thus, a worker was given his "walking papers" because of repeatedly flaunting one of the safety rules. Doubtless, there was a message in this action. However, it is safety management by force, rather than by education and reinforcement. True, according to OSHA policy, use of safety devices, including HPDs, can be made a condition of employment. That is a bureaucratic way of saying, "Wear these or be canned!" It is a fact that negative reinforcement achieves results quickly, but positive reinforcement is longer lasting. Education by example may be necessary as a last resort, but it is included here not as a recommendation, but as a contrast to the following educational format.

Education by Example—Positive

The plant manager in the previous story was wearing a hard hat. He was also wearing HPDs. So, his action against the employee was from the position: "Look, I'm wearing safety devices—you were supposed to wear them too!" When management encourages use of safety devices, then uses them, there is a powerful lesson provided, even though no words are spoken. It is not uncommon to begin a plant "walk through" by being offered safety devices—hard hat, protective eyeglasses, dust mask, gloves, or other items deemed necessary for safe transport through the areas to be visited. But HPDs are not as often issued. When reminded that earplugs or ear muffs should be used also, the manager assures the guest that the period in the plant will be brief and that hearing damage will not occur. That is probably true, but wearing HPDs into the processing areas where workers are required to wear them provides a subtle but positive reinforcement of the stated safety policy. It is well to suggest that management wear the most obvious hearing protectors available so it is readily apparent that they are setting the example. (Even if it comes to putting small flashing lights on them or painting them phosphorescent orange!) This is reinforcement to the worker educational format(s) selected.

Varying Formats

None of the described formats will offer the perfect educational climate for all workers to learn reasons behind the hearing conservation program. It is well to vary the format from time to time in order to maintain interest in the presentation.

FITTING AND ISSUING HEARING PROTECTIVE DEVICES

Technical data and concepts regarding HPDs can be found in Chapter 10. However, the personal contacts with personnel regarding their use of HPDs falls under the aegis of "education" and will be discussed here.

Selection of HPDs

It would be impractical to attempt a comprehensive HPD selection process for each and every employee. Thus, some method must be developed wherein adequate protection is afforded but with an efficient selection process. Often, as an adjunct to the sound measurement and analysis of plant noise, specific HPDs are recommended in the report (Chapter 6). Using HPD specifications, a computer program can be written which will match various octave-band sound levels with those specifications and then select devices appropriate for that sound environment. If left to the purchasing department, the least expensive HPDs will be obtained, regardless of their specifications or effectiveness. Thus, it is necessary for some quantification in order to have reasons not to be satisfied with the least-expensive products available.

It is not advisable to select one HPD for use throughout the facility. Even if it is the most effective HPD available, no single HPD will satisfy the protective needs of an entire workforce. Some persons refuse to insert anything into the ears, thus, restricting the selection to a single ear plug would cause problems with a portion of the staff. Earmuffs are sometimes hot and/or heavy, causing some workers to reject them. Finger dexterity problems limits the use of malleable plugs in some cases. Canal caps are uncomfortable for long-term use and would be rejected if required full time. Ear canal sizes vary quite widely and very few HPDs are sufficiently flexible to fit all ears.

For some employees, the color might be wrong. So, there are numerous compelling reasons to have a selection of HPDs available, both with respect to style and brand of HPDs. Depending upon the variability of sound levels contained in the work environment, it is well to suggest that approximately five different devices be selected and obtained for distribution to workers.

Issuing HPDs

One method of issuing HPDs would be to take a page from the book of the Army drill sergeant by shoving a preselected HPD into the hands of each employee with the stern order: "Here, take these and use them. Don't *ever* let me see you without them on!" It doesn't require a degree in industrial psychology to realize that method will not achieve the desired results.

One of the first things one learns in a journalism class is the five points for an article: What? Where? When? How? and Why? That format is appropriate for this topic as well:

1. *What?* Hearing protectors come in a variety of types and styles. They should be shown to the employee to foster an understanding of the range of choices available.
2. *Where?* HPDs are worn in, at, or over the ears. They are not intended to cover the throat, as can be seen sometimes on contruction sites or airport tarmacs where workers have their earmuffs resting around their necks.
3. *When?* The protectors being issued are to be worn at all times there is noise in the working environment. It is often possible to modify that statement to indicate that HPDs are to be used during a certain set of operations. On occasion, the only time protectors are required is in the operation of a specific noisy machine, for example, a grinder. The workers should be told that the protective devices (earmuffs or canal caps are

most efficient) will be located on a hook above the grinder right next to the safety goggles. Further, depending upon hearing conservation program policy, it might be permissible for the employees to have a duplicate set of HPDs to take home for use when engaging in noisy activities off the job.

4. *How?* Proper placement or insertion of protectors is the primary guarantee that the devices will perform up to their specifications. Simply thrusting ear plugs out to the worker, and expecting that person to insert them appropriately is making a faulty assumption. As shown in Chapter 10, there are specific methods for the placement of these devices and that information must be made clear to the employee.

5. *Why?* Employees must have an appreciation of *why* HPDs are required. That is one of the functions of the employee education program. In the individual contacts at the time HPDs are issued, it is necessary to reinforce information disseminated during the educational programs.

If the worker participates in the final selection of HPDs for his or her personal use, there is more of a chance that person will be willing to undergo the adjustments required to use the protectors appropriately. Personal experience has shown that it works best to offer each employee a choice of three HPDs. Once they make their selection, there is a form of commitment to use them. Contrast that action to simply having some type of earplug or earmuff foisted upon them.

It should be made clear that to be effective, the HPDs must fit snugly, but that the comfort of the employee is important. So, if after a period of time, the HPDs are increasingly uncomfortable, or if there seems to be a sore spot where the devices come into contact with the ears, a refitting should be undertaken.

It can be readily seen from the foregoing discussion that the refitting and issuing of HPDs is a time-intensive process for the hearing conservationist or for that member of the staff assigned this duty. Yet, time and effort relegated to this stage of the hearing conservation program significantly increases the overall prognosis for success of the program.

ENCOURAGEMENT

There is the force of law requiring that certain actions must be taken to effect a hearing conservation program in industry and in some other environments. The authority to require these actions has been given to the hearing conservationist to use whenever and wherever it is deemed necessary. However, personnel management protocol suggests that other approaches be considered in addition to the authoritative attitude some might attain.

Supervisory personnel in the plant are on the "front line" of the HPD enforcement program. Without their cooperation in the use of HPDs and their insistence that others wear them, it will be difficult to achieve the level of HPD usage that is required for successful protection of employees.

Worker encouragement takes many forms:

- Noting that someone is using their hearing protectors and reinforcing that practice in the presence of other workers
- Making sure that supervisors are using their HPDs
- Encouraging management to always use HPDs when in noisy areas of the plant
- Making custom-fitted HPDs a sign of distinction by using them as an award for "employee of the week"
- Offering earplugs to workers' family members who are exposed to high-level sound
- Noting if HPDs are particularly unflattering to a worker (too large, or incorrect in other ways) and offering to exchange them for another style of HPD

DEALING WITH RESISTANCE TO HPDs

In practically every plant, certain persons will resist some of the policies. Those pockets of resistance threaten the quality and success of those policies. It is important that such resistance be identified and the scope of its effect be recognized. In some cases, short of threatening termination of their jobs, it is nearly impossible to obtain the willing cooperation of some workers. A method to dissolve these pockets of resistance has been used with considerable success by Dr. Robert Bertrand in Montreal. The essence of this program is as follows:

1. Identify the "ring leader," that person who seems to be at the forefront of those who are resisting use of HPDs.
2. Issue a challenge to this person. State that if he or she can demonstrate that HPD are of no value, that they won't be required. Note here that the conservationist is playing with a "stacked deck," but the worker is not aware of it.
3. For three consecutive days, test the worker's hearing at the beginning and at the end of a work shift when HPDs are *not* used.
4. The next three days, conduct the pre- and postshift hearing tests and issue HPDs while closely monitoring their use.
5. Go over the results of the testing, demonstrating that *on that worker* there were threshold shifts on the days HPDs were not worn, whereas there were no (or smaller) shifts on the days HPDs were in use.

Using this protocol, it has been found that demonstrating the value of HPDs using the individual's own hearing will enlist support of the program and faithful use of protectors. In fact, the worker will likely turn around and begin to lead others into the use of HPDs rather than lead the resistance against their use. Of course, the plan is not without risk. If not conducted appropriately with close monitoring of HPD use, no effect or a reverse effect might be noted, which throws the entire project into disarray.

CONCLUSION

Most other segments of hearing conservation have been augmented by the burgeoning high technological advances using microchip-based instrumentation and analytic procedures. However, much of employee education has not progressed past the level of accomplishment attained a decade ago. New ideas and development of fresh programming (e.g., Ver Hoef, 1987) can bring the educational efforts into line with other advances in the hearing conservationists armamentarium.

REFERENCES

Ades, H. W., Graybiel, A., Morrill, S., Tolhurst, G., & Niven, J. (1958). Non-auditory effects of high intensity sound stimulation on deaf human subjects (Joint Report #5). Dallas: University of Texas Southwestern Medical School, & Pensacola, FL: U.S. Naval School of Aviation Medicine.

Anonymous (1985). *Hearing conservation: A programmed learning review for industrial workers.* Knoxville, TN: Hear Now Publishers.

Berger, E. H. (1986). *Annotated listing of noise and hearing conservation films and videotapes (E-A-R 82-10/HP).* Indianapolis: Cabot Corporation.

Federal Register. (1983, March). Occupational noise exposure; hearing conservation amendment; final rule. 48(46), 9738–9785.

Ver Hoef, N. (1987). *Preserve your hearing* [Video Tape]. Des Moines, IA: Audiology Associates, Inc.

SUGGESTED READINGS

Feldman, A. S., & Grimes, C. T. (1985). Employee training programs in occupational hearing conservation. In A. S. Feldman & C. T. Grimes (Eds.), *Hearing conservation in industry.* Baltimore: William & Wilkins.

Gasaway, D. C. (1985). How to successfully educate, indoctrinate, and motivate workers.

In D. C. Gasaway (Ed.), *Hearing conservation: A practical manual and guide*. Englewood Cliffs, NJ: Prentice-Hall, Inc.

Mellard, T. J., & Geier, S. R. (1984). Personal hearing protection and employee education programs. In M. H. Miller & C. A. Silverman (Eds.), *Occupational hearing conservation*. Englewood Cliffs, NJ: Prentice-Hall, Inc.

12

The Comprehensive Hearing Conservation Program

Andrew P. Stewart

*T*he necessary components of a comprehensive hearing conservation program (HCP) have been known and discussed in the literature for years. Many authors have described the activities that must be carried out in an industrial setting in order to adequately protect employees' hearing from the effects of long-term high level noise exposure. In recent years, research on the validity and effectiveness of these components has been undertaken. Although empirical evidence has been scanty to date, and documentation of the importance of program components has consisted primarily of anecdotal evidence, available data and other information support the inclusion of traditional components and have expanded our knowledge of what makes a hearing conservation program work.

This chapter discusses the features of a comprehensive HCP. Previous chapters have described the individual features in detail, and that information will not be repeated here. The reader is referred to earlier chapters for in-depth coverage of individual program components. This chapter will synthesize the information, interrelate it where possible, and provide an approach to putting together a comprehensive HCP in industry. No one approach will work in all settings. Individual needs and strengths of companies and available personnel resources, both professional and nonprofessional, will necessitate the modification of the ''classical'' program model to meet local needs. This is completely appropriate, so long as all program components are included and an appropriate melding takes place which results in an effective program that protects the hearing of all noise-exposed employees.

It is important at this point to differentiate between an HCP that meets the requirements of the Occupational Safety and Health Administration's (OSHA) Hearing Conservation Amendment to the Occupational Noise Exposure Regulation 1910.95 (OSHA, 1983), and an effective HCP (see

Appendix A). A program that contains the minimal programmatic elements required by the noise standard may be construed as being comprehensive from a legal standpoint, but it may be far from effective. Currently, no specific requirements are included in the amendment to assess or guarantee effectiveness of program operation. An employer is in compliance with OSHA regulations if he or she simply incorporates all of the required program components into the HCP. An effective program, on the other hand, should incorporate various activities and documentations that are not included in the OSHA noise standard. As used in this chapter, the term *comprehensive* is meant to include all of the additional activities and efforts that should result in an effective HCP.

WHAT IS AN EFFECTIVE HEARING CONSERVATION PROGRAM?

An effective HCP is one that accomplishes the goals established for it. The primary goal of an industrial HCP must be the prevention (or, at least, limitation) of permanent hearing loss associated with exposure to industrial noise (Royster, Royster, & Berger, 1982). Other goals may be formulated in addition to this primary goal, such as compliance with OSHA regulations, reduction of employee stress and absenteeism, reduction of work place accidents due to plant noise levels, and reduction of the company's liability to worker compensation claims for occupational hearing loss. However, the primary goal remains the prevention of noise-induced permanent threshold shift (NIPTS) caused by workplace noise exposure.

But, practically speaking, how is program effectiveness to be measured? The OSHA amendment does not really address the issue, although it implies that the absence of "standard" threshold shifts in employees constitutes an effective program. Historically, two approaches have been used to discuss effectiveness in HCPs. The earliest approach was to define a program as effective if it was complete; that is, if it contained

all of the elements thought to be necessary for an effective HCP. Many authors (e.g. Maas, 1972; Sataloff & Michael, 1973; Olishifski & Harford, 1975; Dear & Karrh, 1979) have used this criterion, stressing the importance of providing a complete HCP to accomplish the goal of preventing worker hearing loss. However, even if a program is complete and includes all of the major elements known to be necessary to HCPs, its success in preventing hearing loss is not assured. Various sources of contamination and error exist in HCPs, which may confound the efforts of program personnel to carry it out effectively (Melnick, 1984). These may be found in all elements of the HCP and include such factors as inaccurate assessment of worker noise exposure, calibration errors in noise measuring equipment and audiometric instruments, improper use and deterioration of hearing protectors, insufficient motivation and intelligence in workers and technicians, and excessive ambient noise levels in testing areas. These and similar problems will prevent an apparently complete HCP from being effective.

The second approach, which arose out of the inadequacies of the first, was to define a program as effective if it worked; that is, if it succeeded in preventing or limiting the occurrence of NIPTS in an employee population. In recent years, several authors (e.g. Pell, 1973; Summar, 1976; Royster et al., 1980; Melnick, 1984) have stressed the need for some type of empirical, objective data to document whether a HCP is successful in this way. They generally use the findings of the annual audiometric evaluations as the source of these data, comparing them to the audiometric data of nonindustrial noise-exposed populations of similar demographic characteristics. The assumption is made that, if the noise-exposed population is not incurring any additional hearing loss beyond that which occurs in the nonindustrial noise-exposed population, the HCP is functioning effectively.

It appears obvious that, although programmatic completeness is an essential element of effective HCPs, criteria based on evaluation of employee audiometric results must also be used to definitively establish the effectiveness of individual HCPs. At the time this chapter was being written, a working group of the American National Standards Institute (ANSI S12-12) was working on the development of such criteria.

FEATURES OF A COMPREHENSIVE HEARING CONSERVATION PROGRAM

The following components of an HCP, which are necessary for its effective functioning, have been widely accepted (Fox, 1965; Maas, 1972; Newby, 1964; Royster et al., 1982):

1. Measurement of work-area noise levels
2. Identification of over-exposed employees
3. Reduction of hazardous noise exposure to the extent possible through engineering and administrative controls
4. Provision of personal hearing protection if other controls are inadequate
5. Initial and periodic education of workers and management
6. Motivation of workers to comply with HCP policies
7. Initial and periodic evaluation of worker hearing levels
8. Professional audiogram review and recommendations
9. Follow-up program for audiometric changes
10. Detailed record keeping system for the entire HCP
11. Professional supervision of the HCP

It is important that proper attention be given to all program components in order to achieve an effective HCP. Often, industrial management neglects certain areas of the program while emphasizing the implementation of other areas. For example, some companies neglect to perform noise exposure measurements in all plant areas and choose instead to include all areas in the hearing protector program. Such "scattergun" approaches seldom work effectively, since they neglect the realities of the work environment and place unnecessary burdens on under-exposed workers, thus reducing the credibility of the HCP in the eyes of the workers. Other companies place emphasis on the completion of annual employee audiograms, while neglecting to perform any follow-up activities based on the results of these tests, thus permitting worker hearing loss to continue unchecked. It is critical that all program elements be activated in every work place, but the exact "blend" of components will vary between companies, depending upon their individual needs. This fact underscores the necessity for the last program element in the previous list: professional supervision and consultation, which will assist a company in determining its exact needs and strengths.

The OSHA Occupational Hearing Conservation Rule

On March 8, 1983, the Hearing Conservation Amendment to the OSHA Occupational Noise Exposure Rule 29CFR 1910.95 was published in the *Federal Register* (OSHA, 1983). This document, the culmination of years of preparation, public comment, and political manipulation, outlined a relatively nonspecific set of guidelines for industry to follow in the structuring of HCPs. Somewhat diluted from its original "final version," which was published on January 16, 1981, it nevertheless provided a useful framework for the establishment of an effective program of hearing conservation activities, which could be "fleshed out" and enlarged upon through professional guidance. It included the essential elements of an effective HCP, while leaving the method of accomplishment of many of the activities involved in the program to the discretion of industrial management, although specifying that at least the audiometric testing program was to be under the supervision of a professional in hearing conservation or a physician.

The general program areas that are

included in and required by the Hearing Conservation Amendment are the following:

1. Noise monitoring
2. Employee notification (time-weighted average [TWA] exposure)
3. Audiometric testing program
4. Qualified audiometric technician support
5. Professional supervision
6. Calibrated test equipment
7. Audiometric test specifications
8. Audiogram review and follow-up procedures
9. Employee notification (Standard Threshold Shift [STS])
10. Personal hearing protectors
11. Fitting, training, enforcement, replacement of protectors
12. Measurement of protector attenuation
13. Employee training
14. Record keeping
15. Access to records
16. Transfer of records

Figure 12-1 is a compliance checklist used by one firm to assist its clients in assessing the completeness of their HCP with regard to the requirements of the Hearing Conservation Amendment. It also serves as documentation of compliance. Figure 12-2 is a form used by the same firm to assess and document its clients' HCPs during annual program reviews.

Noise Monitoring

The initial step in a company's HCP is to measure the sound environment to determine if employee overexposures exist. The OSHA regulation contains minimally specific guidelines as to the proper procedures for accomplishing these measurements, although the earlier (1981) version was much more specific. Monitoring must be done in order to determine which employees have 8 hour time-weighted average (TWA) exposures of 85 dBA or higher and must therefore be included in the HCP. Employees with TWA exposures above 90 dBA must

also be identified and required to use personal hearing protectors if engineering or administrative noise controls are not feasible. A further purpose of noise monitoring is to identify the level of attenuation that must be provided by hearing protectors used in various work areas. The monitoring must include all noise levels to which employees may be exposed between 80 and 130 dBA. The method of monitoring, whether area measurements with a sound level meter or personal sampling with a dosimeter, is left up to the employer, with the specification that area monitoring is permissible unless it would be inaccurate due to employee mobility during the workday or the intermittency of the noise sources. Each employee is to be notified of his or her 8 hour TWA noise exposure level, but the manner of notification is unspecified.

Since accurate determination of overexposure can be affected by such factors as worker mobility (which may vary significantly from day to day) and varying production process conditions, it is essential that measurements be made under ''worst case'' conditions. Often, management wishes to eliminate as many employees as possible from the HCP for economic reasons (a valid approach if this can reasonably be done with due concern for and understanding of the auditory risk factors involved); this becomes the primary emphasis of noise monitoring. They must be reminded that the true purpose of noise monitoring is to attempt to specify as accurately as possible the actual, effective exposure of each employee, in order to determine the degree of care that must be exercised in helping employees to protect themselves adequately from auditory insult. ''Splitting hairs'' during noise monitoring does a disservice to both workers and management, and may well leave the company open to later legal action, which may be more costly than including a few more workers in the HCP.

It must be remembered that the OSHA risk criterion does not guarantee protection against NIPTS; it is simply a statistic that suggests when the general population of noise-exposed workers is at risk for NIPTS

ELB/MONITOR, INC.

OSHA HEARING CONSERVATION STANDARD COMPLIANCE CHECKLIST	Date:		Page 1 of 2
COMPANY:	LOCATION:		

PURPOSE
This hearing conservation program checklist is intended for use by companies in conducting annual hearing conservation program evaluations to assure continuing effectiveness.

REFERENCE
Refer to OSHA Standard 29 CFR 1910.95(c)-(p), Hearing Conservation Amendment, for details concerning the standard requirements.

NO.	29 CFR 1910.95 REQUIREMENT	REF NO.	YES	NO	COMMENTS
	PROGRAM				
1	Employee noise exposures equal or exceed 85 dBA, 8 hour time-weighted average (action level)	(c)(1)			
	MONITORING				
2	Where 85 dBA, 8 hour TWA may be exceeded, monitoring conducted	(d)(1)			
3	Representative personal sampling used for mobile workers	(d)(2)			
4	Continuous, intermittent, and impulsive sound levels from 80-130 dBA measured	(d)(3)			
5	Repeat monitoring when noise exposures increase	(d)(4)			
	EMPLOYEE NOTIFICATION				
6	Each employee at or above 85 dBA, 8 hour TWA notified	(e)			
	OBSERVATION				
7	Affected employees or their reps given opportunity to observe monitoring	(f)			
	AUDIOMETRIC TESTING				
8	Available to all employees exposed at or above action level	(g)(1)			
9	Tests performed by professional or certified technician	(g)(2)			
10	Technician responsible to audiologist, otolaryngologist, or physician	(g)(3)			
11	Audiograms meet standard Appendix C requirements	(g)(4)			
	BASELINE AUDIOGRAM				
12	Establish within 6 months of first exposure at or above action level or within one year if mobile van used. Hearing protection worn if more than 6 months elapsed before baseline	(g)(5)(i)			
13	There is a fourteen-hour period without workplace noise exposure before test (ear protection may be used)	(g)(5)(ii)			
14	Employee informed to avoid high noise levels before baseline	(g)(5)(iii)			
	ANNUAL AUDIOGRAM				
15	Provided for all employees exposed at or above action level	(g)(6)			
	EVALUATION OF AUDIOGRAM				
16	Each annual test compared to baseline for validity and standard threshold shift	(g)(7)(i)			
17	Retest conducted within 30 days (option) if standard threshold shift occurred	(g)(7)(ii)			
18	Audiologist, otolaryngologist, or physician reviews problem audiograms and determines need for further evaluation	(g)(7)(iii)			
	FOLLOW-UP				
19	Employees with standard threshold shift notified in writing within 21 days	(g)(8)(i)			
20	Unless a physician has determined that the shift is not work-related, have • Employees not using protectors provided with them, fitted, trained, and required to use them • Employees using protectors refitted and retrained • Referred the employees for clinical evaluations when necessary • Told employees with non-work related medical problems of the need for otologic examination	(g)(8)(ii)			
21	Employees informed of retest results	(g)(8)(iii)			
	REVISED BASELINE				
22	Are baselines revised per OSHA criteria	(g)(9)			
	STANDARD THRESHOLD SHIFT				
23	Criteria: 10 dB or more average change for 2000, 3000, and 4000 Hz for either ear; aging allowance made (option)	(g)(10)			
	AUDIOMETRIC TEST REQUIREMENTS				
24	Each ear tested, pure tone, air conduction with frequencies of 500, 1000, 2000, 3000, 4000, and 6000 Hz.	(h)(1)			

(continued)

Figure 12-1. A compliance checklist for assessing completeness of a hearing conservation program.

COMPANY:		LOCATION:				
NO.	29 CFR 1910.95 REQUIREMENT		REF NO.	YES	NO	COMMENTS
25	Audiometers meet ANSI S3.6-1969 standard		(h)(2)			
26	Pulse tone and self-recording audiometers meet requirements specified in standard Appendix C		(h)(3)			
27	Test rooms meet requirements in Appendix D		(h)(4)			
	AUDIOMETER CALIBRATION					
	• Functional checks before each day's use					
	• Annual acoustic calibration					
	• Exhaustive calibration every two years					
	HEARING PROTECTORS					
28	Available to all exposed employees and replaced as necessary		(i)(1)			
29	Worn by employees when					
	• 90 dBA, 8 hour TWA exceeded					
	• 85 dBA, 8 hour TWA exceeded when					
	— no baseline after 6 months					
	— standard threshold shift (STS) experienced		(i)(2)			
30	Employees select protectors from variety		(i)(3)			
31	Employees trained in care and use		(i)(4)			
32	Initial fitting and correct use enforced		(i)(5)			
	ATTENUATION					
33	Protector attenuation evaluated		(j)(1)			
34	Attenuation to 90 dBA (85 dBA for standard threshold shift)		(j)(2)			
35	Attenuation adequacy re-evaluated when necessary		(j)(3)			
	TRAINING PROGRAM					
36	Provided all employees at or above action level		(k)(1)			
37	Repeat training annually		(k)(2)			
38	Information provided up-dated when appropriate		(k)(3)			
39	Training includes					
	• Effects of noise on hearing					
	• Purpose, advantages, disadvantages, attenuation, selection, fitting, use and care of protectors covered					
	• Purpose and procedure of audiometric testing		(k)(4)			
	EMPLOYEE AND OSHA ACCESS					
40	Copy of standard available to employees and reps		(l)(1)			
41	Information provided by OSHA available to employees		(l)(2)			
42	OSHA provided training materials upon request		(l)(3)			
	RECORDKEEPING					
43	Exposure records (2 years)		(m)(1)			
44	Audiometric tests (duration of employment)		(m)(2)			
	RECORD ACCESS					
45	Provided employees, former employees, reps, and OSHA		(m)(4)			
	RECORD TRANSFER					
46	Records transferred to successor employer		(m)(5)			

NOTES: (Use separate sheet as necessary)

REVIEW CONDUCTED BY: Date:

FOR ASSISTANCE CONTACT: ELB/Monitor, Inc., Chapel Hill, N.C., (919) 493-4471

Figure 12–1 (*continued*)

ELB/MONITOR INCORPORATED
HEARING CONSERVATION PROGRAM ANALYSIS

Report No.	Prepared By:	Report Date
D. P. No.:	Employer:	Location:

SUMMARY DATA

YES	NO		YES	NO		YES	NO	
___	___	Conducted Survey	___	___	Ear Protection Fitted	___	___	Administrative Controls
___	___	Educational Program	___	___	Audiometric Testing	___	___	Feasible Engineering Controls
___	___	Ear Protection	___	___	Qualified Technician			

SECTION I

Appropriate Documentation?   Appropriate Recommendation Made?

SURVEY DATA

	DATE	SURVEYED BY	METER DESCRIPTION	CALIBRATION BEFORE	CALIBRATION AFTER
(1)					
(2)					
(3)					
(4)					

SECTION II

Audits?   Fitting Technique?   ≥ 85 dBA?
Otoscopic Check?   Records?   Fit ✓ annually?
% Observed Using?   Discipl. Policy?   Distribution Proc.?

HEARING PROTECTION

	TYPE USED	MANUFACTURER	ATTENUATION	FITTED BY
(1)				
(2)				
(3)				

SECTION III

Technique?   Otoscopic Check?   Follow-up?
Pre-empl. Tests?   Bio Calib.?
Annual Tests?   List Check?

AUDIOMETRIC TESTING

TECHNICIAN	CERTIFICATION DATE	CERTIFICATION BY	AUDIOMETER TYPE	CALIBRATION DATE	BOOTH TEST	AUDIOGRAMS REVIEWED BY
(1)						
(2)						
(3)						

SECTION IV

EDUCATIONAL PROGRAM

	TECHNIQUE USED
Initial Date _____	
Last Date _____	___ Meetings ___ Posters
Annual Review _____	___ Film ___ Pamphlets

Figure 12-2. A consultant's form for evaluating completeness of a hearing conservation program.

(continued)

ENGINEERING PROGRAM DATA

(A) Consultation: DATE	EQUIPMENT MANUFACTURER	ASSOCIATION CONTACTS	INSURANCE CARRIER	SUBJECT OR ASST. OBTAINED
(1)				
(2)				
(3)				
(4)				

(B) Engineer Training Program DATE	LOCATION	PROGRAM		HOURS
(1)				
(2)				
(3)				

(C) Feasible Engr. Controls				dB(A)	
DATE	LOCATION	EQUIPMENT	DESCRIPTION	BEFORE	AFTER
(1)					
(2)					
(3)					
(4)					
(5)					
(6)					

(D) Engineering Controls Utilized

_____ Mufflers	_____ Noise Control Curtains	_____ Purchasing Controls
_____ Vibration Damping	_____ Pipe/Dust Wrapping	
_____ Acoustical Foam	_____ Machinery Enclosures	_____ Routine Maintenance
_____ Plastic Gears	_____ Personnel Enclosures	

(F) Additional Engineering Data

Figure 12-2 (*continued*)

(Gasaway, 1984a). Therefore, the results of noise monitoring, especially personal worker sampling, must be interpreted with great care. If a worker is exposed to noise levels exceeding 90 dBA for occasional periods during the workday, but his or her TWA exposure level is below 90 dBA, he or she should not be considered to be automatically protected from sustaining NIPTS. Individuals have often demonstrated the audiometric indications of NIPTS after a working lifetime of occasional high noise exposures, due to their personal susceptibility to the effects of such exposure (Gasaway, 1985a). In the comprehensive, effective HCP, workers are required to protect themselves from high noise exposures regardless of their duration of exposure or their official TWA exposure level.

In addition to possible costly legal outcomes, eliminating certain employees from the HCP can have other effects. It becomes very difficult administratively to monitor the use of hearing protectors in an area where some workers are required to use them and others are not. Likewise, worker resentment and skepticism about the value of the HCP may develop. For these reasons, many employers choose to require the universal use of hearing protectors by employees in a given work area, rather than selectively requiring their use by certain workers. If this decision is made, more costly personal dosimeter sampling of employees can usually be eliminated in favor of less expensive and less time-consuming area measurements with a sound level meter. If noise levels in excess of the selected permissible exposure limit (whether 85 or 90 dBA) are measured, all workers in the area are required to use hearing protectors at all times during the workshift.

Engineering and/or Administrative Controls

Although not included as a part of HCPs in the OSHA Hearing Conservation Amendment, engineering and administrative controls are still required by the original provisions of the 1970 Noise Exposure Regulation (Williams-Steiger Occupational Safety and Health Act, 1970). They must be included in a discussion of the comprehensive hearing conservation program.

Controlling workplace noise at the source by quieting machinery (engineering controls) or limiting employee exposure time to noise (administrative controls) should be considered the primary means of protecting workers from auditory damage because of the unquestionable effectiveness of these practices (Erdreich, 1985; Suter, 1984). However, these controls are often technologically or economically infeasible. Recent court decisions and OSHA interpretation of these decisions (U.S. Department of Labor, 1983) have relegated this type of noise control "officially" to a secondary position. Engineering controls are now required by OSHA only if worker TWA noise exposures are "so high that hearing protectors alone may not reliably reduce noise levels. . .to the levels specified in Table G-16 or G-16a of the standard. . .(that is) when employee exposure levels border on 100 dBA. . .(and) the costs of engineering and/or administrative controls are less than the cost of an effective hearing conservation program" (U.S. Department of Labor, 1983, p. 2). This implies, of course, that these types of controls are no longer to be enforced for the vast majority of U.S. workplaces.

Despite this situation, companies that are committed to an ongoing program of engineering noise controls have managed to achieve significant reductions of employee noise exposure levels at reasonable cost (Bailey, Hart, & Stewart, 1980). This has reduced the need in many cases for dependence on personal hearing protection, which has consistently been demonstrated to be a less-effective means of protecting workers from NIPTS because of problems with the "real-world" use of hearing protectors (Berger, 1980a,b).

Administrative noise controls involve adjusting work schedules to rotate noisy operations among different workers for shorter exposure periods, or limiting the amount of time during which a noisy machine may operate. They also involve decisions and policies made by employers to prohibit the purchase of machinery that cannot be demonstrated by the manufacturer to conform to maximum noise output levels set by trade and manufacturing associations. All companies should have such purchasing controls in effect and should make every possible effort to administratively control worker exposure.

Audiometric Testing Program

The goal of an occupational HCP is to prevent significant occupationally related NIPTS in all employees working in otohazardous noise. An audiometric testing program, supported by a well-structured, goal-directed program of follow-up activities, is the only way to monitor whether or not this goal is being met. In addition, the testing program will benefit employee health by identifying other types and sources of hearing loss, such as otologic problems and hearing loss due to off-the-job noise exposure.

The OSHA regulation requires that management provide a baseline audiogram for all employees whose 8 hour TWA noise exposure level is 85 dBA or higher, at no cost to the employees. This test, which must be preceded by a period of at least 14 hours without exposure to workplace noise, must be performed within 6 months of an employee's first exposure to noise of 85 dBA TWA or above (within 1 year if mobile audiometry is employed). Management should be advised that, whereas adhering to the 6/12 month policy is acceptable under OSHA criteria, it may not be protective of the employee's hearing (not a true indication of his or her actual pre-exposure thresholds) or of the company's liability under state or federal worker compensation statutes (which may designate that baseline or pre-employment audiograms be performed within a more restricted period after hire date). The regulation permits the 14

hour quiet period to be achieved by wearing hearing protectors, which is certainly inadvisable in view of the well-documented inadequacy of this method in assuring that the wearer is protected from temporary threshold shift (TTS) from noise exposure (Berger, 1980a,b). No mention is made in the regulation of the need for review of baseline tests by a professional, although a supervising professional should review these tests as well as annual monitoring tests, so that proper recommendations can be made for follow-up referral in cases of probable auditory pathology or pre-existing compensable hearing loss.

An annual monitoring audiometric evaluation must be made available to all employees whose 8 hour TWA exposure is 85 dBA or higher, at no cost to the employees. The regulation contains no specifications for hearing protection from workplace noise prior to monitoring tests, but the program supervisor may wish to recommend this to avoid the effects of potential TTS contamination. Some have suggested that it is inappropriate to take any extra steps to protect workers before monitoring tests, as this may disguise hearing protector use problems and give a false picture of hearing protector program adequacy. Testing employees during the workshift and when they have been wearing their usual hearing protectors helps to evaluate the degree of protection being provided by the protectors (Royster & Royster, 1984).

The regulation permits the supervising professional to adjust the baseline test if subsequent tests indicate significantly better hearing sensitivity or if a persistent threshold shift has occurred. This practice is advisable as it has the beneficial effect of ensuring that later audiograms are compared to an accurate reference audiogram. Of course, if the baseline test is adjusted because of a persistent threshold shift, it is important to keep track of the initial threshold results and the number of times the thresholds have shifted and to take appropriate action to halt the degradation in hearing.

Qualified Audiometric Technician

This is clearly the key person involved in any audiometric testing program, since he or she interacts directly with employees at the critical times of the initial hearing test/hearing protector fitting session and the annual monitoring testing and refitting sessions (Royster, 1985). This is often the only time during the year when employees interact with a trained person who can properly advise and encourage them about hearing loss and hearing protector use, and it is essential that these be productive encounters. Technicians are required by the OSHA regulation to be certified through a course approved by the Council for Accreditation in Occupational Hearing Conservation (CAOHC) or at least to have demonstrated competence to the program supervising professional in their ability to test hearing accurately and care for and check the calibration of their audiometer. It should be noted that, in a spasm of illogicality, OSHA personnel responsible for the final wording of the Hearing Conservation Amendment included a clause that allows a technician testing with a microprocessor audiometer not to become "certified." It is important that the supervising professional set higher standards for audiometric technicians than those set by OSHA. The training, commitment, and motivation of these persons to ensure the prevention of NIPTS in every one of their employees is, without question, one of the keys to the success of an HCP. Technicians should also be urged to take a refresher course in occupational hearing conservation at least every 5 years, although this is not required by the regulation.

Professional Supervision

The regulation states in very explicit terms that audiometric technicians are to be responsible to a supervising professional. This person is defined as an audiologist, an otolaryngologist, or a physician ("qualified" physician in the original draft of the regulation). The supervisor, although this is not stated in the regulation, should be a professional well versed in the theoretical and practical aspects of noise-induced hearing loss and HCPs. The supervisory responsibility of this person extends beyond the mere supervision of audiometric techniques (as is clear from the wording of the preamble to the Hearing Conservation Amendment) to include judgments concerning audiogram interpretation, significance of audiometric results and case history findings, significance of threshold shifts, work-relatedness of threshold shifts, stability of threshold shifts, and referral criteria. Logically, this person's responsibility should extend also into other phases of the HCP, including employee training, hearing protector use, and noise monitoring (Feldman, 1985). No requirement exists in the regulation for on-site supervision, but it is difficult to imagine how any useful or meaningful supervision could take place without, as a minimum, an annual visit by the supervisor to the plant.

Calibrated Test Equipment

The regulation requires that audiometer calibration be assessed at regular intervals to ensure its conformity to the ANSI audiometer specifications (ANSI S3.6–1969, R.1973) for limited range audiometers. Biological calibration is required on every day of testing (including a threshold test on a person with known, stable hearing thresholds and a listening check), with an acoustic calibration check required for deviations of 10 dB or more. Acoustic calibration (a limited check of sound pressure level [SPL] output and attenuator linearity) is required, in any event, once a year, with an "exhaustive" (more extensive than acoustic) calibration check required for deviations of 15 dB or more. Exhaustive calibration is required for all industrial audiometers every 2 years. It is recommended that, in view of the potential legal use of audiometric data gathered in industrial settings, an exhaustive calibration check should be made annually as a minimum.

The importance of accuracy in equipment calibration and operator technique cannot be over-emphasized in industrial hearing testing. Important, potentially costly decisions are made on the basis of these tests. For example, accurate pre-employment audiograms are absolutely essential because the company may be called upon at a later date to prove that a worker's hearing loss pre-dated his or her employment with the company. This would be necessary if the employee filed a compensation claim alleging noise-induced hearing loss stemming from workplace exposure. Likewise, accurate monitoring tests are essential because they are measured against the baseline test to discover if any changes have taken place which may be associated with exposure to workplace noise. These tests will provide the only obtainable information to determine if a worker may be abnormally sensitive to noise exposure or may not be adequately protecting himself or herself from workplace noise. Accurate audiometer calibration is the minimum essential condition for obtaining this type of information.

Environment for Testing

The regulation requires that audiometric testing be performed in an area that meets the background noise criteria specified in Table D-1 of the Hearing Conservation Amendment. There is general agreement among professionals that these criteria (which evolved from the 1960 version of ANSI criteria document S3.1–1960) are not stringent enough for accurate testing down to 0 dB HTL at all the specified test frequencies ranging from 500 to 6000 Hz. In a 1983 addendum to their "Guide for Conservation of Hearing in Noise," the American Academy of Otolaryngology–Head and Neck Surgery ([AAO-HNS], 1982) recommended levels that were 10 dB more stringent than the OSHA levels at all frequencies. It is recommended that HCPs attempt to comply with these stricter levels. Another proposed background noise criterion is Hirschorn and Singer's 1973 recommendation of an A-weighted level of 43 dBA, which their study found allowed accurate threshold testing to 0 dB HTL at all frequencies from 500 to 6000 Hz, when MX-41/AR cushions were used.

This criterion could serve as a quick measure for a technician to make with a simple sound level meter whenever it appears that background levels have increased such that employees' thresholds have uniformly worsened by 10 or 15 dB for 500 Hz. It is not recommended that circumaural earphones be used at any time in industrial testing because of their relatively poor attenuation for lower frequencies, which prevents them from being an acceptable substitute for a true sound-treated enclosure where low-frequency noise is present.

Audiogram Review and Follow-up Procedures

The key to obtaining the maximum benefit from an industrial audiometric program is to maximize the amount of personal follow-up activity that takes place after the annual test has been given. Ideally, the employee should be counseled immediately after the test as to its results and whether or not it indicates that his or her hearing has shifted since the baseline test. Further counseling may need to be delayed until instructions are received from the program supervisor after analysis of the test, if the tester is not an audiologist or qualified physician. However, the importance of some type of immediate feedback to the employee cannot be overemphasized. As mentioned earlier, this may be the only time during the year when the employee interacts with the best-trained member of the plant hearing conservation team, and this makes it an ideal time for reinforcement of correct hearing protector use, counseling about the effects of noise exposure, and discussion of other topics concerning hearing and ears. Information about hearing status should be inherently interesting to an employee and should provide motivation to take better care of hearing if it has been allowed to deteriorate through improper hearing protector use. It has been suggested that the degree of personal attention shown to the employee at the time of the annual hearing test represents the company's sincerity regarding the overall HCP (Royster, 1985).

The OSHA regulation requires that "problem" audiograms be evaluated by a professional, although the initial determination of whether or not the employee has experienced a significant change in hearing (STS) may be made by the technician. Although acknowledging the vital role of the technician in performing and communicating the results of the annual hearing test to the employee, the careful professional program supervisor will probably wish to evaluate all audiograms performed by the technician, as well as auditory and otologic histories obtained from employees. The legal hazards of failing to identify a potentially pathologic condition and so notifying the employee make such precise review and program supervision advisable. Likewise, careful review of data and judicious referral for follow-up examination by a physician or audiologist is essential, both for the employee's benefit when referral is clearly indicated, and to prevent the recommendation of unnecessary referrals in cases where they would be superfluous.

The following are some considerations for audiologists or physicians who intend to review industrial audiometric data.

Experience

The best interpreters of industrial test data and the best advisors of appropriate follow-up actions are professionals who understand the industrial situation from first-hand experience, having visited many different types of workplaces before undertaking industrial audiologic or otologic practice. Audiologists and physicians who contract with a company to review its audiometric data should, if possible, visit the plant first to become acquainted with its personnel and its problems.

Promptness

Nothing more effectively kills an employee's motivation to take steps to protect his or her hearing than not to be notified of hearing test results until a month or two after the test occurred. Of course, one hopes that the in-plant technician explained the test results to the employee immediately after the test, but one cannot always be certain of this. Such explanations are not always full and correct and may not provide the proper recommendations for follow-up action.

Information

Information provided to the employee must be as complete and factual as circumstances permit. Granted, this is not a review of a personally administered test, and the professional may or may not have access to detailed and accurate otologic history information. But it is vital that what the employee is told about his or her hearing be factual, as complete as possible, and motivational. The report must contain enough information so that the employee understands his or her hearing status and what should be done about it, without using professional jargon or being excessively wordy. Remember that many employees in noisy industries don't read very well—in fact, a surprising number are functionally illiterate.

Whether or not a computer-generated report is used, what is given to an employee after a professional data review must include as a minimum: (1) a description of the employee's hearing status that accurately describes the audiogram and what it means, and (2) a statement about changes in thresholds since the baseline test. Coding the audiogram "A," "B," or "C" won't help the employee to understand the results, and informing the employee only when there has been a significant change in his or her hearing according to OSHA's criterion is not as useful as noting gradual, smaller changes that may precede a legally significant change. Remember that the fundamental role of monitoring audiometry is to identify NIPTS long before the individual would notice a hearing impairment.

Recommendations

Recommendations, to be useful and effective in motivating the employee to take follow-up action, must be specific and practical. Over-referral is just as bad as under-referral, as it will undermine employees' confidence in the factualness and usefulness of future recommendations. Recommendations for referral should be given in all cases of potential medical problems (e.g., asymmetrical loss and low-frequency loss) as well as in cases of unusually large threshold shifts. Recommendations should also be given for the refitting and more careful use of hearing protectors. Of course, the reviewer must consult beforehand with company management so that management will know what may be recommended to their employees and why, and so that they will co-operate fully with recommended follow-up activities.

Regarding the audiometric frequencies that should be considered in determining whether the employee has exhibited a significant change in hearing, the OSHA regulation is specific in its direction that only changes at 2000, 3000, and 4000 Hz be considered. An average change for the worse of 10 dB or more for these frequencies is termed a Standard Threshold Shift (STS), and requires that certain follow-up activities take place. The question of which are the most appropriate frequencies to use in deciding when an employee has had a notable change in his or her hearing is currently under study and is a matter of some controversy (Royster & Royster, 1982b). In 1983, the AAO-HNS issued a set of referral criteria for HCPs which incorporated both audiometric and medical criteria. They recommended otologic referral for employees who exhibit shifts greater than 15 dB for the average of 500, 1000, and 2000 Hz, or greater than 20 dB for the average of 3000, 4000, and 6000 Hz. Some companies use these criteria to determine what they term "medical" threshold shifts as opposed to "OSHA" shifts, assuming that they represent a greater degree of auditory change than would normally be expected from early exposure to workplace noise. The AAO-HNS criteria are displayed in Table 12–1.

In summary, it is important that an industrial audiometric program be structured

TABLE 12–1.

AA0–HNS Otologic Referral Criteria for OHC Programs.

Audiological Criteria
Baseline audiogram
  Avg. loss >25 dB for .5, 1, 2, 3 kHz, either ear
  Avg. difference between ears
    >15 dB for .5, 1, and 2 kHz
    >30 dB for 3, 4, and 6 kHz

Annual audiogram
  Change for the worse from baseline, either ear,
    >15 dB for .5, 1, and 2 kHz
    >20 dB for 3, 4, and 6 kHz

Medical Criteria
History of any of the following: Ear pain; drainage; dizziness; severe persistent tinnitus; sudden, fluctuating, or rapidly progressive hearing loss; fullness or discomfort in ears.

Cerumen accumulation or foreign body visible in ear canal

according to a plan based upon clearly defined goals, reasons for tests of various types, and criteria and procedures for follow-up action (Gasaway, 1985b). The plan must include well-trained and highly motivated technicians and be closely supervised by competent professionals. Findings must be communicated clearly and promptly to employees and they must be motivated to carry out recommended follow-up actions only in this way will the testing program provide a reliable, effective assessment of the overall company hearing conservation effort.

Personal Hearing Protection

Perhaps the most difficult area of the HCP in which to achieve effective employee participation and co-operation is the hearing protective device (HPD) program. In no other area does the success of the activities undertaken to protect workers' hearing depend as greatly on the understanding, motivation, goodwill, and co-operation of employees as in the HPD program. For this reason, many companies experience an enormous degree of difficulty in making this part of their HCP work and frequently report frustration and failure, despite their best efforts. Well aware of management frustration in this area, one experienced hearing conservation practitioner listed four goals for HPD programs: (1) convince workers and management that noise is a real threat to hearing, (2) teach each employee the skills for proper use and care of HPDs, (3) establish behaviors and attitudes to ensure that HPDs will always be properly worn, and (4) raise workers' consciousness so that they will avoid excessive noise, both on and off the job (Gasaway, 1984b).

According to the OSHA regulation, HPDs must be made available to employees with an 8 hour TWA noise exposure of 85 dBA or higher, and their use must be made mandatory for employees whose TWA exposure exceeds 90 dBA. Although use of HPDs is ordinarily optional for TWAs between 85 and 90 dBA, it is required for workers in that range who have experienced an STS. HPD attenuation must be sufficient to lower worker exposure to at least 90 dBA, or 85 dBA if they have experienced an STS. Workers must be able to select their HPDs from among a variety of appropriate types, must be trained in their proper use and care, and must be required to use them in areas of overexposure. The regulation is very clear in its intent that management supervise employees' use of HPDs and succeed in having them used at all times by overexposed persons. Workers who experience an STS must have their HPDs refitted and be retrained in their proper use and care.

Because of the problems involved with selective use of HPDs by employees in different plant areas, many employers choose to make the use of HPDs mandatory above 85 dBA rather than above 90 dBA. Despite research evidence suggesting that these employees are at somewhat less risk of material hearing impairment, this practice can be supported from the standpoint of more effective implementation of the HCP.

Royster and colleagues (1982) have suggested some characteristics of successful HCPs. All, in one way or another, pertain directly to the HPD program. They include:

(1) the sincere and active support of top management in a company; (2) strict enforcement of the requirement that all employees use HPDs in designated areas; (3) the existence of a key individual who is personally responsible for the success of the HCP; (4) active lines of communication among all management levels of a company; and (5) the available HPDs are all potentially effective in preventing occupational hearing loss. What they are implying in all of these characteristics is that plant management must be sincerely interested in the HCP and expend a significant effort, as opposed to token compliance, to make the HCP work. The program must be adequately funded, staffed by competent technicians and professionals, well advertised within the company through frequent educational programs, and adequately enforced. Without such sincere efforts, the entire HCP, but most especially the HPD program, will fail to realize its potential.

The comprehensive HCP will be structured so that a strong interrelationship exists between the HPD program and the other parts of the HCP, particularly the audiometric testing program. Results of annual hearing tests should be used to assess an individual's success in the proper use of HPDs, and extensive employee counseling should occur on an individual basis when an employee exhibits possible occupationally related hearing loss (Royster, 1985). These counseling sessions are an ideal time for intensive personal re-education, further indoctrination about the company's HCP policies, and remotivation where needed (Gasaway, 1984b; Harris, 1980). In addition, the annual hearing test session is a useful occasion to perform an inspection of the fit and condition of the employee's HPDs, to replace them if necessary, and to provide re-education concerning the need for personal vigilance in this area. Often, employees will have substituted another type or size of HPD for one originally issued to them, especially if distribution of replacement HPDs is not under the explicit control of the trained person who initially fits and issues the HPDs to employees (Royster & Royster,

1985). Results of annual hearing tests are also a means of providing feedback (positive as well as negative) to supervisors and managers about the overall success of the company's hearing conservation efforts, especially the HPD program, and will indicate particular plant areas where greater efforts in requiring the proper use of HPDs by employees are needed.

A successful HPD program also requires that periodic audits of employee compliance be made throughout the plant. Each supervisor should be required to personally audit his or her workers' use of HPDs on a daily basis and report violations or problems to management and the medical department so that corrective action can be taken. A clear, firmly enforced disciplinary policy for failure to comply with the HPD use policy must be in place and used when necessary. Of course, positive reinforcement and persuasion are preferable means of obtaining HPD acceptance by workers, but the ultimate tool of punishment must also be available and used if the former techniques are unsuccessful.

Another major problem with HPD programs in industry is relying on the Noise Reduction Rating (NRR) as a means of selecting the appropriate HPDs for a group of workers. It is well known (Berger, 1980a,b) that laboratory attenuation measurements significantly overestimate the actual industrial performance of HPDs, due to the fact that laboratory subjects insert HPDs much more carefully than workers do. Therefore, the resultant NRR seriously overestimates the "real-world" attenuation that employees obtain from HPDs (Berger, 1983). Likewise, comfort and employee acceptance must be considered, in addition to the amount of attenuation needed, when arriving at a decision as to which HPDs are going to be offered to employees (Royster & Holder, 1982). Managers, professionals, and technicians should "use-test" HPDs for at least several days in the work environment before deciding to make them available for employee use, in order to assure their comfort and ease of proper use.

Employee Training

The care devoted by a company to the employee training phase of its HCP will determine, in large part, the success of the HCP. Because occupational hearing loss is almost totally asymptomatic during its earlier stages and occurs so gradually even in its later stages, most individuals are unaware of its progression. For this reason, many industrial employees are openly skeptical and tend not to co-operate with the policies and practices of the HCP. Strong, ongoing employee training programs, geared to workers' educational levels and learning needs (Stapleton & Royster, 1985), as well as to management's needs and concerns, are therefore essential to the success of the HCP.

The OSHA regulation calls for annual training of all workers exposed at or above the action level (85 dBA TWA), although it makes no specifications about how or when the program is to be carried out. It does, however, specify the minimum contents of the program:

1. The effects of noise on hearing
2. The purpose of hearing protectors; the advantages, disadvantages, and attenuation of various types, and instructions on their selection, fitting, use and care
3. The purpose and procedures of audiometric testing

The regulation does not specify a need for documentation of employee participation in annual training, but this would seem to be essential for record keeping purposes.

The effectiveness of the HCP is measured by its ability to minimize the amount of NIPTS experienced by workers. In order to achieve this goal, a company must make both managers and workers aware of the dangers associated with exposure to excessive noise and teach them to understand and accept the steps being taken by the company to protect employees from NIPTS. This is particularly true with regard to managers and supervisors, since their commitment to the goals of the HCP will determine their workers' potential commitment.

Although the regulation contains no directives regarding the training of managers and supervisors, their education is extremely important, for they are the ones who must implement the program. How well it works will depend on their knowledge, understanding, commitment, and willingness to actively support the HCP and enforce its requirements. First-line supervisors are especially important, for they must answer employees' questions, enforce HPD use on a day-to-day basis, ensure that machine modifications made to reduce sound levels are not removed or modified, and see to it that hearing test policies and schedules are followed. Supervisors' co-operation is the key to the ongoing effectiveness of the HCP; therefore, special efforts must be made to elicit this co-operation through advanced training programs that provide more information than may be given to workers.

In addition to the minimum program contents listed, annual training programs for noise-exposed workers should also include the following: the goals of the HCP; the reasons and need for the program, based on the results of noise surveys; the function of the ear, and how noise exposure may damage this function; and the responsibilities of workers and management in the program. The inclusion of these additional topics will make the program more comprehensive and therefore more valuable, to employees; with more understanding they will be more motivated to protect their own hearing.

Education should not stop after a group meeting, however. It must continue throughout the year, particularly at the time of each employee's annual hearing test and whenever new or replacement HPDs are issued. These are times to remind workers of the hows and whys of HPD use and to motivate them to continued or improved co-operation with the HCP, based on the results of their audiometric test (Gasaway, 1984c). Any apparent threshold shifts should be discussed in order to determine if inconsistent use of HPDs may be the cause, and the worker should be motivated to better protect his or her hearing. Re-instruction in proper HPD insertion and use techniques should accompany these interaction sessions with workers.

Record Keeping

Record keeping is often a weak link in the chain of activities that constitutes an effective occupational HCP. Management and technicians usually give more attention to setting up the program, staffing, arranging for the mechanics of it, buying the necessary equipment and supplies, etc. than they do to documenting the activities and results of the program. The OSHA regulation is not much help in deciding what records to keep. It specifies only four: noise exposure measurements (to be retained for 2 years), audiograms (to be retained only for the duration of employment), measurements of background sound pressure levels in the audiometric test room, and audiometer calibration data (no specified retention period for the latter two records). Maintenance of only these records will not provide sufficient documentation of HCP activities to adequately protect an employer's liability to unjustified worker compensation claims, nor will it provide enough information about HCP activities to adequately protect employees' hearing.

Table 12–2 contains a list of records that should be kept by plant personnel and by professional supervisors in occupational HCPs. The list could easily be increased, but these are considered to be the essential records for adequately documenting the program and and providing information to protect employees' hearing.

Much of what is done in the name of record keeping in HCPs is neither complete nor carefully done. In many cases, this occurs through a lack of understanding of the dual purposes of occupational health record keeping: functional and historical. Records serve a functional purpose in that they provide a "road map" for activities to be performed in each segment of the HCP

TABLE 12-2.
Record Keeping in OHC Programs

Records Required by OSHA:
  Employee noise exposure measurements (Retain for 2 years)
  Employee audiometric tests, to include
    Name and job classification
    Date of test
    Examiner's name
    Date of last acoustic or exhaustive calibration of audiometer
    Employee's most recent noise exposure assessment (TWA)
    Background sound pressure levels in test room (Retain for duration of employment}
  Contents of employee training program

Other Necessary Records (In-Plant)
  Medical history (otologic), including employee's signature and date
  Previous noise exposure history (military and employment)
  Recreational noise exposure history—type, duration, frequency
  Audiometer: make, model, SN of audiometer; calibration standard
  Audiometer calibration—biological, acoustic, exhaustive
    Date
    Values
    Instrument ID data
    Technician's name
  Additional Employee Audiogram Data: age, sex, SSN, time of test, recent noise exposure, use of hearing protectors, employee's signature
  Otoscopic inspection of employee's ear canals
  Medical/audiologic referral history
  Hearing protector data
    Initial and subsquent fitting dates
    Type
    Size for each ear
    Dates of training in use and care
    Record of occasions when employee was found not to be satisfactorily using protectors
  Technician certification evidence
  Documentation of employee training—dates, signatures

Records to Be Maintained by Audiologist
  OHC technicians trained/retrained: dates, locations, test scores
  Audiometer calibration data: plant, clinical and/or mobile van equipment
  Test room background sound levels: plant, clinical and/or mobile van rooms

Noise exposure measurements for individual plants, to include
    Date and time
    Equipment identification data
    Evidence of calibration
    Measures obtained
    Examiner's name
    Recommendations to company
  Reports of findings and recommendations during OHCP evaluation visits
  Audiogram review/interpretation/recommendation data for individual plants
  Copies of all correspondance with individual plant representativss
  Records of telephone or face-to-face conversations with individual plant repressntatives, other OHCP consultants (physicians, IHs, etc.), and individual employees
  Diagnostic audiological evaluation data and recommendations for individual employees
  Evaluation of individual plant and corporate audiometric data bases, interpretations, and recommendations
  Legal correspondance and transcripts
  System for determining when individual employees are scheduled for annual, or more frequent, hearing tests
  Industrial audiograms (to include 3 and 6 kHz) performed in clinic or mobile van for individual plants
  Otologic histories, previous noise exposure histories, and recreational noise exposure histories of employees tested in clinic or mobile van
  Otoscopic inspection findings for above employees
  Hearing protector data (as in "In-Plant" list) for above employees
  Microfilming of audiograms

Information Needed by Employers After Audiogram Review
  Audiometric data and ID data for individual employees (digital or graphic)
  Individual employee notification forms, listing
    Results of recent hearing test
    Interpretation
    Recommendations
  Otologic history data, hearing protector use status, noise exposure status for individual employees
  Display of previous audiograms (at least baseline) for each employee

(continued)

TABLE 12-2 (continued)

Information Needed by Employers After Audiogram Review (continued)

Individual calculations, including

Changes from baseline audiogram

Code designation of type and degree of indicated loss to facilitate follow-up activity

Indication whether a significant threshold shift, according to OSHA or other criteria, has occurred after appropriate age corrections have been applied

Indication of degree of average loss for speech and high frequencies, together with changes of these degrees from baseline audiogram

Other code designations (e.g., Early Loss Index of NSC)

Percent of potentially compensable hearing loss, for current and baseline tests

Specific recommendations for follow-up activity for individual employees, including retests, diagnostic evaluations, hearing aid evaluations, medical referral

Technical data, including audiometer ID, examiner ID, test environment sound levels, etc.

Scheduled dates for subsequent tests for each employee

Lists of employees with notable findings

and the forms to record results of those activities. Therefore, if a form is omitted or inaccurately completed, some activity that may be critical to the prevention of NIPTS in a worker may go undone. Second, records serve a historical purpose in that they provide documentation of what was actually done. They create a ''paper trail'' that may assist HCP personnel in years to come to recreate a history of employee performance for legal or health maintenance purposes. Again, completeness and accuracy are vital to the protection of the employee and the company (Gasaway, 1985c,d).

However, in record keeping, as in all phases of the HCP, it is important to remember that the interactions to be recorded are between people, and not forms (Gasaway, 1985d). The prevention of NIPTS is a dynamic, interactive process in which people communicate with people; it is the effectiveness of that communication that determines how successful the HCP is in preventing hearing loss. No degree of record keeping has ever prevented NIPTS from occurring. It is the use of the records that accomplishes this task, as well as the commitment and the coordinated efforts of the personnel involved in the HCP.

APPROACHES TO ACHIEVING A COMPREHENSIVE HEARING CONSERVATION PROGRAM

There exist a number of different approaches to achieving a comprehensive HCP. The choice of a particular approach will be based on a company's individual needs and will be successful to the extent that the methods chosen fit the company's management philosophy and all personnel within the company are committed to the goals of the HCP.

HCP Policy

The first step in structuring a comprehensive HCP is to develop a detailed HCP policy statement. This statement should formally express the company's goals for the HCP and its commitment to achieve those goals. It should delineate each department's role in the HCP and the particular responsibilities of certain important individuals in each department, as well as the overall responsibilities of all company employees. The HCP policy is useful for formalizing the company's commitment to the program at the very outset and for informing all affected personnel of their part in the HCP. It is also valuable as documentation for OSHA of the company's compliance with the noise regulation. Likewise, it provides new employees with an overview of the entire program and the part that their department plays in it. Figure 12–3 displays a sample HCP policy.

Approaches to Structuring a Program

There are three basic approaches to the provision of HCP activities within a company: (1) a totally in-house program, (2) a partly in-house/partly consultant service program, and (3) a totally consultant program. The choice of any of these will be based on a number of different factors, such as available in-house personnel, level of training and experience of personnel, costs involved, management support, size of the company, number of plant locations, company operating procedures, and available outside consultants and resources. No one approach is necessarily better than another; the company's particular needs and preferences will be the final determinant after consultation with appropriate professionals.

Totally In-House Program

This method is often selected by larger corporations with extensive professional support within the company. It involves delegating responsibility for each HCP task to the appropriate department, such as medical, safety, industrial hygiene, and engineering. This method works well if suitable professionals are available and if they are all well trained in hearing conservation. For example, a number of companies throughout the U.S. have hired audiologists and otologists for the purpose of providing the audiogram review and referral portion of the program within the company. Of course, if this approach is taken, extensive computer capability must also exist within the company to facilitate in-house review of large numbers of audiograms. This approach also requires that the data processing staff be authorized and willing to devote the time, energy, and computer storage space necessary to set up an effective audiogram review and reporting system. If this capability is available, on-line transmission and storage of audiometric data from remote plant sites is often a possibility, resulting in large savings in record keeping time and costs.

The totally in-house approach, although possible for larger corporations, is quite expensive in terms of overhead costs, and for that reason has not been widely used in the U.S.

Combination of In-House and Consultant Service Program

This appears to be the most commonly selected approach to structuring HCPs. Companies using this approach have in-house personnel involved in the program (often the nurse, personnel manager, or an occupational hearing conservationist [OHC]), but some of the services are provided by outside consultants. Commonly provided by the latter are noise surveys, mobile audiometric testing, audiogram review, and referral of employees. More commonly provided by in-house personnel are employee education, HPD fitting and training, and (often) in-plant audiometric testing. Record keeping is usually a jointly shared responsibility. Often employee education and hearing protector fitting and training sessions are provided by the firm supplying mobile audiometric testing services. HCP supervision and consulting services will normally be performed by an outside consultant, usually an audiologist or physician who also provides mobile audiometric testing or data review and referral services to the company. Noise control engineering is often provided by outside consultants when this combined approach is used.

Numerous combinations of responsibility for HCP activities are found in various companies and industries. For example, sometimes the plant occupational hearing conservationist performs pre-employment and retest audiograms and an outside consultant provides mobile testing for annual audiograms. This approach works well in very large plants, where it would be extremely time-consuming for the OHC to perform hundreds or thousands of annual audiograms, but the testing service could complete them easily in several days

HEARING CONSERVATION PROGRAM POLICY

1. *General Policy*

    It is the policy of the company to maintain an effective hearing conservation program to protect the hearing of employees and visitors exposed to high noise levels.

2. *Provisions of the Policy*

    a. *Applicable Law*

    This policy, with the resulting programs, is to be in compliance with the Occupational Safety and Health Act of 1970, Paragraph 1910.95 (Occupational Noise Exposure), as amended March 8, 1983 (48 FR 9738).

    b. *Employees and Areas Involved*

    (1) A continuing, effective hearing conservation program will be instituted and maintained in all facility areas in which employee noise exposures equal or exceed an 8-hour time-weighted average sound level of 85 dB measured on the A scale. For the purposes of this policy, the above areas are considered high noise level areas.

    (2) High noise level areas as defined in 2b(1) above will be clearly marked in each company facility.

    c. *Hearing Protective Devices*

    (1) The occupational noise standard states that hearing protective devices must be used by employees exposed to a time-weighted average sound level of 90 dB or greater. It further states that they must be used by employees exposed to a time-weighted average of 85 dB or greater who have experienced a permanent significant threshold shift. However, it is the policy of this company that hearing protective devices must be worn at all times by anyone in facility areas with a sound level which equals or exceeds 85 dB (high noise level area). This applies to regular employees, visitors, and walk-through traffic. The use of hearing protective devices in high noise level areas is mandatory and is a condition of employment.

    (2) *Original Issue and Replacement*

    The company will furnish the original approved hearing protective devices to each employee working in a high noise level area. Thereafter, the employee will be responsible for replacement due to loss or destruction beyond normal aging and deterioration of the devices. Employees who own or acquire personal hearing protective devices may use these devices provided they are approved in each individual case by the medical department.

    (3) *Maintenance of Supply*

    The replacement supply of hearing protective devices will be maintained in the plant medical department for issue or sale to employees. In case medical facilities are not available or open, each plant facility will provide a supply source of hearing protective devices for issue or sale to employees. The distribution of hearing protective devices from these supply sources will be under the direct supervision of the local medical department facility.

    d. *Safety and Health Department Participation*

    (1) *Noise Exposure Monitoring*

    The Safety and Health Department will make the initial determination of individual employee noise exposures and area noise levels. These measurements will be repeated every two years and within sixty days of a change in production processes, equipment, or personnel which may result in new noise exposures equal to or exceeding a time-weighted average of 85 dB.

(continued)

Figure 12–3. Ajax Textile Corporation Hearing Conservation Program Policy.

(2) *Protective Device Requirement*

The Safety and Health Department will notify each company facility of its noise level measurements and employee noise exposure measurements, identifying areas in which the use of hearing protective devices is required.

(3) *Record Keeping*

The Safety and Health Department will be responsible for maintaining records of employee noise exposures and area noise levels.

e. *Medical Department Participation*

   (1) *Audiometric Testing*

   (a) Baseline audiometric testing will be conducted on all new employees, transferring employees, and employees returning to work after sick leave or military leave.

   (b) Monitoring audiometric testing will be conducted annually on all employees with a noise exposure which equals or exceeds a time-weighted average of 85 dB. Retesting will be performed when deemed necessary by the company audiologist or medical director.

   (2) *Hearing Protective Devices*

   The Medical Department will fit and issue proper protective devices to all employees needing them. Except as noted in 2c(3) above, replacement hearing protective devices will be issued only by the Medical Department.

   (3) *Employee Training*

   The Training Department, in consultation with the Medical Department, will institute a training program in hearing conservation for all employees who are exposed to noise at or above a time-weighted average level of 85 dB. This program will be repeated annually.

   (4) *Record Keeping*

   The Medical Department will be responsible for maintaining audiograms, records of issue of hearing protective devices, and medical records on audiologic and otologic referrals.

   (5) *Medical Surveillance*

   The Medical Department will monitor individual facility hearing conservation programs and act as consultants to plant management and counselors to employees on matters of hearing conservation and hearing problems.

f. *Management Participation*

   (1) *Responsibility for Enforcement*

   (a) Plant management will enforce the mandatory use of hearing protective devices and take disciplinary action as outlined in Policy No. 116 as needed.

   (b) Plant management will ensure that appropriate employees participate in the audiometric testing program and the hearing conservation training program.

   (2) *Noise Reduction*

   Plant management will institute a noise awareness program in each facility with high noise level areas and will take steps to reduce noise levels to the extent economically feasible, subject to approval of divisional lines of authority.

g. *Corporate Participation*

Corporate Engineering, Research and Development, and Technical Services will initiate and continue programs which are considered appropriate and feasible to study methods of reducing high noise levels at the source.

Figure 12-3 (*continued*)

or weeks. Often, large corporations use a different approach or combination of approaches in different plants, depending on the availability of in-house and consultant personnel. In these cases, however, (and, in fact, in all situations where the combined approach is used) it is essential that there be a key individual in each plant and the corporate office who has overall responsibility for the entire HCP. This person must be knowledgeable in all areas of the program and must work to integrate the various functions so that the HCP effectively protects all employees' hearing. Constant attention and follow-up activity is needed, as well as frequent communication with outside consultants to assure that activities are proceeding in a timely fashion and that the goals of the HCP are being met.

There are many advantages to in-house staff provision of audiometric testing, HPD fitting and training, and employee education:

1. More flexible scheduling of activities throughout the year
2. Better management contact with workers throughout the year, insofar as frequent contacts for testing, etc. provide increased opportunities for informal training and motivation
3. Ease of performing pre-employment audiograms and recommended retests
4. More continuity in follow-through on consultants' recommendations
5. More consistent monitoring of HPD use by workers and easier correction of problems encountered
6. Employee association of one or two key persons with the HCP, making it more likely that they will seek help with HPD problems
7. Constant presence in the plant of at least one well-trained person to give attention to the HCP and problems that may develop

Advantages are to be found also in using consultant services, especially mobile testing services:

1. The ability to test, train, and check the HPDs of large numbers of workers in a relatively short time
2. Third-party professional involvement, with its attendent legal advantages
3. The opportunity to draw attention to the importance of the HCP through well-advertised, yearly activities involving all employees
4. Opportunities for key company personnel to maintain currency of understanding of HCP concepts through contact with consultants
5. Avoidance of overhead costs associated with in-house personnel, equipment, and space devoted to the HCP
6. The use of well-trained professionals in the provision of HCP services

Totally Consultant Program

This method is sometimes selected by companies (usually smaller companies) wishing to avoid the expense and personnel acquisition problems attending in-house HCPs. These companies lack the necessary staff support to provide HCP services to workers and, instead, contract with outside consultants to provide all needed activities. These may be provided on-site (in the plant), as with mobile hearing testing and employee training sessions, or they may take place off-site in the consultant's office or clinic. Generally, only hearing testing, HPD fitting, employee training sessions, and audiogram review and referral lend themselves easily to off-site programming, and having even these services provided away from the plant entails numerous problems related to insurance, transportation costs, and expenses associated with increased worker time away from the job. Activities associated with noise surveys and engineering noise controls naturally must take place on-site.

Because of these types of problems, as well as the expenses involved in a totally consultant operated program, this approach

TABLE 12-3.

Role of the Audiologist on the Interdisciplinary Team in Occupational Hearing Conservation

Noise Exposure Monitoring
  Provide it directly.
  Train company personnel to perform it.
  Determine employee TWA exposures.
  Recommend exposure monitoring schedule/ criteria.
  Select/calibrate sound measuring equipment.

Engineering and Administrative Controls
  Determine effects on HCP; advise management of these.
  Recommend engineering consultants to management.

Personal Hearing Protection
  Advise on selection of appropriate HPDs.
  Study nature of employee noise exposures and attenuation characteristics of HPDs in order to do the above.
  Advise on methods of performing in-field attenuation tests, or perform them personally.
  Provide HPD fitting, or training and supervision of technicians in HPD fitting.
  Train employees in HPD use and care, or train technicians in methods of doing this.

Audiometric Testing, Audiogram Review, and Referral
  Train and supervise OHC technicians.
  Select/calibrate audiometric test equipment.
  Select/document suitable test areas.
  Provide or recommend mobile testing services.
  Review and interpret in-plant audiograms.
  Develop and implement pass-fail and referral criteria.
  Provide follow-up diagnostic audiological testing.
  Establish/monitor proper record keeping system.
  Refer employees to appropriate medical services.
  Counsel hearing-impaired and "shifting" employees.
  Retrain/recertify OHC technicians.
  Consult with management regarding overall OHCP, including their liability under OSHA regulations and comp. laws.
  Testify as expert witness in legal forums.

Employee/Management Education
  Provide it directly.
  Develop or recommend appropriate training materials.
  Train company personnel to provide employee education.
  Write articles for publication in company newsletters or trade journals.
  Speak before industry groups.
  Maintain up-to-date knowledge of laws and regulations.

Record Keeping
  Advise on required and recommended records at plant.
  Establish and maintain an adequate personal system for all OHC services to a company, whether direct, supportive, or consultative.

to HCPs is rather infrequently used. In addition, the lack of a key responsible person at the plant to oversee program activities and have the HCP as a primary interest and concern make this approach less effective and, therefore, inadvisable for most companies.

The "Team Approach"

Because hearing conservation is clearly a multidisciplinary field and requires the contribution of many different persons, both professional and nonprofessional, a "team approach" is always needed to achieve an effective HCP (ASHA, 1985; Harford, 1978; Miller, 1976; Olishifski, 1975; Sataloff, 1973). This is the position that has been officially adopted by the Council for Accreditation in Occupational Hearing Conservation (CAOHC) and recommended to its course directors as the philosophy to adopt with OHC training courses (CAOHC, 1985).

Although the importance of interdisciplinary activity to the success of HCPs seems obvious, few professions have addressed the issue of the differing roles of various professionals in HCPs. The American Speech-Language-Hearing Association (ASHA), one of the professional associations of audiologists working in HCPs, did

TABLE 12-4.

Responsibilities of Members of the OHCP Team

Audiologist (see Table 12-3)

Acoustical Engineer
Assess feasibility of noise control engineering efforts
Advise company engineers "how to do it" or implement controls personally
Assess success of efforts
Specify acceptable noise output levels for machinery being considered for purchase

Industrial Engineer
Do time studies for TWA noise exposure calculations and administrative noise control decisions

Industrial Hygienist
Perform noise surveys and dosimetry
Calculate TWA noise exposures of employees
Advise concerning need for engineering noise controls or personal hearing protection
Assess attenuation effectiveness of HPDs
Analyze incidence of noise-induced hearing loss within departments and plants
Assist in plant education program

Otolaryngologist
Diagnose and treat ear and hearing problems of employees
Advise HCP technician concerning referral criteria
Review employee audiograms
Provide diagnostic otologic reports to company
Recommend employee restrictions from further noise exposure when warranted

Industrial Physician
Supervise in-house HCP if on staff
Provide medical consultation to plant HCP if not on staff
Provide initial examination of employees with ear or hearing problems
Make referrals to otolaryngologist or audiologist
Review and follow-up on employee reports from specialists

Safety Specialist
Integrate HCP into total plant safety program
Assist in plant education program
Supervise hearing protective device program
Audit HPD compliance among employees
Inform management of HCP status on periodic basis
Coordinate employee education program

Occupational Health Nurse/OHC Technician
Perform employee hearing tests
Train employees in OHC concepts
Communicate TWA exposures to employees
Provide employee motivation to prevent noise-induced hearing loss
Perform otoscopic examinations, if trained to do so
Take otologic case history
Fit HPDs and train employees in use and care of HPDs
Check fit and condition of HPDs on periodic basis
Ensure that audiogram review takes place
Communicate referral recommendations from audiogram review to employees
Arrange for employee referrals
Follow-up on recommendations of referral source
Assist employees in solving HPD problems

Data Management Specialist
Program computer to review audiograms according to recommendations of professional reviewer
Provide computer reports as needed
Design programs to assist with administration of HCP
Maintain and retrieve statistics on HCP activities

Plant Manager
Supervise and manage the overall HCP
Coordinate HCP activities with other plant activities and production goals
Support other HCP team members as needed
Enforce company HCP policies

Superintendent/Supervisor
Supervise and manage the HCP in their area
Support HCP technical staff as needed
Enforce company HCP policies in their area
Audit use of HPDs by employees in their area
Assist employees in solving HPD problems

Safety Committee
Audit HPD use in company
Discuss and make recommendations for solving HCP problems
Arrange for employee training and motivation programs in OHC as needed

Personnel Manager
Ensure adequate staffing so that company HCP policies can be carried out
Support goals of HCP in dealings with employees
Communicate HCP policies to new employees
Assist in employee motivation tasks undertaken by nurse or HCP technician

address the issue of its members' role in the overall HCP when it published a position statement in 1985 (ASHA, 1985). Table 12–3 contains a list of HCP activities recommended for qualified audiologists by ASHA.

But the team approach cannot be limited to professional contributions. Clearly, vital responsibilities must be carried out by in-plant personnel also. These responsibilities will in some ways be even more important than those of outside consultants since they are concerned with the day-to-day activities of workers on the job. These activities will vary from company to company, but certain general functions can be listed for specific plant members of the HCP team. These are listed in Table 12–4 along with functions that would usually be assigned to other professional team members.

SUMMARY

This chapter has provided information concerning the development of a comprehensive HCP in industry. It has defined the goals of an HCP and discussed problems involved with demonstrating the effectiveness of a particular HCP in meeting those goals. It has described the general features of a complete HCP and has related those features to the requirements of the OSHA Noise Regulation and its Hearing Conservation Amendment. It has outlined the difference between a complete, effective HCP and one that only meets OSHA requirements. It has listed and explained various approaches to achieving a comprehensive HCP, pointing out reasons why and situations when one or another approach might be preferable to another. Finally, it has discussed the "team approach" to occupational hearing conservation, in which each team member, nonprofessional as well as professional, has his or her interrelated role to play in creating a successful HCP.

Although an effective HCP cannot be guaranteed by the presence of each of the elements of a comprehensive HCP discussed in this chapter, any program that does not include all of the elements will doubtlessly be ineffective. The actual degree of effectiveness experienced by a company in its HCP depends: (1) on the commitment of company management to the adequate implementation of each program element; (2) on the enthusiasm and interest shown by the company's hearing conservation technician in each employee's hearing health, (3) on the competence and dedication of each member of the HCP team, both professional and nonprofessional; and (4) on the motivation of each worker to protect his or her own hearing. Hearing conservation can work in industry, but it must be vigorously and continuously supported by management and must include the technical assistance of professional supervisors to guarantee that each aspect of the HCP is functioning with maximum effectiveness. If this is accomplished on a wide scale in industry, as seems possible since the implementation of the OSHA Hearing Conservation Amendment, eventually, occupational hearing loss with its needless costs in human and economic terms will be a thing of the past.

REFERENCES

American Academy of Otolaryngology—Head and Neck Surgery (1983). *Otologic referral criteria for occupational hearing conservation programs.* Washington, DC: AAO-HNS.

American Academy of Otolaryngology—Head and Neck Surgery Foundation, Inc. (1982). *Guide for conservation of hearing in noise* (rev. ed.). Rochester, MN: Custom Printing.

American Speech-Language-Hearing Association (1985). The audiologist's role in occupational hearing conservation. *Asha, 27,* 41–45.

Bailey, J. R., Hart, F. D., & Stewart, N. D. (1980). *Noise control technology assessment* (Report submitted to the Occupational Safety and Health Administration). OSHA Docket #OSH–011 500-7.

Bearce, G. R. (1975). Hearing conservation—a call for action. *Sound Vib, 9,* 24–28.

Berger, E. H. (1983). Using the NRR to estimate the real world performance of hearing protectors. *Sound Vib, 17,* 12.

Berger, E. H. (1980a). The performance of hearing protectors in industrial noise environments.

EARLog #4. E-A-R Division, Cabot Corporation.

Berger, E. H. (1980b). Hearing protector performance: How they work-and-what goes wrong in the real world. *EARLog #5.* E-A-R Division, Cabot Corporation.

Chung, D. Y., Gannon, R. P., Roberts, M. E., & Mason, K. (1982). Hearing conservation based on hearing protectors. In P. W. Alberti (Ed.), *Personal hearing protection in industry.* New York: Raven Press.

Cluff, G. L. (1980, September). Limitations of ear protection for hearing conservation programs. *Sound Vib,* 19–20.

Council for Accreditation in Occupational Hearing Conservation (1985). *CAOHC Manual* (2nd ed.). Springfield, NJ: Association Management Corporation.

Dear, T. A., & Karrh, B. W. (1979). An effective hearing conservation program—federal regulation or practical achievement? *Sound Vib, 13,* 12–19.

Erdreich, J. (1985). Alternatives for hearing loss risk assessment. *Sound Vib, 19,* 22–23.

Feldman, A. S. (1985). Federal regulations dealing with occupational noise. In A. S. Feldman, & C. T. Grimes (Eds.), *Hearing conservation in industry.* Baltimore: Williams & Wilkins.

Feldman, A. S. (1976). Industrial hearing conservation programs. In D. Henderson, R. Hamernick, D. Dosanjh, & J. Mills (Eds.), *Effects of noise on hearing.* New York: Raven Press.

Fox, M. S. (1965). Industrial and audiometry. In A. Glorig (Ed.), *Audiometry: Principles and practices.* Baltimore: Williams & Wilkins.

Gasaway, D. C. (1985a). Why the conservation of hearing is so vital. In *Hearing conservation: A practical manual and guide.* Englewood Cliffs, NJ: Prentice-Hall.

Gasaway, D. C. (1985b). Purpose of audiometric tests important in defining procedures. *Occ Health Saft, 54,* 61–67.

Gasaway, D. C. (1985c). Documentation: The weak link in audiometric monitoring programs. *Occ Health Saft, 54,* 28–33.

Gasaway, D. C. (1985d). Using documentation to enhance monitoring efforts. In *Hearing conservation: A practical manual and guide.* Englewood Cliffs, NJ: Prentice-Hall.

Gasaway, D. C. (1984a). Safety in numbers misleading in assessing auditory risk criteria. *Occ Health Saft, 53,* 58–66.

Gasaway, D. C. (1984b, September). Foundations for an effective hearing protection program. *Ind Hyg News,* 44–48.

Gasaway, D. C. (1984c). Motivating employees to comply with hearing conservation policy. *Occ Health Saft, 53,* 62–66.

Glorig, A. (1973). Industrial hearing conservation. In *Noise-Con 73.* Washington, DC: National Conference on Noise Control Engineering.

Gosztonyi, R. E. (1975). The effectiveness of hearing protective devices. *J Occ Med, 17,* 569–580.

Harford, E. R. (1978). Industrial audiology. In D. M. Lipscomb (Ed.), *Noise and audiology.* Baltimore: University Park Press.

Harris, D. A. (1980). Combating hearing loss through worker motivation. *Occ Health Saft, 49,* 38–40.

Hicklish, D. E., & Challen, P. J. (1966). A serial study of noise exposure and hearing loss in a group of small- and medium-size factories. *Ann Occ Hyg, 9,* 113–133.

Hirschorn, M., & Singer, E. (1973). The effect of ambient noise on audiometric room selection. *Sound Vib, 7,* 18–22.

Karrh, B. W. (1973). Effective hearing conservation programs. In *Noise-Con 73.* Washington, DC: National Conference on Noise Control Engineering.

Lipscomb, D. M. (1974). *Noise: The unwanted sounds.* Chicago: Nelson-Hall.

Maas, R. (1972). Industrial noise and hearing conservation. In J. Katz (Ed.), *Handbook of clinical audiology.* Baltimore: Williams & Wilkins.

Maas, R. (1970). The challenge of hearing protection. *Ind Med, 39,* 29–33.

Maas, R. (1961, September). Hearing protection—whose failure? *Natl Saft News,* 20–25.

Mellard, T. J., Doyle, T. J., & Miller, M. H. (1978). Employee education—the key to effective hearing conservation. *Sound Vib, 12,* 24–28.

Melnick, W. (1984). Evaluation of industrial hearing conservation programs: A review and analysis. *Amer Ind Hyg Assc J, 45,* 459–467.

Melnick, W. (1978). Industrial hearing conservation. In J. Katz (Ed.), *Handbook of clinical audiology* (2nd ed.). Baltimore: Williams & Wilkins.

Michael, P. (1972, June). Hearing conservation. *Mining Congress J,* 74–82.

Miller, M. H. (1976). The audiologist's role in occupational hearing conservation. *Asha, 18,* 846–849.

Miller, M. H., & Silverman, C. A. (Eds.). (1984). *Occupational hearing conservation.* Englewood Cliffs, NJ: Prentice-Hall.

Newby, H. A. (1964). *Audiology* (2nd ed.). New York: Appleton-Century-Crofts.

Occupational Safety and Health Administration (1983, March 8). Occupational noise exposure; hearing conservation amendment; final rule. *Federal Register 46*, 9738–9785.

Olishifski, J. B., & Harford, E. R. (1975). *Industrial noise and hearing conservation.* Chicago: National Safety Council.

Pell, S. (1973). An evaluation of a hearing conservation program—A five-year longitudinal study. *Amer Ind Hyg Assc J, 34*, 82–91.

Pell, S. (1972). An evaluation of a hearing conservation program. *Amer Ind Hyg Assc J, 33*, 60–70.

Pell, S., Dear, T. A., & Karrh, B. W. (1982). The long-term effectiveness of a hearing conservation program based upon personal hearing protection. In P. W. Alberti (Ed.), *Personal hearing protection in industry.* New York: Raven Press.

Royster, J. D. (1985). Audiometric evaluations for industrial hearing conservation. *Sound Vib, 19*, 18–24.

Royster, L. H., & Holder, S. R. (1982). Personal hearing protection: Problems associated with the hearing protection phase of the hearing conservation program. In P. W. Alberti (Ed.), *Personal hearing protection in industry.* New York: Raven Press.

Royster, L. H., Lilly, D., & Thomas, W. G. (1980). Recommended criteria for evaluating the effectiveness of hearing conservation programs. *Amer Ind Hyg Assc J, 41*, 40–48.

Royster, L. H., & Royster, J. D. (1985). Are we educating the right people? *Sound Vib, 19*, 4.

Royster, L. H., & Royster, J. D. (1984). Making the most out of the audiometric data base. *Sound Vib, 18*, 18–24.

Royster, L. H., & Royster, J. D. (1982a). Methods of evaluating hearing conservation program data bases. In P. W. Alberti (Ed.), *Personal hearing protection in industry.* New York: Raven Press.

Royster, L. H., & Royster, J. D. (1982b). *Comparing the effectiveness of significant threshold shift criteria for industrial use* (Final Report submitted to OSHA, U.S. Department of Labor, Docket #OSH-011, Submission #366).

Royster, L. H., Royster, J. D., & Berger, E. H. (1982). Guidelines for developing an effective hearing conservation program. *Sound Vib, 16*, 22–25.

Sataloff, J., & Michael, P. (1973). *Hearing conservation.* Springfield, IL: Charles C. Thomas.

Spindler, D. E., Olson, R. A. & Fishbeck, W. A. (1979). An effective hearing conservation program. *Amer Ind Hyg Assc J, 40*, 604–608.

Stapleton, L., & Royster, L. H. (1985). The education phase of the hearing conservation program. *Sound Vib, 19*, 29–31.

Summar, T. M. (1976). An effective hearing conservation program in industry today. In *Inter-Noise 76.* Washington, DC: Proceedings of the International Conference on Noise Control Engineering.

Summar, T. M. (1969). Industrial hearing conservation (A report on a longitudinal study). *Natl Saft News, 100*, 52–54.

Suter, A. H. (1984). Noise control: Why bother? *Sound Vib, 18*, 5.

U.S. Department of Labor (OSHA). (1983, November 9). *Guidelines for noise enforcement* (Instruction CPL 2-2.35). Washington, DC: Office of Health Compliance Assistance.

Vassallo, L., & Sataloff, J. (1978). Procedures in hearing measurement and protection. In D. M. Lipscomb & A. C. Taylor (Eds.), *Noise control handbook of principles and practices.* New York: Van Nostrand Reinhold.

Walker, J. L. (1972). A successful program of hearing conservation. *Ind Med, 41*, 11–14.

Williams-Steiger Occupational Safety and Health Act (1970). Public Law 91-596. (Title 29 CFR, Ch 17). Washington, DC.

Related Hearing Conservation Activities

Hearing conservation has been identified primarily with the industrial scene in recent years. However, prior to most industrial hearing conservation activities, public schools and the military were developing programs to identify persons with sub-normal hearing in order to provide requisite follow up. Those pioneering efforts are honored in this section with discussions of current activities in the context of past history.

One chapter is included in this section to present an additional area of professional service for persons skilled in hearing conservation efforts. Community noise abatement and control efforts utilize many of the concepts discussed earlier in this text, but there are special approaches peculiar to the community noise problem. The last chapter in this section addresses that topic.

13

CHAPTER

Hearing Conservation in an Educational Setting

Deborah A. Arthur

A lthough industrial programs are comparatively new, hearing conservation programs in an educational setting have a historical base of over 60 years, most specifically in the area of identification. The purpose of this chapter is to present a hearing conservation program inclusive of those areas of particular concern to children in a school environment.

In determining the need for an industrial hearing conservation program, the Subcommittee on Noise of the Committee on Conservation of Hearing of the American Academy of Opthalmology and Otolaryngology issued the following conditions as indications for implementation of a program in an industrial setting (taken from the 1982 revision of the *Guide for Conservation of Hearing in Noise*):

1. Difficulty in communicating by speech while in noise
2. Head noises or ringing in the ears after working in noise for several hours

3. A temporary loss of hearing that has the effect of muffling speech and changing the quality of other sounds after several hours of exposure to noise

The component of that hearing conservation program should include (Melnick, 1985):

1. Assessment of noise exposure
2. Control of noise exposure
3. Measurement of hearing

Details of these components of the hearing conservation program are found elsewhere in this text (see Chapters 1 & 12). Those elements of the programs are named here to emphasize the point that there are many more similarities in industrial and educational hearing conservation programs than there are differences. Two unique characteristics of educational hearing conservation programs are (1) the setting in which the program is conducted, and

(2) the fact that noise is not necessarily the primary etiological factor.

STATE AND FEDERAL REGULATIONS

Industrial hearing conservation programs are federally regulated and monitored, as are educational programs. Public Law 94–142, passed by the Congress and signed into law on November 29, 1975, provided for free, appropriate, public education for the handicapped. Incorporated within this legislation were specifications regarding the rights and protection of handicapped children and their parents. The federal government provides funds to state education agencies (SEA) who subsequently pass them on to the local education agencies (LEA) after an approved "plan of services" is presented to the SEA. The local school districts are monitored by representatives of the SEA on a regular basis (every 3 years), and local programs are subject to citations and possible withholding of future funds if found to be out of compliance with their approved plan. The responsibility of the LEA to provide those audiology services is clearly outlined in federal regulations (*Federal Register*, August 23, 1977). One of the glaring problems is that declining enrollment in schools (American Speech-Language-Hearing Association [ASHA], 1983) has eroded the funding base for many state and local school systems. At the same time, costs are accelerating and administrators are faced with the need for delivery of services that are both cost effective and accountable.

The concern in an industrial setting is for prevention of hearing loss due to noise exposure, measurement of hearing, and noise surveys to monitor worker noise exposure. In an educational setting, the focus is on early identification of hearing problems, then referral for intervention (medical and/or professional) and habilitation/rehabilitation, if needed. The two most overlapping areas between industrial and educational forms of hearing conservation activities are identification and referral for medical or professional management.

A recent survey of audiological service programs across the United States (Wall, Naples, Buhrer, & Capodanno, 1985), indicated that a system of state-mandated hearing screening programs was in effect in 35 states. Federal legislation required programs in 19 other states.

GOALS OF A HEARING CONSERVATION PROGRAM

Identification of hearing impairment is a primary concern in the establishment of a comprehensive hearing conservation program. Frequently though, the hearing screening process becomes the focus and, in some cases, the only component of the program.

Although one of the goals of an industrial hearing conservation program is to identify hearing losses as the result of excessive noise exposure in the work environment, the goal of the educational hearing conservation program is directed more toward prevention of educationally significant hearing losses as the result of undetected sensory neural loss or transient conductive pathology. In addition to being an incomplete approach, screening alone can result in poor public relations within the medical and professional community. The inclusion of a school-based program should be aimed at minimizing the handicapping effects of a hearing impairment through early identification and intensive broad-based management of each child. The goals of an efficient and effective program should include provisions for:

1. *Prevention*—Providing information on causes and effects of hearing loss to the public on an ongoing basis. This information can be provided in the form of brochures, teacher workshops, parent and student counseling, health fairs, and public service announcements.

2. *Identification*—Including ongoing identification programs utilizing pure tone audiometry and acoustic immittance measures. These procedures should provide for periodic screening for ages 0 to 21 and should establish a mechanism for comprehensively testing preschoolers.

3. *Assessment*—Establishing the nature and degree of hearing impairment and its effect on communicative function and educational performance.

4. *Habilitation and Rehabilitation*—Selecting appropriate amplification for use in the educational environment, assisting in instructional services by participating in the multidisciplinary team, ensuring the provision of auditory and visual (speechreading) training, and providing a systematic continuum of support.

5. *Follow-up, Referral, and Record Keeping*—Including medical referral, audiometric/tympanometric monitoring, and record keeping that includes notification of parents, professionals, and physicians of the child's hearing condition/status on an ongoing basis. (American Speech-Language-Hearing Association, 1983.)

PREVALANCE OF HEARING LOSS IN SCHOOL-AGE POPULATION

The reported incidence of hearing loss in school-age children has varied widely, dependent on type, degree, method of assessment, and age of children tested. Eagles et al. (1973) reported that 5 percent of school children have hearing levels (HL) at one or both ears, at one or more frequencies, beyond the range of normal hearing. Ross and Giolas (1978) stated that 30 to 50 hearing-impaired children exist per 1,000 students, for a total of approximately two to three million hearing-impaired school-age children. The numbers in these studies include middle ear infections resulting in conductive hearing loss, children with sensory loss, and those with central auditory processing disorders.

Ross and Giolas (1978) further suggested that a child with a hearing loss of 15 dB HL or more in the speech frequencies at the better ear is at risk for that loss having an adverse effect on some aspect of their educational, vocational, or social competency.

Recognizing and accepting the fact that hearing impairment is a reality is a necessary first step. For a number of years, statistics cited only the incidence of sensory hearing loss, generally severe to profound in nature. The most common cause of hearing loss in children is middle ear diseases. By far, the most frequent of these diseases is otitis media (middle ear effusion) with its associated complications. It is a widely known and statistically supported fact that the incidence of otitis media among young children is very high and occurs most frequently in children under the ages of 8 to 10 (Shepherd, Davis, Gorga, & Stelmachowicz, 1981). Figures have shown 76 to 95 percent of all children experience at least one episode of otitis media by 6 years of age (Howie, Ploussard, & Sloyen, 1975). According to McCandless (1979), 90 percent of all ear problems in children result in some form of otitis media. The deleterious effects of otitis media, whether episodic or persistent, on the development of speech and language has been documented by Holm and Kunze (1969). They further showed that these same children experienced significant difficulties in other academic areas.

In the majority of the studies reported in the literature, focus has been on the identification of hearing loss at the earliest possible age, with emphasis put on kindergarten through third grade and all those children considered high risk (to be discussed in a later section). Although intervention is crucial at an early age, there has been evidence for a number of years that sensorineural hearing loss in the high frequencies increases dramatically with age and is becoming more common in high school students (Lipscomb, 1972; Woodford & O'Farrell, 1983). In three separate studies conducted at the University of Tennessee in 1968, modified hearing screening tests were given to 1,000 sixth, ninth, and twelfth graders to determine the screening failures to high frequency (above 2000 Hz) tones. The results, using the high frequency screening criterion of 15 dB (ISO, 1964), revealed 3.8 percent of the sixth graders failed, followed by a jump to 11 percent for the ninth graders and a similar level of 10.6 percent for the twelfth graders. This study was followed (Lipscomb, 1972) by a similarly conducted technique with two successive groups of college freshmen. The results were even more startling, with 32.9 percent failing the screening criteria in the first group, followed by 60.7 percent in the second group. Consistently throughout, the incidence of high frequency hearing loss has been greater in males than in females. It is important to identify these very young hearing-impaired children before they enter the educational process, but at the same time, can we ignore those students who are just beginning to enter career training and the job force? When asked, the majority of these secondary and college students were not aware of the existence of a hearing deficit. As Lipscomb (1972) and Woodford and O'Farrell (1983) have pointed out, these youngsters are entering the work force with retirement-age ears.

IDENTIFICATION AUDIOMETRY

The audiometric identification program in the schools is intended to identify those children with educationally significant hearing loss, either permanent or transient. Identified children are then referred for medical or professional management. It must be remembered that, in children, even a mild (15 dB HL in the better ear) or fluctuating (middle ear effusion) hearing loss can affect speech/language development and the fundamentals of learning. It is therefore necessary that the screening procedures used are sensitive to middle ear disease as well.

The American Speech-Language-Hearing Association (1985a) has stated in its screening guidelines that the use of only pure tone air-conduction screening is inefficient in identifying many of the middle ear disorders. Brooks (1979) reported on the interference of environmental noise in pure tone screening. On studies carried out in England, a median ambient noise level of 44 dB HL with frequent peak levels exceeding

60 dB HL were reported. As can be seen by Table 13-1, the median level exceeds the maximum acceptable level at 500 Hz and the peak levels exceed all frequencies.

Application of pure tone audiometry in the detection of middle ear pathology has been shown to be only 50 percent accurate (Brooks, 1979). This same figure was reported by Melnick, Eagles, and Levine (1964) and was not improved, even with a changed criteria (lowering) of screening levels. These researchers further suggested the addition of otologic input to the screening program. This, of course, is not feasible due to the limited availability of physicians and the inevitable high expense. In 1973, Bluestone, Beery, and Paradise published an article on the superiority of immittance testing over traditional pure tone audiometry in the identification of middle ear disease. Notable in this study was that 34 of the 84 children tested by pure tone audiometry had pure tone averages (PTAs) better than 25 dB HTL (ANSI-69). Unfortunately, 16 of those 34 had serous otitis media, verified by otoscopy and subsequent myringotomy. It is interesting to note that, although the purpose of the identification program is to find middle ear disorders needing medical attention and hearing problems that are educationally handicapping, a very small percentage of programs combine immittance and pure tone screening.

It is recommended in the 1984 ASHA screening guidelines that identification

TABLE 13-1.

Minimum Allowable Octave Band Ambient Noise Levels

Test Frequency	Allowable Ambient Noise Levels
500	41.5
1000	49.5
2000	54.5
4000	62

Adapted from American Speech-Language-Hearing Association (1985). Guidelines for identification audiometry, *ASHA, 27,* 49–52.

programs in the schools should incorporate acoustic immittance measures with pure tone audiometry as a means of identifying persons who are in need of audiological and medical services (ASHA, 1985a).

PROCEDURES

The use of speech stimuli for screening, as in the fading numbers test and group pure tone testing (Reger & Newby, 1947), has been predominantly replaced by individual sweep frequency pure tone screening. As stated earlier, educational hearing screening programs test in the school where there is a poor acoustic environment. Further, hearing screening tests are designed for accomplishing rapid and efficient identification of hearing impairment, especially with young children.

ASHA published its original identification audiometry guidelines in 1974 and revised them in 1984 (published in 1985). These guidelines stipulate that identification audiometry should be conducted or supervised by an audiologist. Wall and colleagues (1985) reported in their survey of audiological practices and procedures in the schools that support personnel popularly are the on-site testing providers, with the vast majority (61 percent) being school nurses. In this same survey, it was found that 74 percent of the state respondents indicated that no licensure or certification requirements are necessary for conducting hearing screening. This, of course, differs vastly from the industrial setting where training and recertification of hearing test examiners is an ongoing process.

As stated previously, concentration in the identification process is on the very young child. ASHA recommends hearing testing of kindergarten through third grade and all "at risk" children, which includes (1) grade repeaters, (2) children in special education, (3) new students or transfer students, (4) those who were absent during a previously scheduled screening, (5) children with speech or language problems, (6) children who failed a threshold test the

previous year, (7) students experiencing health problems related to hearing impairment, and (8) those involved in coursework that places them at risk for noise exposure. ASHA recognized that some school systems may elect to screen routinely after third grade, but stated that more data appear to be needed to determine the merit of this practice (ASHA, 1985b). The 1983 survey of audiological procedures and practices (ASHA, 1983) indicated that approximately 60 percent or more of the surveyed school systems screen through third grade and a high proportion, approximately 50 percent, screen fifth and seventh grades as well. The figures decrease dramatically as age increases, to a low of 10 percent of twelfth graders screened. Lipscomb (1972) has shown the need for reassessment of the presently recommended age criteria with his study of high frequency hearing loss in secondary and college freshmen students. It has been postulated that the combination of high intensity recreational sounds and educational/vocational training environmental sounds have evidenced a very tangible effect on our young people.

The relatively poor acoustic environment (Brooks, 1973) where educational hearing screening takes place does not allow for presentation levels below 20 dB HL (ANSI-1969). The ASHA recommended frequencies are 1000 Hz, 2000 Hz, and 4000 Hz with the inclusion of 500 Hz when the ambient noise in the testing environment does not exceed the recommended levels (see Table 13-1). ASHA does not recommend the inclusion of 3000 Hz, as a substitute for 4000 Hz, nor the addition of 6000 Hz, due to interactions between earphones and ears (Villehur, 1970). In Table 13-2, the reported test frequencies and levels used by the respondents show that better than 90 percent follow ASHA guidelines.

The failure criterion in an educational screening program initially appears quite rigid, with failing to respond to the screening level at any frequency at either ear necessitating retest. The rescreening process should be accomplished in the same session and should involve a repositioning of the

TABLE 13-2.
Test Frequencies and Levels Used for Hearing Screenings (N-551)

Level	250 N	250 %	500 N	500 %	1000 N	1000 %	2000 N	2000 %	4000 N	4000 %	6000 N	6000 %	8000Hz N	8000Hz %
<20 dB	3	0.54	18	3.27	29	5.26	30	5.45	28	5.08	11	2.00	16	2.90
20 dB	33	5.99	124	22.51	256	46.46	259	47.01	185	33.58	68	12.34	60	10.89
25 dB	78	14.16	252	45.74	217	39.38	210	38.11	288	52.27	86	15.61	88	15.97
30 dB	12	2.18	36	6.53	14	2.54	14	2.54	17	3.09	10	1.82	6	1.09
>30 dB	2	0.36	4	0.76	2	0.36	2	0.36	1	0.18	1	0.18	1	0.18
no level given	2	0.36	7	1.27	6	1.09	6	1.09	7	1.27	1	0.18	2	0.36
TOTAL	130	23.59	441	80.08	524	95.09	521	94.56	526	95.47	177	30.31*	173	31.39

Taken from Wall, Naples, Buhrer & Capodanno (1985). A survey of audiological services within the school system. *ASHA*, 31–34.

*sic

earphones and careful reinstruction of the student. Wilson and Walton (1974) reported that a 52 percent reduction in failures can be achieved by rescreening. Expedient referral for the threshold assessment by an audiologist is the next step in the identification process and should be done with the informed consent of the parents. At this time, a medical referral can also be made, especially if a middle ear disorder is likely. Frequently, this is the point where audiological services in the schools break down. There is no assured mechanism of follow up and referral. The school system needs to have designated personnel to monitor the referral process and ensure recommendations have been followed. In the survey (Wall, 1985) conducted by the American School Health Association, it was reported that 80.4 percent of school systems provided some form of counseling regarding referral and management to the parents and/or school personnel. A quite large number (90.8 percent) stated they have an established follow up mechanism for tracking the children once referred. Data are incomplete regarding how effective the tracking mechanism actually is. The primary referral designees reported in the survey were 60.6 percent to the family physician or otologist and 38.2 percent to an audiologist. This figure is interesting because 97.8 percent of the respondents stated they solely used pure

tone air conduction screening and only 30.13 percent indicated the incorporation of acoustic immittance measures.

For a number of years the literature has documented the incidence of high frequency hearing loss in secondary students (Hull, Mielke, Timmons & Willeford, 1971; Lipscomb, 1972). The incidence reported is greater in males than in females and, assuming that in our society males are typically exposed to high levels of noise to a larger extent than are females, the possibility of overexposure to noise as an etiological factor is entirely feasible (Woodford & O'Farrell, 1983). The popularity of high intensity recreational sound sources such as motorcycling, sport shooting, and live rock music has increased in recent years. Investigations by Lipscomb (1974), Woodford (1973), and Lipscomb (1972), have attributed the increased incidence of these supposedly noise-induced losses to gunfire and loud music. It has been further documented (Woodford, 1980; Woodford & O'Farrell, 1983) that industrial arts programs in schools are also a potential source of excessive noise exposure. Woodford (1980) and Woodford & O'Farrell (1983) collected and reported data on sound levels in public schools shops and school band practice rooms (see Table 13-3).

It is obvious that the sound levels experienced in these environments are just

TABLE 13-3.

Typical Levels of Environmental Sounds in Secondary Schools

Noise Source	Sound Level dBA
Band	90–112
Carpentry/Building/Construction	92–114
Auto Shop (Body and Maintenance	94–106
Welding	72–108
General Shop	92–109

Adapted from Woodford, C., & O'Farrell, M. (1983). High frequency loss of hearing in secondary school students: An investigation of possible etiologic factors. *Language, Speech, and Hearing Services in Schools, 14,* 22–28.

as hazardous as many of those reported by the industry. Because of this, it was of great concern when Garstecki (1978) found that, of the surveyed educational audiologists, it was reported that *no* time was spent by audiologists in the schools in initiating and maintaining hearing conservation programs. OSHA protects the hearing of workers in industry from excessive noise exposure, but students and employees in the school system are not afforded protection under federal regulation. Schneider (1985) stated:

> Although Occupational Safety and Health Act hearing conservation regulations do not apply to schools, the educational audiologist has the obligation to make students and teachers in noisy industrial environments aware of the hazards. Education as to effects of noise, hearing loss, noise control and ear protection is necessary. (p. 818)

The lack of knowledge and information about hearing loss and the hazards of intense noise exposure may be the primary targets of the audiologist when establishing a hearing conservation program. Teachers and students alike are in need of education regarding both topics. The problem of education is of major concern in both the industrial and the educational setting. Woodford and O'Farrell (1983) feel that if time were spent on educating students and teachers as to the effects of intense noise

exposure on hearing and they were then provided with ear protectors, losses of hearing in many individuals would be prevented and some problems eliminated. In their survey of secondary vocational shop programs (Woodford & O'Farrell, 1983), 69 percent of respondents indicated having difficulty getting students to wear hearing protection. Their explanation as to why included: (1) lack of knowledge regarding the hazard, (2) inconvenience or laziness, (3) discomfort, and (4) forgetfulness. For those programs reporting a success in the use of hearing protectors, the reasons mainly were confined to (1) a class requirement, (2) grades were cut if they didn't wear them, or (3) an explanation was given and an example was set for wearing them.

It certainly appears as if the same problems plague the audiologist in the educational setting as in the industrial setting. Although federal mandates regulate the monitoring, assessment, education, and management of the industrial employee, it is still amazing that American young people are not afforded the same protection under similar guidelines. Whereas hearing identification programs are well established in the U.S. school systems, the remainder of the hearing conservation program remains sparse, at best. In a very recent survey of public school hearing conservation programs in Oregon (Pelson, 1987), the major problems identified continue to be training of screening personnel; lack of audiological consults; no reorganized testing protocol; and minimal actual followup and retest after medical management. In the educational setting we seem to be able to identify our problems, but do not appear to be effectively remediating them.

REFERENCES

American National Standards Institute. (1969). Specifications for audiometers. ANSI S3.6–1969. New York: American National Standards Institute, Inc.

American Speech-Language-Hearing Association. (1975). Guidelines for identification audiometry. *ASHA, 17,* 94–99.

American Speech-Language-Hearing Association. (1984). Proposed revision of guidelines for identification audiometry. *ASHA, 26*(2), 47–50.

American Speech-Language-Hearing Association. (1983). Audiologic services in the schools position statement. *ASHA, 25*(5), 53–60.

American Speech-Language-Hearing Association. (1985a). Guidelines for identification audiometry. *ASHA, 27,* 49–52.

American Speech-Language-Hearing Association. (1985b). A survey of audiological services within the school system. *ASHA, 27*(1), 31–34.

Bluestone, C., Beery, Q., & Paradise, J. (1973). Audiometry and tympanometry in relation to middle ear effusion in children. *Laryngoscope, 83,* 594–604.

Brooks, D. (1979). Impedance in screening. In J. Jerger & J. Northern (Eds.), *Clinical impedance audiometry.* Mass: American Electromedics Corporation.

Brooks, D. N. (1973). Hearing screening—A comparative study of an impedance method and pure tone screening. *Scandinavian Audiology, 2,* 67–72.

Davis, J., Shepherd, N., Stelmachowicz, P. & Gorga, M. (1981). Characteristics of hearing impaired children in the schools: Part II—Psycho-educational data. *Journal of Speech and Hearing Disorders, 46,* 130–137.

Eagles, E., Wishik, S., Doerfler, L., Melnick, W., & Levine, H. (June, 1963). Hearing sensitivity and related factors in children. *Laryngoscope,* (Monograph Suppl.), 73.

Federal Register. (1977). Monday, August, 23, 42, (163).

Garstecki, D. (1978). A survey of school audiologists. *ASHA, 20,* 391–296.

Holm, V. A., & Kunze, L. H. (1969). Effect of chronic otitis media on language and speech development. *Pediatrics, 43,* 833–839.

Howie, V. M., Ploussard, J. H., & Sloyen, J. (1976). Natural history of otitis media. In D. J. Lim, C. D. Bluestone, W. H. Saunders, & B. H. Senturia (Eds.), *Recent advances in middle ear effusion. Annals of Otol. Rhin. Laryngol. 85* (Supp. 25) 18–19.

Hull, F., Mielke, P., Timmons, R., & Willeford, J. (1971). The national speech and hearing survey: preliminary results. *ASHA, 11,* 231–237.

International Standards Organization. (1964). Standard reference zero for the calibration of pure-tone audiometers. ISO Recommendation R 389. New York: American National Standards Institute.

Lass, N. J., Carlin, M. F., Woodford, C. M., Campanelli-Humphreys, A. L., Judy, J. M., & Hushion-Stempla, E. A. (1985). A survey of classroom teacher's and special educators' knowledge of and exposure to hearing loss. *Language, Speech, and Hearing Services in Schools, 16,* 211–216.

Lipscomb, D. M. (1972). The increase in prevalence of high frequency hearing impairment among college students. *Audiology, 11,* 231–237.

Lipscomb, D. M. (1974). Dangerous playthings. *Noise: The unwanted sounds.* Chicago: Nelson-Hall.

McCandless, G. (1979). Impedance measures. In W. F. Rintleman (Ed.), *Hearing assessment.* Baltimore: University Park Press.

Melnick, W., Eagles, E. L., & Levine, H. (1964). Evaluation of a recommended program of identification and audiometry with school children. *Journal of Speech and Hearing Disorders, 29,* 3–13.

Melnick, W. (1985). Industrial hearing conservation. In J. Katz (Ed.), *Handbook of clinical audiology* (3rd ed.), pp. 721–740). Baltimore, MD: Williams & Wilkins.

Pelson, R. O. (1987). Public school hearing conservation in Oregon. *Language, Speech, and Hearing Services in Schools, 18,* 245.

Reger, S. N., & Newby, H. A. (1947). A group pure-tone hearing test. *Journal of Speech and Hearing Disorders, 12,* 61–66.

Ross, M., & Giolas, T. G. (1978). *Auditory management of hearing impaired children.* Baltimore: University Park Press.

Schneider, D. (1985). Educational audiology. In J. Katz (Ed.), *Handbook of clinical audiology* (3rd ed.). Baltimore: Williams & Wilkins, 818.

Shepherd, N., Davis, J., Gorga, M., & Stelmachowicz, P. (1981). Characteristics of hearing impaired children in the schools: Part I—Demographic data. *Journal of Speech and Hearing Disorders, 46,* 123–129.

Villehur, E. (1970). Audiometer-earphone mounting to improve inter-subject and cushionfit reliability. *Journal of Acoustical Society of America, 48,* 1387–1396.

Wall, L., Naples, G., Buhrer, K., & Capodanno, C. (1985). A survey of audiological services within the school system. *ASHA, 27,* 31–34.

Wilson, W. R., & Walton, W. K. (1974). Identification audiometry accuracy: Evaluation of a recommended program for school-age children.

Language Speech Hearing Services in Schools, 5, 132–142.

Wilson, W. R., & Wilson, W. K. (1978). Public School Audiometry *Pediatric audiology,* F. Martin (Ed.). Englewood Cliffs, NJ: Prentice-Hall, Inc.

Woodford, C. (1973). A perspective on hearing loss and hearing assessment in school children. *Journal of School Health, 43,* 572–576.

Woodford, C. M. (1980, July). Notes on audiology in the public schools. *Hearing Aid Journal,* 5–9.

Woodford, C., & O'Farrell, M. L. (1983). High frequency loss of hearing in secondary school students: An investigation of possible etiologic factors. *Language, Speech, and Hearing Services in Schools, 14,* 22–28.

14

CHAPTER

Occupational Hearing Conservation in the Military

Donald C. Gasaway ■

A lthough today the need for an occupational hearing conservation program (OHCP) to conserve the hearing of military and civilian employees in the services is well recognized, such was not the case until during and following World War II. Review of the contents of two bibliographies that included reports on hearing and noise problems among military personnel published prior to mid-1950 identified problems resulting from blasts or explosions, barometric stresses encountered within hyperbaric and hypobaric environments, gunfire, and noise exposures associated with military hardware and vehicles (*An annotated bibliography on noise*, 1955; Stevens, Loring, & Cohen, 1955). Prior to this period, effective types of hearing protection devices were almost nonexistent (Anderman, 1960; Ruedi & Furrer, 1948).

The span of time required to develop organized and structured "hearing conservation programs" within the military was directly linked to the evolution of specialists in fields of audition, speech science, psychoacoustics, electroacoustics, and bioacoustics. Many of these specialists came from other fields of science: electronics, physiology, psychology, otology. To a great extent, the field of audiology evolved as a direct result of needs and problems associated with rehabilitation of military personnel who acquired losses in hearing as a result of exposures to loud noises and other types of auditory stresses (Anderman, 1960; McLaughlin, 1978; Ruedi & Furrer, 1948; Sedge, 1986).

The types of exposures to loud noises encountered within the military, then, as now, cover a wide range of auditory stresses. Some exposures are so intense that a single event for an unprotected person could result in severe, permanent noise-induced hearing impairments (Ades et al., 1955). The gamut of noise types include construction, repair, refurbishing, retrofitting and refitting of ships, tanks, armored vehicles, aircraft, rockets, missiles, ballistic weapons, and the myriad support systems associated with military hardware and devices.

EVOLUTION OF HEARING CONSERVATION IN THE MILITARY SERVICES

Although meager, fragmented, and somewhat informally established and followed, the earliest existence of occupational hearing conservation efforts can be traced to the 1920s and 1930s, where blasts, explosions, and gunfire were recognized as major causes of hearing loss (Anderman, 1960; *An annotated bibliography on noise*, 1955; Ruedi & Furrer, 1948; Stevens et al., 1955). Many early military aviators within both the Army (Army Air Corps) and the Navy revealed significant amounts of hearing loss and otologic problems associated with climatic and barometric extremes, which contributed to middle ear disorders among those flying in open cockpits of early types of aircraft (*An annotated bibliography on noise*, 1955; Gasaway, 1986; Stevens et al., 1955).

The turbulent years of World War II brought many newer and even more hazardous types of noise exposures, and mobilization of U.S. soldiers, sailors, and airmen resulted in many men and women being exposed to noises that caused thousands to lose substantial amounts of hearing.

Clinicians working in military hospitals and rehabilitation centers during the early 1940s encountered many who exhibited serious hearing disorders. This growing problem could not be ignored, as more and more reportings of the seriousness of damaged hearing among military members found their way into professional journals. Expanding documentation, in conjunction with briefings and formal reports that were presented to leaders within the military services by concerned medical staff personnel, confirmed the serious nature of an otherwise preventable form of impairment.

During the mid and late 1940s and throughout the 1950s, the number of people dedicated to conserving the hearing of military and civilian employees began to grow. Aural rehabilitation centers or services were established as a direct result of the large numbers of hearing-impaired military members who served during World War II. These centers were initially established by the U. S. Army at Walter Reed General Hospital, Washington, D.C. (later moved to Deshon Army Hospital in Butler, Pennsylvania), Borden Army Hospital in Chickasaw, Oklahoma, and Hoff Army General Hospital in Santa Barbara, California (Sedge, 1986). The U. S. Navy created an aural rehabilitation facility at the U. S. Navy Hospital at Philadelphia (Ruedi & Furrer, 1948). Shortly thereafter, the Veterans Administration staffed similar facilities to deal with those who were eventually released from active duty (Anderman, 1960; Ruedi & Furrer, 1948). Much of the emphasis and expertise needed to formulate a hearing conservation protocol within an occupational health program in order to prevent occurrences of noise-induced hearing losses developed directly from the activities and concern of the dedicated clinicians and therapists who worked at these medical diagnostic, treatment, and rehabilitation centers. The resources dedicated to auditory treatment and rehabilitation became strongly established during this period and would continue in the coming years, but a major change in operating philosophy was taking place; the emphasis would change from rehabilitation to prevention.

In 1948, the major approach to dealing with hearing loss associated with excessive noise exposures within the military services was based on a philosophy of prevention and conservation. Regulatory documents specifying various elements of what eventually became concerted and organized comprehensive programs of hearing conservation within the three services (the U.S. Air Force became a separate service in 1947) were first issued by the Air Force in 1948; followed by the Navy in 1955, and the Army in 1956. These early regulations and instructions were issued by Medical Headquarters in each of the three services. The most significant contributions of each consisted of defining noise exposures as a hazard, directing use of hearing protection, and requiring audiometric monitoring of people who encountered potentially hazardous noises. For example, one of the earliest regulations dealing with hearing conservation was Air Force Regulation 160–3, issued on October 21, 1948, which specified that personnel working in high-level noises would wear hearing protection, noise measurements would be performed to determine degrees of risk, exposure periods would be minimized, and audiometric monitoring would be performed on people engaged in testing and operating turbojet and rocket engines (*Precautionary measures against noise hazards*, 1948). Guidance provided to field commanders and medical, health, and safety personnel within all three services was very limited in the first of these regulatory directives, and usually consisted of only one or two pages. However, each revealed that the major approach was one of prevention of noise-induced hearing losses.

These first hallmark directives were followed by revisions that substantially expanded and standardized hearing conservation program elements within each of the services. The initial 1948 document issued by the Air Force was followed with major revisions in 1953, 1956, 1973, and 1982; the 1955 Navy directive was followed by revisions in 1959, 1970, 1979, and 1983/1984; and the 1956 Army guidance was followed by revisions in 1965, 1972, and 1980. Modern regulatory directives of all three services serve as a monument to the knowledge, dedication, and bureaucratic skills of many people within the structure of the military medical services. They accepted the challenge of turning the recognition of a problem into movement toward the solution. These directives are usually referred to as ''Regulations'' by the Air Force, ''Technical Bulletins'' by the Army, and either ''Bureau of Medicine,'' ''Navy Medical Command Instructions,'' or ''Operational Naval Instructions'' by the Navy; thus, AFR, TB, and either BuMed, NAVMEDCOMINST, or OPNAVINS.

The U.S. Coast Guard, a sister service of the U.S. Navy, actually comes under the U.S. Department of Transportation and has also issued regulatory guidance concerning hearing conservation in the form of a ''Circular'' (*Recommendations on control of excessive noise*, 1982). Normally, the Coast Guard obtains medical and health services from the U.S. Department of Public Health, and thus, ''borrowed'' an audiologist for a short time from the department during the 1970s to help with problems of noise and hearing losses. Basically, the Coast Guard Circular deals with the same component parts of an overall program as found in the three military services.

The comprehensive nature of the current series of military hearing conservation programs contained in each respective service directive may not be fully appreciated by persons who are currently involved in occupational hearing conservation activities outside of the military. Although the military structure of personnel activities for both uniformed and civilian employees gives a broader scope to ''mandatory'' compliance with the elements of the individual programs, all are designed to rely on self-motivation, education, and self-discipline. Obtain copies of the three service's hearing conservation guidance materials and digest their contents. You will find the information also has non-military applications.

A national resource developed from military hearing conservation programs is the number of audiologists who gained unique experiences dealing with various aspects of military hearing conservation programs. They continued to provide such services to the civilian sector following separation from active duty. Additionally, thousands of medical, health, and safety technicians carry their experiences and skills with them as they enter or return to the civilian side of occupational health and safety. Industrial hygienists, bioenvironmental engineers, occupational health nurses, otolaryngologists, and safety officers have within their ranks many who benefited by knowledge obtained in military hearing conservation programs. In fact, many of the recognized leaders in civil occupational health and safety fields gained a wealth of hands-on experience during their military service.

There is no doubt that knowledge of audition, clinical audiology, aural rehabilitation, anatomy and physiology of auditory systems, and differential diagnostics provide a valuable foundation for any audiologist who will eventually work primarily with military occupational hearing conservation. Those who used their basic skills and training, accepting and embracing the preventive medicine philosophy needed to properly apply their unique formal training, are the ones who have contributed major impetus to military hearing conservation programs.

The following list identifies basic types of position descriptions to which audiologists have been or may be assigned in the military:

- Clinical Audiology
- Aural Rehabilitation
- Psychoacoustics
- Bioacoustics
- Occupational Hearing Conservation
- Academic Training in Areas of Clinical Audiology
- Academic Training in Areas of Occupational Hearing Conservation
- Speech Science and Communications

Generally, the majority of military and civilian audiologists entering the military services are primarily prepared to deal with clinical-type activities. Some are assigned strictly to clinical activities (usually at large military medical centers), others to hearing conservation activities where their expertise regarding industrial or occupational audiology must be developed. Others are assigned to dual roles that combine clinical and occupational audiology. A few are brought into the military to perform research and development activities, where their knowledge of acoustics, noise, hearing conservation, psychoacoustics, speech science, and related areas will be applied to existing and future research efforts unique to operational military requirements. The roles to which audiologists are assigned in the military are continually changing as time and experience reveals the various ways in which their professional knowledge and skills can enhance accomplishment of an occupational hearing conservation program. Although some who enter the military have voiced displeasure with nonclinical duties, most have devoted their energies and skills in a manner that has been mutually rewarding.

There is no question that the primary area of need within the military is related to conserving the hearing of people who encounter potentially hazardous noises associated with military activities. Although a noise-induced hearing loss involves the ear, one clinically trained in audiology may not be particularly qualified to deal with the problems and activities associated with occupational or preventive medicine. Herein lies a potential problem for some audiologists. Occupational hearing conservation programs are based on philosophies associated with preventive, not clinical, medicine.

COMPARISONS BETWEEN ARMY, NAVY, AND AIR FORCE PROGRAMS

Although there are differences in details between hearing conservation programs conducted by the three services, there are many practices common to all. Even when the same procedural guidance exists within a given element or area of activity, ways to accomplish the task may vary considerably. This variance may be visible throughout the overall service programs or may be evident between commands and individual bases. Accepting and embracing procedures that have been standardized for conducting service-wide hearing conservation tasks are essential to achieving success with these programs.

When overall control of a given task, such as monitoring audiometry, is relocated to only one specially trained functional group of people, there is greater conformance to standardized procedures.

The following discussion illustrates similarities and differences that exist between service programs.

Common Elements

The functional elements that compose a program of hearing conservation are very common. Differences can be found between the three services in the underlying criteria, procedural methods, and the skills and training of the people who perform the different tasks, but a visit to almost any military establishment with an ongoing hearing conservation program should reveal these component elements:

- Educational, motivational, supervisory, and disciplinary aspects of training and enforcement
- Noise measurements and assessments of potentially hazardous exposures
- Monitoring audiometry
- Clinical referrals for people identified with ear and hearing problems
- Noise control measures
- Personal hearing protection
- Record keeping and records management
- Hearing data registry interfaces
- Disposition and follow-up actions for managing people who exhibit changes in the status of their hearing
- Methods for evaluating effectiveness of effort
- Command supervision of the conduct of the overall program
- Formal training of those who perform hearing conservation duties

These elements are not mutually exclusive; several coexist and interrelate.

Education, Motivation, Supervision, and Discipline Aspects of Training and Enforcement

All three services have similar programs in these areas of concentration. Here, the elements include structured programs for those who routinely encounter potentially hazardous noise. Although the services have been active in educating, motivating, and indoctrinating members for many years, it was not until the early 1970s that these elements assumed more formally structured approaches to the tasks (*Hazardous noise exposure*, 1982; *Hearing conservation*, 1980; *Hearing conservation and noise abatement*, 1983; *Occupational noise control and hearing conservation*, 1984). Personnel who undertake these tasks have been drawn from several specialties, including: occupational health nurses, bioenvironmental engineers, industrial hygienists, occupational health officers (some trained as veterinarians), safety officers, and medical technicians (aeromedical, environmental, ear, nose, and throat technicians, veterinarians, and general). Initially, audiologists were not primary educators within these programs.

The Army was at the forefront in designating audiologists to provide education and motivation for uniformed and nonuniformed personnel. This action was the result of two studies that revealed 20 to 30 percent of all combat arms (armour, infantry, and artillery) personnel with more than one and one-half years of duty showed significant losses in hearing and, among those with 15 years or more service, 50 percent exhibited significant hearing impairments (Walden, Worthington, & McCurdy, 1971; Walden, Prosek, & Worthington, 1975). The impact of these findings convinced Army leaders to turn to audiologists to reverse this trend and by mid-1970, 58 positions for audiologists were established (Sedge, 1986). Although some audiologists in the Navy and Air Force have been actively providing such training, their roles in this area of activity have yet to be organized into service-wide programs.

Each of the services developed educational/motivational films and other types of training aids for use in their respective programs. In fact, all three services can lay claim to occupational hearing conservation program (OHCP) training and motivation films dating back to the 1950s and 1960s, some of which are recognized as classics.

Areas of supervisor responsibility for enforcing hearing conservation conformance among subordinates are basically the same for all three services. The Uniformed

Code of Military Justice applies to all three uniformed services where disciplinary actions may be employed. For civil service personnel, the three services follow the guidelines of the Department of Labor (OSHA). Although disciplinary actions have been used as a method for achieving compliance, all three services have clearly recognized that this approach is less desirable than achieving compliance through education, motivation, and supervisory influences. A most effective approach to such tasks as conforming to mandatory use of hearing protection is to have the requirement to properly wear hearing protection stated in job/position descriptions as a "condition of employment," especially for nonuniformed employees.

Noise Measurements and Assessments of Potentially Hazardous Exposures

All three services have been involved in accomplishing noise measurements of potentially hazardous types of noise exposures since the late 1940s. The Army and Air Force have even established noise data registries where a multitude of noise measurements are on file and copies of pertinent noise summaries and surveys can be requested by field personnel for evaluating similar noise sources at their facilities. Generally, these data are classified according to specific types of weapon or support systems or equipment and machinery to be found in the appropriate military service. The Air Force noise data repository even provides applications of auditory risk associated with given noise measurements and gives at-the-ear exposures expected from wearing approved types of hearing protection devices for the major types of aircraft, ground support, and maintenance activities associated with a given weapon system (Cole, 1975). Generally, results of noise assessments may be found in a designated medical-environmental health function at each base or facility where potentially hazardous noises are known to exist; in similar medical-environmental health functions at major command headquarters; or

at a designated service-wide repository for such data.

All three services require assessments of individual noise exposures. The results of these measurements may be attached to or included in "shop environmental health folders" where the contents may be inspected to ensure that each person who routinely works in noise has been included in such risk assessments. Typically, the task of taking noise measurements, including dosimetry assessments, are assigned to occupational health specialists, industrial hygienists, or bioenvironmental engineers (*Hazardous noise exposure*, 1982; *Hearing conservation*, 1980; *Hearing conservation and noise abatement*, 1983; *Occupational noise control and hearing conservation*, 1984). Some audiologists, especially in the Army, may also routinely perform such assessments (*Hearing conservation*, 1980).

The Army (*Hearing conservation*, 1980) currently uses a risk criterion for intermittent and/or steady-state noises which differs from that employed by the Air Force and Navy (*Hazardous noise exposure*, 1982; *Hearing conservation and noise abatement*, 1983). Although all three services employ a "trading" relationship of 4 dBA for each halving and doubling of "allowable" durations of exposure, the starting point or boundary for 8 hours/day assessments is 85 dBA for the Army and 84 dBA for the Navy and Air Force. This may not appear to be a significant difference, however, the durations of allowable exposure existing between these two criteria becomes dramatically different for noise levels in the lower range of exposure risks. Table 14–1 contains the durations (in minutes/8 hr) of allowable unprotected exposure associated with these two sets of auditory risk criteria. Comparisons reveal significant differences in durations of assessment between the two for A-weighted noise levels, especially at and below 90 dB. For example, whereas the Army criterion for 90 dBA would allow an exposure duration of 202 minutes/day, the Navy/Air Force criterion would permit an exposure duration of 107 minutes/day for the same sound. This represents a difference of 32 minutes/day

TABLE 14–1.

Auditory Risk Limits for Two Department of Defense Criteria Which Trade Allowable Durations of Exposure for A-Weighted Levels*

| dBA for 8 Hours/Day | Allowable Minutes/8 hours[a] | | Minute(s) Difference |
	84 dBA Air Force/Navy	85 dBA Army	
115[b]	2.2	2.7	.5
114	2.7	3.2	.5
113	3.2	3.8	.6
112	3.8	4.5	.7
111	4.5	5.3	.8
110	5.3	6.3	1.0
109	6.3	7.5	1.2
108	7.5	8.9	1.4
107	8.9	10.6	1.7
106	10.6	12.6	2.0
105	12.6	15.0	2.4
104	15.0	17.8	2.8
103	17.8	21.2	3.4
102	21.2	25.2	4.0
101	25.2	30.0	4.8
100	30.0	35.7	5.7
99	35.7	42.4	6.7
98	42.4	50.5	8.1
97	50.5	60.0	9.5
96	60.0	71.4	11.4
95	71.4	84.9	13.5
94	84.9	101.0	16.1
93	101.0	120.0	19.0
92	120.0	143.0	23.0
91	143.0	170.0	27.0
90	170.0	202.0	32.0
89	202.0	240.0	38.0
88	240.0	285.0	45.0
87	285.0	339.0	54.0
86	339.0	404.0	65.0
85	404.0	480.0	76.0
84	480.0	571.0	91.0
83	571.0	679.0	108.0
82	679.0	807.0	128.0
81	807.0	960.0	153.0
80	960.0	960.0+[c]	—

*Entries are minutes per day
[a]Based on doubling/halving of allowable time using 4 dB "trades."
[b]Maximum intermittent or steady-state allowed for unprotected exposures.
[c]960.0+ indicates greater than 16 hours/day.

between the two. The equivalent allowable durations of exposure differences between the two criteria become even more dramatically evident when levels of noise exposure, such as at and below 85 dBA, are considered. For this level, 85 dBA, the Army criterion would allow an exposure duration of 480 minutes/day and the Navy/Air Force, 404 minutes/day; a difference of 76 minutes/day.

All three military services have established auditory risk criteria that are more stringent than the OSHA-1970 (reaffirmed, 1983) limits (*Federal Register*, 1983). Another point of considerable importance is that all three services, although they functionally apply these two criteria to assess at-the-ear exposures and to prioritize noise-control versus personal hearing protection approaches, adopted an individual exposure guidance which states that personnel will wear hearing protection during any and all durations of exposure once the level exceeds the lower boundary of 85 dBA for the Army and 84 dBA for the Navy and Air Force.

Differences that exist between risk limits of these two criteria assumes importance, especially for lower levels of at-the-ear exposures, because of two operational considerations. First, the number of people encountering lower levels of unprotected exposure, such as below about 90 dBA, increases dramatically. Decisions regarding enlistment in a comprehensive OHCP become paramount. Second, the at-the-ear exposures for those wearing hearing protection moves the degrees of auditory risk to lower levels and alters assessments of auditory risk for protected hearing. Here, the different exposure criterion levels between the Army and the Navy/Air Force come into play. This second consideration assumes greater significance than one might casually recognize because the requirement to wear hearing protection when levels of unprotected exposure equal or exceed either 85 dBA (Army) or 84 dBA (Navy and Air Force) is followed by all three services. Therefore, assessments of at-the-ear risk limits involve applying safe versus unsafe exposure limits

against the attenuated value achieved through use of hearing protection; one which, for high levels of noise, would move the at-the-ear level to the lower ranges of allowable durations. Since it is within this lower range where the most significant differences exist between the services, it follows that a potentially significant variance can be found between the Army and the Navy/Air Force approaches to assigning degrees of auditory risk.

Generally, all three services follow the same risk limit assessments for impulse and impact type noise exposures. The boundary beyond which these types of noises are considered a threat to unprotected hearing is 140 dB (peak-unweighted). All have embraced an approach that gunfire of any type is a threat to unprotected hearing and hearing protection is to be worn during such exposures (*Hazardous noise exposure*, 1982; *Hearing conservation*, 1980; *Hearing conservation and noise abatement*, 1983; *Occupational noise control and hearing conservation*, 1984).

Monitoring Audiometry

All three services have energetically adopted audiometric monitoring as a primary method for evaluating the status of hearing of uniformed and nonuniformed personnel included in their respective hearing conservation programs. Although the Air Force established a Hearing Conservation Data Registry (Repository) in October 1956, where duplicate copies of audiometric monitorings were sent, it was not until October 1974 that automation (computer handling and storage) of the contents of such records was routinely established. Prior to this date, studies of data entries contained on such records were accomplished using manual methods. A few computer data files on selected samples were established and limited studies were performed by the Air Force prior to October 1974. At present, all three services have established similar data registries: the Army in 1980 and the Navy in 1983. Studies completed by the services using such data have served to strengthen the roles of audiometric monitoring and have provided each service with valuable insights concerning the effectiveness of their programs.

All three services depend on the people who conduct audiometric monitoring in the field to utilize the findings on each person who is examined. Under supervision of a medical-health professional, technicians actually conduct the audiometric examinations, compute the presence or absence of significant threshold shift (STS), determine the appropriate initial disposition and follow-up actions that will be followed, and then, depending on the supervisors' directions, follow through to ensure the results obtained are used to actually manage each person. Immediate interpretations of audiometric findings (discovery of STS) can significantly improve the overall cost-effectiveness associated with performing these monitorings (Gasaway, 1985a). The most successful of these monitors appear to be those who have been personally trained by those most knowledgeable in performing the tasks associated with such monitorings. This assurance has been achieved to some degree as military leadership acknowledged compliance with the OSHA act that required that technicians and others who perform audiometric monitorings receive training which, at a minimum, equals or exceeds guidance provided by the Council for Accreditation in Occupational Hearing Conservation (CAOHC) (Hearing conservation, 1978).

All three services initiated training and certification courses following a CAOHC meeting in Waterville, Maine, during the summer of 1973. Thousands of "Hearing Conservationists" (CAOHC certified) have subsequently completed training at formal courses that were developed by each of the services and coordinated and approved by the Council for Accreditation. Although some technicians and professionals receive only the initial basic 22 hours of instruction required for CAOHC certification, others receive 36 or more hours of formal academic

and practical training (Hazardous noise exposure, 1982; Hearing conservation, 1980; Occupational noise control and hearing conservation, 1984).

The people resources that evolved from these training programs have proven the logic and value of achieving hands-on education of persons who actually perform audiometric examinations and follow-up actions according to the findings revealed during monitoring. After receiving a standardized curriculum of training with audiologists and others in the safety-health professions serving as instructors, these "at the bench" people have performed admirably. This type of standardized training also enhances adherence to policies, criteria, and procedures established by each of the services, improving hearing test-retest consistency.

As a result of adopting commonality of approaches to recording audiometric data (standardized forms), using specific defined "purposes of the test," and adhering to standardized follow-up and disposition actions, the role and value of routine audiometric monitorings have significantly contributed to the achievement of goals established for the services programs.

The basic types of audiometric examinations used within the services include the following "purposes of the test" categories:

- Reference or baseline hearing examination
- 90 day follow-up on personnel newly assigned to work in noise
- Annual monitoring examination
- Initial follow-up, 15 and 40 hours out of noise, examinations conducted on those exhibiting threshold shifts on either the 90 day or annual evaluation (Navy includes a 5 day follow-up)
- Detailed follow-up tests (Intervals of 3 and 6 months for Air Force and 2 and 4 months for Navy; following confirmation of the presence of significant shifts)
- Termination hearing test (completed at the end of noise-exposure employment)

Reference audiograms are obtained in all three services and recorded on a standard Department of Defense (DOD) form (DD 2215). An out-of-noise period preceding the accomplishment of the reference is 48 hours for Air Force (*Hazardous noise exposure*, 1982) and 14 hours for Army and Navy (*Hearing conservation*, 1980; *Occupational noise control and hearing conservation*, 1984). Use of hearing protection to achieve this noise-free period is specified by the Navy but not addressed by either the Air Force or Army directives. Pre-employment audiograms are generally not acceptable for reference purposes by the Air Force but are allowed by the Army and Navy (Note: The 1983 Navy OPNAVINST directive did not allow pre-employment audiograms to be used for references, but they are allowed by the 1984 Navy NAVMEDCOMINST directive).

All three services perform 90 day re-evaluations on newly assigned personnel and a mandatory out-of-noise period preceding this examination is not required, except as would occur from normal use of hearing protection. Results of the 90 day examinations are recorded on a standardized DOD form (DD 2216).

Annual audiometric examinations are mandatory and also reported on the DD 2216. Annual examinations do not require a mandatory noise-free period preceding their administration, except as obtained by normal use of hearing protection (*Hazardous noise exposure*, 1982; *Hearing conservation*, 1980; *Occupational noise control and hearing conservation*, 1984).

Initial out-of-noise follow-up examinations, consisting of 15 and 40 hours of noise-free activity, are specified by all three services when significant threshold shifts (STSs) are observed on either the 90 day or annual examinations (*Hazardous noise exposure*, 1982; *Hearing conservation*, 1980; *Occupational noise control and hearing conservation*, 1984). These sequential out-of-noise follow-up examinations are performed to confirm and validate findings of STS and are also recorded on the DD 2216. Both the Army and the Air Force define STS as shifts of 20 dB or greater at 1000, 2000, 3000, or 4000 Hz,

either ear, noted between the current 90 day or annual and the reference (*Hazardous noise exposure*, 1982; *Hearing conservation*, 1980). The Navy uses a dual STS finding; either 15 dB or greater at 1000, 2000, 3000, or 4000 Hz, either ear, or an average of 20 dB at 2000, 3000, or 4000 Hz, either ear (*Occupational noise control and hearing conservation*, 1984). Entries for either or both of these follow-up validation re-examinations are contained on the DD 2216, which is designed for recording either 90 day or annual examinations.

Detailed or stringent follow-up examinations may be performed to ensure that a person who is allowed to return to work in noise does not show further shifts in hearing. Although all three services employ such follow-up examinations, only the Air Force uses a standard form (AF 1671) for recording these monitorings. Intervals for performing detailed or stringent follow-up examinations are at 3 and 6 month intervals for the Air Force (*Hazardous noise exposure*, 1982); 2 and 6 months for the Navy (*Occupational noise control and hearing conservation*, 1984); and the Army leaves the scheduling periods to the discretion of medical staff; usually a physician or audiologist (*Hearing conservation*, 1980).

Termination audiograms (accomplished when individuals retire, separate, or are removed from hazardous noise duties) are performed by all three services. The Air Force and Navy perform termination audiometric examinations only on military and civilian employees who were included in the hearing conservation program (*Hazardous noise exposure*, 1982; *Occupational noise control and hearing conservation*, 1984). The Army requires all military personnel to receive termination examinations and only civilian employees who were included in the OHCP (*Hearing conservation*, 1980). All services record results of termination audiometric examinations on form DD 2216.

Audiologists are generally involved in most of the routine occupational audiometric monitoring programs within the Army and Navy; if not in the day-to-day testing activity, they are at least available to supervise audiometric monitoring activities.

The Air Force Logistics Command is the only command within the Air Force using audiologists to accomplish routine occupational audiometric monitorings. In this command, large numbers of people at six major bases are involved in retrofit, reconstruction, and maintenance of Air Force aircraft and support equipment.

The roles of audiologists in these most critical monitorings can be significant. When audiologists are directly involved in day-to-day monitoring, they ensure adherence to required follow-up and disposition actions. Further, they contribute to personal persuasion and motivation of participants in such examinations.

Although mobile audiometric testing facilities at select locations have been employed by the Army and Navy, the Air Force has located their examination facilities at fixed sites.

The types of audiometers used to monitor the status of hearing of those included in OHCPs encompass the full spectrum of commercially available devices; manual, self-recording, microcomputer-controlled audiometers, and even clinical/diagnostic instruments. The three services recognize the ease with which the current generation of microcomputer-controlled instruments, with their associated software, can enhance routine monitorings and the handling of hearing data. All three services have rigorously explored adoption of existing commercially available computer-controlled units. They have expanded their investigations to include specially designed units that can be customized to meet defined psychoacoustic parameters of auditory testing as well as incorporate data management protocols. The potential for these sophisticated units will probably lead to expanded adoption of microcomputer systems as standard devices for performing audiometric evaluations, especially when the data being obtained during each audiometric negotiation may be fed into an automated data handling computer located at each service-wide hearing data registry. The results of each monitoring performed with microcomputer systems can also be designed to print the test data directly on the entry spaces of standardized DOD or service forms to be filed in individual medical-health folders.

Electroacoustic calibration of audiometers is required by all three services. The Air Force requires electronic calibrations at 6 month intervals for all instruments used in their OHCP (*Hazardous noise exposure*, 1982). The 1983 OPNAVINST specifies that such calibrations will be performed at 12 month intervals for audiometers located in shipboard spaces and at 24 months for those located on land-based facilities (*Hearing conservation and noise abatement*, 1983), whereas the 1984 NAVMEDCOMINST only requires a 12 month interval for all audiometers (*Occupational noise control and hearing conservation*, 1984). The Army requires calibrations at intervals of 12 months for manual and microcomputer-controlled instruments and at 6 months for self-recording type audiometers (*Hearing conservation*, 1980).

Biological checks of the status of audiometers are required by all three services and a standard DOD form (DD 2217) is used to record findings. The Air Force and Navy (*Hazardous noise exposure*, 1982; *Occupational noise control and hearing conservation*, 1984) require biological checks to be performed on each day of testing whereas the Army (*Hearing conservation*, 1980) requires such a check only once a week.

Ambient noise levels in audiometric testing environments are also required to be within acceptable limits by the Army and Navy. Certification or validation of acceptability is specified as "every 2 years" by the Navy (*Occupational noise control and hearing conservation*, 1984); "periodically" by the Army (*Hearing conservation*, 1980); but is not clearly indicated by the Air Force (*Hazardous noise exposure*, 1982). Both the Army and Navy directives contain specific maximum allowable ambient noise levels for use in performing compliance checks.

One unique requirement of audiometric monitorings used in the military involves structuring the tasks to deal with a highly transient workforce. The active duty component of each service represent a large population of transient people who receive periodic and routine occupational monitorings. Obtaining realistic and meaningful longitudinal audiometric examinations of these members demands the most stringent standardization between audiometers, their consistent adherence to calibration and repair standards, and standardized methods for administering, determining, and reporting threshold findings. This logistic problem has been, and will continue to be, a major concern for hearing conservation personnel in all three services. Test-retest reliability between sequential measurements (plus or minus 5 or 10 dB) may be brought to smaller acceptable ranges in some programs, such as locations where civilian employees remain stable for an entire work life, but adopting a plus or minus of 5 dB for the range of acceptable variance may be too restrictive to apply to the general military population.

Clinical Referrals of People Identified with Ear and Hearing Problems

Fortunately, all three services have a rather enviable structure for dealing with audiologic and otologic referrals when consultation work-ups are needed. All three services have otolaryngologists, audiologists, and other clinical professionals at strategic locations who can provide comprehensive consultations. Clinical referral is critical to the successful disposition of those who exhibit ear and/or hearing problems. Failure to adhere to standardized procedures and follow-up actions that have been carefully structured into a service's hearing conservation effort may negate the ultimate effectiveness of referral actions. Therefore, professionals who provide consultations must know and accept standardized follow-up actions and dispositions dictated in the appropriate hearing conservation regulations. The primary goal to be achieved by consultation referrals is to differentiate between noise and other causes of the altered status of hearing; that is, differentiating between noise as the primary culprit

and other singular or contributing (superimposed) clinical causes of an observed change in the status of hearing.

Referral criteria are specified in each of the service OHCP directives. The Air Force employs referrals when STS persists or changes during monitorings (*Hazardous noise exposure*, 1982). The Navy directs referral actions when hearing threshold levels (HTLs) exceed 25 dB at 500, 1000, 2000, or 3000 Hz, either ear, or when a unilateral loss exceeds 20 dB at 500, 1000, or 2000 Hz, or is greater than 40 dB at 3000, 4000, or 6000 Hz, either ear (*Occupational noise control and hearing conservation*, 1984). The Army directs referrals when average HTLs at 500, 1000, or 2000 Hz equal or exceed 25 dB and/or when an abrupt loss is more than 55 dB at 3000, 4000, or 6000 Hz, either ear (*Hearing conservation*, 1980).

Noise Control Measures

Reducing noise exposures to relatively safe levels of unprotected exposure may be facilitated by using noise-control engineering approaches, either during the initial procurement or by altering and changing the noise emitted by existing machinery, equipment, or vehicles. Although all three services can and do have existing military standards and specifications for new equipment, this approach may not be rigorously adopted and enforced (*To provide proper compensation for hearing impairments*, 1978). Many vehicles, such as tanks and armored vehicles, ships, and aircraft, already have limited acceptable noise exposure standards (Van Cott & Kinkade, 1972) which, if obtained during initial procurements, would help reduce the noise associated with their operation. However, many procurements receive waivers for noise since it is believed that strict adherences to lower noise levels exceed technological and/or economic feasibilities.

Personal Hearing Protection

Dependence on personal hearing protection is a main line of defense against overexposure to noise. Although engineering controls are discussed in each service directive as a preferred technique for controlling excessive noise and are considered important and necessary, the services have adopted and energetically enforced the use of hearing protection.

Most hearing protection devices (HPDs) used by the military services are obtained from federal supply channels where large quantities are purchased periodically and made available through normal world-wide supply and distribution networks. Each type of HPD procured under the federal supply system must be approved by one or more of the services and assigned a specific alpha/numeric code; called a federal supply number (FSN). A specific assigned FSN is then common across the services. The total list of FSNs applicable to HPDs, especially insert and semi-insert devices, are periodically (usually annually) reviewed by each service and assigned approved or restricted status for their respective procurement. For example, a specific HPD may be approved by the Army and Navy but restricted for procurement by Air Force requestors. Procurements of insert and semi-insert devices are considered medical items (since they either insert into or come into contact with the ear canals) and are assigned FSN codes (Class 7755) generally restricted to qualified requestors only within designated medical units. Circumaural muffs (Class 4240), headsets, and communication-muffs (Class 8415) are assigned FSNs and considered "line items" and are available to nonmedical units. Inserts, semi-inserts (ear-caps), circumaural muffs, and noise-attenuating communication devices (used alone as headsets or fitted in helmets) are the basic types of devices included as hearing protectors (*Hazardous noise exposure*, 1982; *Hearing conservation*, 1980; *Hearing conservation and noise abatement*, 1983; *Occupational noise control and hearing conservation*, 1984).

Although the medical side of the military services apparently issue HPDs, the variety of circumaural muffs procured may vary considerably. Once a given muff manufacturer has received general approval for a muff or communication device, the devices actually obtained are based on awarding procurement purchases according to competitive bids submitted by manufacturers.

Insert devices include premolded types, such as the single-flange (V-51R) and the 3-flange plugs (supplied by Plasmed, North Health Care, etc.), and a few types of user-molded devices, such as the Silaflex (Flents), wax-impregnated cotton stopples (Flents), and expandable vinyl foam plugs (E-A-R Division of Cabot Corp. or North Health Care "Decidamp"). Since the current approved premolded plugs are sized, each service's regulation states that all users should receive individual sizings and fittings by appropriately trained medical-health personnel. Each service directive also stipulates that people supplied with user-molded devices be individually instructed to ensure adequate usability.

The Army (*Hearing conservation*, 1980) and Air Force (*Hazardous noise exposure*, 1982) directives state that insert devices should be chosen by medical and health professionals and, where feasible, user choice should be considered. The Navy (*Occupational noise control and hearing conservation*, 1984) stipulates that industrial hygienists should guide such selections.

Many people in the military depend on electroacoustic communication devices to achieve a certain degree of hearing protection. These people usually are carefully monitored audiometrically to ensure optimum protection (*Hazardous noise exposure*, 1982; *Hearing conservation*, 1980; *Occupational noise control and hearing conservation*, 1984). When listening to communications in very high noise levels, many must also wear insert devices beneath their communication units.

Hearing protection devices that accommodate alterations in barometric pressure associated with hypobaric and hyperbaric environments are a common requirement within the military. Expandable vinyl foam plugs accommodate such changes and have improved the ease with which alterations in environmental pressure changes can be tolerated (Stork & Gasaway, 1977).

Two basic schemes exist within the services for accounting for values of attenuation provided by HPDs (*Hazardous noise exposure*, 1982; *Hearing conservation*, 1980; *Hearing conservation and noise abatement*, 1983; *Occupational noise control and hearing conservation*, 1983). Although the manufacturer's published Noise Reduction Rating (NRR) has received some degree of acceptance (*Hearing conservation and noise abatement*, 1983; *Occupational noise control and hearing conservation*, 1984; *Recommendation on control of excessive noise*, 1982), the method used by the Air Force consists of taking the dB difference noted between the C- and A-weighted noise measurements of a given type of exposure and the dB difference (C minus A = X) so noted is then used as an *index* under which different values of equivalent at-the-ear A-level attenuation are determined (*Hazardous noise exposure*, 1982). Generally, the greater the dB difference noted between a given dBA and dBC, the smaller the quantity of hearing protection attenuation. Conversely, the smaller the dBC and dBA difference, the greater the amount of attenuation. Use of either approach, the NRR or the C-A index, employs values of mean attenuation at different frequencies obtained in laboratory settings which have been reduced by two sigmas (*Hazardous noise exposure*, 1982). Since the A-weighted at-the-ear level is the primary unit of assessment used to determine degrees of risk for protected hearing by all three services, any procedure that is used to compute the at-the-ear exposures among those wearing protection is translated to this value; that is, "equivalent" at-the-ear dBA.

Each service has auditory research facilities to evaluate hearing protection devices, although the extent of involvement fluctuates depending on interservice agreements, authorizations to perform such studies, and availability of operational fundings. These facilities for the Army are at Ft. Rucker, Alabama (Army Aeromedical Research

Laboratory), for the Navy at Pensacola, Florida (Naval Aerospace Medical Institute), and for the Air Force at Wright-Patterson AFB, Ohio (Aeromedical Research Laboratory). Formal or informal overview of the hearing protection phases for each respective service is allocated to the Bioacoustics Division at Aberdeen Proving Ground, Maryland (Army); the Naval Environmental Health Center at Norfolk, Virginia (Navy); and the Consultants Division of the USAF Occupational, Environmental, and Health Laboratory at Brooks AFB, Texas.

Record Keeping and Records Management

Anyone who has been involved with military programs of hearing conservation recognizes the numerous forms on which audiometric monitorings and other types of appropriate data are entered to document findings. A few OHCP monitoring forms have been substantially standardized between the services. Table 14-2 contains a listing of major forms used to record OHCP data and information. Those identified as DD represent forms that are common for all three services. These forms resulted from tri-service agreements reached between the services during an ad hoc meeting held in Annapolis in 1977. The Air Force forms identified with AF represent standardized forms used only within this branch of service.

Forms used in support of OHCP activities include entries dealing with a wide range of occupational hearing conservation activities and monitorings. Audiometric examinations represent separate forms for recording reference audiograms (all three use DD 2215; see Figure 14-1); annual and 90 day examinations (all three use DD 2216; see Figure 9-1); 15 hours and 40 hours follow-up examinations (also recorded on DD 2216 and used by all three); and detailed or stringent follow-up results (AF Form 1671 which is only standardized for use by the Air Force). All three services use a standardized form for recording results of biological checks of audiometers (DD Form 2217). Results of individual noise surveys where

TABLE 14–2.

Standardized Forms Used by Military Services to Record Occupational Hearing Conservation Data and Information

Type of Information/Data	Air Force	Navy	Army
Audiometric monitoring data:			
Reference audiogram	DD 2215	DD 2215	DD 2215
Hearing conservation data (routine)	DD 2216	DD 2216	DD 2216
Detailed/stringent audio follow-up	AF 1671	none	none
Otologic/audiologic referral	AF 1672	SF 513	SF 513
Biologic audiometer check:			
Biologic threshold checks	DD 2217	DD 2217	DD 2217
Noise measurement/assessment:			
Noise survey (overall levels)	DD 2214	sample	DD 2214
Evaluation of individual noise exposure	AF 1621	none	none
Noise dosimetry assessment	AF 2756	none	none
Engineering noise survey (octave-band)	AF 1622	none	none

NOTE: DD refers to Department of Defense tri-service standardized forms; AF refers to standardized form used only by Air Force; SF refers to standard clinical consultation form not specifically designed for OHCP referrals; and sample refers to a sample of the form contained in the Navy directive but not standardized for Navy use.

only overall levels are recorded are standardized for all three services (DD Form 2214), but noise dosimetry monitorings (AF Form 1621) and detailed engineering or octave-band analyses (AF Form 1622) are only standardized for use by the Air Force. The Air Force also has a standard form (AF Form 1672) for otologic/audiologic referral actions. Records or forms used for recording ambient noise levels in audiometric areas, electro-acoustic calibration data, and other types of OHCP data and information are left to the devices of end-users.

The most important concept recognized for adoption of standardized forms regarding documentation of monitoring audiometry, within and between the services, is the critical need to provide consistent records against which an evaluator can determine changes in hearing occurring over time (Gasaway, 1985a). Sequential records on an individual, especially since noise-induced hearing impairment usually develops longitudinally over time, provides the most efficient means of determining and validating cause-effect relationships between overexposures to noise and the hearing condition.

Hearing Data Registry Interface

All three services now have operating hearing conservation data registries or repositories. Duplicate copies of entries on various HCP forms are received, scrutinized for quality control, and entered into computer storage. Protocols have also been established to return records with omissions or incorrect entries. This not only enhances the value of the data that will ultimately enter computer files but, more importantly, ensures that the person involved in an HCP monitoring is dealt with properly. These data are then used to perform studies of collated entries and even for recalling of individual record files. Effective use of such valuable data has armed each of the services with a viable tool with which to study and evaluate the pertinence and validity of the contents of regulatory documents designed to conserve hearing. Strong and weak elements of almost all aspects of an overall OHCP can be pinpointed and corrective actions can be taken or recommended immediately when deficiencies are identified. The potential for operational management

REFERENCE AUDIOGRAM

(THIS FORM IS SUBJECT TO THE PRIVACY ACT OF 1974 - Use Blanket PAS - DD Form 2005)

ZIP CODE/APO

DOD COMPONENT		SERVICE COMPONENT	
☐	A—ARMY N—NAVY F—AIR FORCE M—MARINE CORPS 1—OTHER DOD ACTIVITY	☐	R—REGULAR V—RESERVE G—NATIONAL GUARD 1—OTHER

PERSONAL DATA

SSN

LAST NAME—FIRST NAME—MIDDLE INITIAL

SEX	DATE OF BIRTH	PAY GRADE, UNIFORMED SERVICES	GRADE, CIVILIAN	SERVICE DUTY OCCUPATION CODE
☐ M—MALE F—FEMALE	year | month | day			

MAILING ADDRESS OF ASSIGNMENT

LOCATION—PLACE OF WORK	MAJOR COMMAND	DUTY PHONE

AUDIOMETRY

☐ 1. REFERENCE ESTABLISHED PRIOR TO INITIAL DUTY IN HAZARDOUS NOISE AREAS
2. REFERENCE ESTABLISHED FOLLOWING EXPOSURE IN NOISE DUTIES
3. REFERENCE RE—ESTABLISHED AFTER FOLLOWUP PROGRAM

HEARING THRESHOLD LEVELS OF TEST FREQUENCIES RE: ANSI S3.6

LEFT EAR						RIGHT EAR					
500	1000	2000	3000	4000	6000	500	1000	2000	3000	4000	6000

DATE OF AUDIOGRAM	DAY OF WEEK	MIL·TIME·DAY	HOURS SINCE LAST NOISE EXPOSURE	ENT PROBLEM AT TIME OF TEST
year | month | day	☐ 1·SUN 4·WED 7·SAT 2·MON 5·THURS 3·TUES 6·FRI			☐ 1—NO 2—YES 3—UNKNOWN

EXAMINER

LAST NAME—FIRST NAME—MIDDLE INITIAL	TRAINING CERT NO. SSN	SERVICE DUTY OCCUPATION CODE	OFFICE SYMBOL

AUDIOMETER

TYPE	MODEL	MANUFACTURER	SERIAL NUMBER	LAST ELECTROACOUSTIC CALIB DATE
☐ 1·MANUAL 2·SELF·RECORDING (automatic) 3·MICROPROCESSOR				year | month | day

PERSONAL HEARING PROTECTION

TYPE USED	EARPLUGS ISSUED	SIZE EARPLUGS	DOUBLE PROTECTION USED	GLASSES WORN (including goggles)	FREQUENCY GLASSES WORN
☐ 1·SINGLE FLANGE (V51R) 5·NOISE MUFFS 2·TRIPLE FLANGE 6·OTHER 3·HAND FORMED EARPLUGS 4·EAR CANAL CAPS	☐ 1·NO 2·YES 3·PREVIOUSLY ISSUED	R L 1·XS 2·S 3·M 4·L 5·XL	☐ 1·NO 2·YES	☐ 1·NO 2·YES	☐ 1·ALWAYS 2·SELDOM 3·N/A

REMARKS

CONTENTS REVIEWED AND VALIDATED BY

NAME OF REVIEWER *(Signature)*	SERVICE DUTY OCCUPATION CODE	AUTOVON	SSN	OFFICE SYMBOL

DD FORM 2215 1 SEP 79   REPLACES AF FORM 1491, MAR 73, WHICH IS OBSOLETE.

Figure 14-1. Form used by Army, Navy, and Air Force to record audiometric data and information designated as reference or baseline.

of a service-wide OHCP is significant and considered as a critical component of each service's overall program. In addition, frequent written and telephone contacts between managers at a registry and those in the field establishes a line of communication and rapport which can solidify and strengthen the overall effort.

Disposition and Follow-up Actions for Managing People Who Exhibit Changes in the Status of Their Hearing

Monitoring the status of hearing to detect changes that may be attributed to overexposures to noise results in identifying people who should receive follow-up actions. Once discovered, follow-up procedures become critically important to ensure no further noise damage is allowed to occur. All three services have specific follow-up policies. These policies are monitored by medical and health people responsible for limiting subsequent encounters with potentially hazardous noises.

All services have adopted and standardized follow-up procedures. Perhaps no group of professionals has acquired a more comprehensive understanding of the alternative courses of actions to follow when managing noise-exposed people than is evident within the massive programs of the services. Although space limitations prevent details, the contents of hearing conservation directives of each service provide useful direction based on insights gained through years of working with the disposition and follow-up of millions of people.

Methods for Evaluating Effectiveness of Effort

Periodically, and by ad hoc studies and evaluations, all three services perform ongoing reviews of their respective programs to identify strengths and weaknesses. The results of program reviews are usually forwarded to the respective medical command. Each Surgeon General's Office is required to heed and respond to the findings, ensuring continued support of existing elements that are apparently working and

to direct or guide actions that can rectify deficient or ineffective elements.

Command involvement is a critical component of any program for continued enhancement of service OHCP. But, the potential value of the evaluations and assessments conducted at each unit should never be overlooked. Findings obtained at functional levels of responsibility foster local involvement and support of the program (Gasaway, 1985b). Data obtained at each unit identifies specific people and areas where apparent deficiencies may be found which consistently capture the attention of commanders and area supervisors (*To provide proper compensation for hearing impairment*, 1978).

Command Supervision of the Overall Program

The regulatory directives of each service specifically specify involvement by commanders and their staff personnel to ensure the success of noise-induced hearing impairment prevention in the military (*Hazardous noise exposure*, 1982; *Hearing conservation*, 1980; *Occupational noise control and hearing conservation*, 1984). A detailed study by the U.S. General Accounting Office (GAO) of the effectiveness of OHCPs at Army, Navy, Marine Corps, and Air Force facilities and compensations for hearing impairment, primarily among civil service workers, pointed out that commanders play a critical role that may not be recognized by them personally (*To provide proper compensation for hearing impairment*, 1978). The GAO report recommended that if the expense of compensation for noise-induced hearing impairment came from monetary funds allocated to the functional operation of their facility, a commander's personal involvement in supporting and enforcing the hearing conservation effort would be more direct and energetic. The GAO reported that some local commanders failed to either accept or acknowledge the importance of their role in furthering and encouraging compliance with the elements of a services hearing conservation effort.

Involvements by local commanders and the major command under which they are assigned will increase when those in positions of power and responsibility can *appreciate* the impact of noncompliance with the dictates of hearing conservation program policies. One way to obtain high level support, outside of the sting of compensation, is for those who evaluate the overall effectiveness of program elements to give specific feedback to commanders at all levels, to promote personal involvement in the successful conduct of their respective programs as a result of pride.

Military programs include a wide spectrum of degrees of command involvement, ranging from total disregard for the program to strong and full support. Unless a commander has been convinced personally that noise-induced hearing loss is a serious problem and he or she accepts the proposition that such occurrences can be almost totally controlled, his or her involvement will, predictably, be less than optimum. Audiologists, otolaryngologists, and other environmental health professionals who have been successful in gaining interest and support of commanders, at base level, command level, or at the level of the Surgeon General's Office, attest to this fact.

Formal Training of Those Who Perform Hearing Conservation Duties

Generally, those who will perform hearing conservation tasks do not enter the service with the comprehensive knowledge and skills required to perform in military settings; they must be taught.

Many audiologists, otolaryngologists, industrial hygienists, bioenvironmental engineers, occupational health nurses, flight surgeons, and other medical professionals who enter active military duty or who become civil service employees possess limited knowledge of the functional conduct of a hearing conservation program. Since these professionals will operate the hearing conservation program, their edification must be assured. The most successful professional personnel have attended special courses and training sessions where the elements of the hearing conservation program were covered in sufficient detail to equip them to carry on the work in a quality fashion. Clearly, 2 to 3 hours of training do not suffice.

Corpsmen and technicians involved with the conduct of elements of the program do receive some degree of formal training, but the extent varies considerably within the same service and between the services. Some of the most motivated players in providing hearing conservation duties are nonprofessional people. They have obviously become concerned and interested during contacts with instructors who provided their course(s) of OHC training. As with all forms of hearing conservation duties, the person who receives reasonably adequate technical training performs in a superior manner when those skills are combined with personal dedication and concern for the people they are serving. When technical knowledge and applied skills have been adequately established in the mind of an individual, the ingredient that can ensure the most energetic and dedicated performance of established technical skills is self-motivation.

THE FUTURE OF MILITARY HEARING CONSERVATION

There is no doubt that the extent of hearing conservation programs within the military services will continue to involve more and more people within the total population of uniformed and civil service workers. As more and more significance is placed on the problem of noise-induced hearing impairments among people, the need to include almost all people who enter active military or civilian duty will increase.

Persons who are intimately involved with hearing conservation programs in the military and who are now working in areas of occupational health within the private sector state that the experience received while in the employ of the military armed them with far greater skills and knowledge than they possessed when they entered the service.

Fulfilling the following *ifs* can enhance the further effectiveness of hearing conservation programs in the services:

- *If* hearing conservation personnel embrace the ideology and philosophy of preventive medicine
- *If* academic institutions that provide training in audiology and otolaryngology expand comprehensive courses in occupational hearing conservation
- *If* audiologists and other health professionals in military hearing conservation programs receive comprehensive initial and follow-up training
- *If* those providing such training are drawn from the most knowledgeable and dedicated personnel available
- *If* microcomputer-controlled audiometers are standardized and consistently used for all routine audiometric monitorings
- *If* use of standardized forms and management procedures is rigorously enforced
- *If* education and motivation sessions for employees are intensified
- *If* commanders who support and enforce compliance with hearing conservation activities receive awards for this accomplishment
- *If* commanders are held responsible for the effectiveness of their programs
- *If* the voices of those dedicated to hearing conservation are heard by commanders, by supervisors, and by hospital and environmental health service commanders
- *If* personal hearing protectors, of all types, are issued or recommended only after each user has received vigorous indoctrinations and training
- *If* employees receive frequent and vigorous scrutiny of their acceptance and proper use of hearing protectors (work area inspections)

- *If* disposition and follow-up actions are established and followed for all personnel
- *If* otolaryngologists and audiologists better understand their unique role in providing consultations

REFERENCES

Ades, H. W. et al. (1953). *Exploratory study of biological effects of noise* (BENOX Report). Chicago, IL: University of Chicago.

An annotated bibliography on noise, its measurement, effects and control. (1955). Pittsburgh, PA: Industrial Hygiene Foundation of America, Inc., Mellon Institute.

Anderman, B. M. (1960). The Veterans Administration audiology program. In H. Davis & S. R. Silverman (Eds.), *Hearing and deafness*, (rev. ed., pp. 477–488). New York: Holt, Rinehart and Winston, Inc.

Cole, J. N. (1975) *USAF bioenvironmental noise data handbook: Organization, content and applications.* (AMRL-TR-75–50, Vol. 1). Wright-Patterson AFB, OH: USAF Aeromedical Research Laboratory.

Gasaway, D. C. (1985a). Using documentation to enhance monitoring efforts. In D. C. Gasaway (Ed.), *Hearing conservation: A practical manual and guide*, (pp. 209-237). Englewood Cliffs, NJ: Prentice-Hall, Inc.

Gasaway, D. C. (1985b). How to successfully educate, indoctrinate, and motivate workers. In D. C. Gasaway (Ed.), *Hearing conservation: A practical manual and guide*, (pp. 85-102). Englewood Cliffs, NJ: Prentice-Hall, Inc.

Gasaway, D. C. (1986). Noise levels in cockpits of aircraft during normal cruise and considerations of auditory risk. *Aviation, Space and Environmental Medicine*, 57(2), 103–112.

Hazardous noise exposure. (1982, April 9), (AFR 161-35). Washington, DC: Department of the Air Force.

Hearing conservation. (1978, June 8). (DOD Instruction Number 6055.3). Washington, DC: U.S. Department of Defense.

Hearing conservation. (1980, March). (TB MED 501). Washington, DC: Department of the Army.

Hearing conservation and noise abatement; (1983, August). (Chapter 18, OPNAVINS 5100.23B). Washington, DC: Department of the Navy.

McLaughlin, R. M. (1978). Audiologists in the United States. In D. L. Lipscomb (Ed.). *Noise and Audiology,* (pp. 361-408). Baltimore, MD: University Park Press.

Occupational noise control and hearing conservation. (1984, April 26). (NAVMEDCOMINST 6260.5). Washington, DC: Department of the Navy.

Federal Register, (1983, March 8). *Occupational noise exposure; hearing conservation amendment; final rule. 48*(46), 9737–9785.

Precautionary measures against noise hazards. (1948, October 21). (AFR 160-3). Washington, DC: Department of the Air Force.

Recommendations on control of excessive noise. (1982, June 2). (Navigation and Vessel Inspection Circular Nu. 12-82). Washington, DC: United States Coast Guard, U.S. Department of Transportation.

Ruedi, L., & Furrer, W. (1948). Traumatic deafness. In E. P. Fowler, Jr. (Ed.) *Medicine of the ear,* (pp. 377–420). New York, NY: Thomas Nelson and Sons.

Sedge, R. K. (1987). Administration of a military-based program of speech-language pathology and audiology. In H. J. Oyer (Ed.), *Administration of Programs in Speech-Language Pathology and Audiology* (pp. 129–156).

Englewood Cliffs, NJ: Prentice-Hall, Inc.

Stevens, S. S., Loring, J. G. C., & Cohen, D., (Eds.). (1955). *Bibliography on hearing.* Cambridge, MA: Harvard University Park Press.

Stork, R. L., & Gasaway, D. C. (1977). *Evaluation of V-51R and "E-A-R" earplugs for use in flight.* (SAM Technical Report No. 77-1). Brooks Air Force Base, TX: USAF School of Aerospace Medicine.

To provide proper compensation for hearing impairments; The Labor Department should change its criteria. (1978). (U. S. GAO Report HRD-78-67). Washington, DC: U. S. General Accounting Office.

Van Cott, H. P., & Kinkade, R. G. (1972). *Human engineering guide to equipment design* (rev. ed.). Washington, DC: American Institutes for Research.

Walden, B. E., Worthington, D. W., & McCurdy, H. W. (1971, 21 Dec). *The extent of hearing loss in the army: A survey report.* Washington, DC: US Army Medical Research and Development Command.

Walden, B. E., Prosek, R. A., & Worthington, D. W. (1975, 31 Aug). *The prevalence of hearing loss within selected US Army branches.* Washington, DC: US Army Medical Research and Development Command.

Community Noise Assessment

David M. Lipscomb ■

Many of the capabilities that allow one to perform competently in hearing conservation activities can be applied to community noise problems as well. Therefore, one who becomes known as a hearing conservationist may be sought out by citizens as officials of communities to assist in dealing with environmental noise problems that arise. This chapter is intended to offer some information on factors that need attention in community noise assessment and control. It is beyond the scope of this text to offer exhaustive coverage of the topic, thus, suggested readings at the end of the chapter provide expanded information for those interested in deeper investigation of the subject.

TYPES OF COMMUNITY NOISE

Transportation Noise

Most forms of transportation produce noise that can impact a community. Usually, single events pose no problem. However,

traffic units in quantity at high speed, an army of poorly maintained trucks, or numerous aircraft flyovers during the span of a day accumulate to produce an unacceptable sound environment. Two concerns face one who engages in community noise assessment: How does one quantify noise? What are the appropriate technical components of an enforceable noise code or regulation for community applications?

Methods of Quantification

This problem is made complex because, in most applications, it is not possible to take single-number sound measures and know precisely all of the applicable acoustical parameters. Variation in sound output by different forms of transportation which fall into the same category, complex interaction of numbers of units at different times of the day, varieties of spectra of sound sources, differences in speed of operation, and age and level of maintenance of noise sources all contribute their unique characteristics to the total acoustic climate. Therefore, various approaches have been developed to deal with each type of sound source. The two most common categories of community noise source, aircract noise and traffic noise, will be elaborated.

AIRCRAFT NOISE. Seriously impacting as many as 25 million citizens of the United States (Sperry, 1978, p. 240), aircraft noise is a candidate for regulation and ongoing noise monitoring. This statistic is all the more significant when one realizes that as the L_{dn} levels increase, even more people are "annoyed" (Sperry, 1978, p. 241).

Sound emanating from aircraft is generally quantified by using the Effective Perceived Noise Level (EPNL) sound descriptor. The EPNL incorporates a number of variables, including sound level of the aircraft; spectrum of the sound source, including tonal quality of the sound; time of day; number of aircraft takeoffs and landings during a day; and scales based on the perceived noise (in dB) or "noy" scale developed by Kryter (1970). The noy scale was modeled after the sone scale, but listeners were asked to judge when they felt a sound was more "noisy" by a certain

amount (e.g., twice as noisy). This is essentially a subjective scale; therefore, interpretation of EPNL values assumes individual variability. An example of the use of EPNL measures is shown in Figure 15-1. The message in this chart is that some aircraft are more noisy than others. It is well to take note that the thin lines representing each aircraft should be expanded to incorporate a range of EPNL values due to individual variations in human response to aircraft noise. Thus, when EPNL measures of aircraft differ by only a small amount, it is not possible to say that one noise source would be universally judged to be more quiet (or noisy) than a comparison aircraft.

The EPNL is an unfamiliar descriptor to most persons active in the hearing conservation arena. To be told that an aircraft is rated for 103 EPNL may mean nothing. As a general rule, EPNL approximates dBA + 13. Therefore, if EPNL = 103, the sound experienced is comparable to an A-weighted sound level of 90 dB. More details on the generation of scales used in EPNL and calculation of EPNL can be found in Kryter (1970) and in Goldstein (1978).

Noise "contouring" is a common method of establishing land use in the

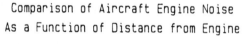

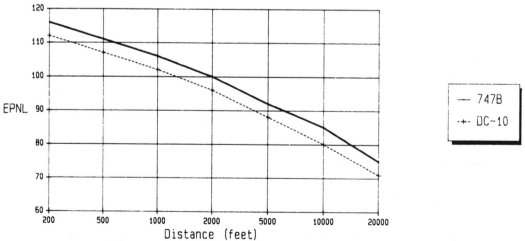

Figure 15-1. Use of EPNL to compare the takeoff noise profiles of aircraft.

vicinity of an airport. The contours are primarily the product of aircract takeoff and flyover patterns as shown in Figure 15–2. Location of the runways determines where the "spikes" of noise contours will be located. This type of mapping assists specialists in land-use planning to determine which areas around an airport should not be inhabited due to noise impact and which areas can be put to some use, usually as manufacturing zones or for other commercial applications.

A predictive contouring method, Noise Exposure Forecast (NEF) (Galloway & Bishop, 1970; Williams, 1978) has been in use since the late 1960s. It utilizes types of information similar to those used to obtain the EPNL, but the NEF is based on *predicted* conditions for some future date. The NEF has been used extensively in determining placement of new airports, regulating the development of land adjacent to airports, and in planned expansions of existing facilities. This descriptor is in the province of the planner, but anyone who may be asked to assess airport noise should not begin that task without first understanding salient features of methods used to evaluate and predict noise impact by airport facilities.

TRAFFIC NOISE. Traffic noise is estimated to place as many as 60 million Americans in sound environments exceeding 60 L_{dn} (Ratering, 1978, p. 249). Therefore, the hearing conservationist might be asked to conduct a traffic noise survey that could range in scope from measures at a single intersection to a community-wide traffic noise assessment. Knowledge of some of the basic descriptors that can be applied to this work along with appropriate application of currently available microchip-based equipment will satisfy most applications. Although not commonly land-use planning experts, hearing conservationists with adequate preparation can make significant contributions in this effort.

Vehicle noise is usually assessed in the short run by using the L_{eq} or L_{dn} descriptors introduced in Chapter 2. Long-term and predictive efforts incorporated the exceedance levels descriptors also discussed

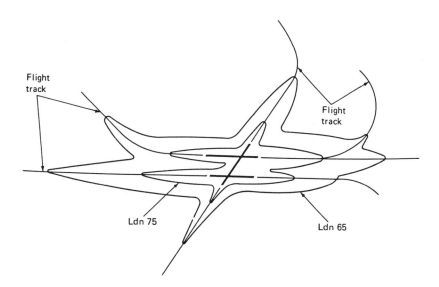

Flight track

Flight track

Ldn 75

Ldn 65

Figure 15–2. Noise isocontours in the vicinity of an airport. Note that the "spikes" in the contours are determined by the location of runways. Reprinted from Sperry, W. C. (1978) Aircraft and airport noise. In D. M. Lipscomb and A. C. Taylor, Jr. (Eds.) *Noise control: Handbook of principles and practices* with permission from Van Nostrand Reinhold.

earlier. In land-use planning for proposed highway constrcution, it is not uncommon for state highway departments to estimate the L_{10} isocontour in areas bordering a proposed highway. For example, any property in Tennessee that is projected to exceed a 70 dB L_{10}, is subject to condemnation proceedings as highway construction progresses.

Time-weighted averages such as L_{eq} or L_{dn} and exceedance measures such as L_{10} are not time-locked. Therefore, it is possible to utilize these descriptors over any period of time that is feasible for consideration. Future impact of traffic noise can be predicted by inserting the estimated levels and mix of units (proportion of passenger cars, trucks, motorcycles, etc.) into the equations for each descriptor. This is facilitated by the observation that most traffic noise over time falls into a "normal distribution," which allows one to convert exceedance levels into equivalent levels or the reverse (Taylor & Lipscomb, 1978, p. 80).

Other descriptors for quantifying community noise levels have been developed. However, it appears that the universal appeal of equivalent and exceedance levels have caused these two types of descriptors to become the predominant ones. For example, there is no reason that one could not apply the EPNL concept to vehicular noise. In fact, descriptors based on the noy scale have been generated for evaluating traffic noise problems. But, these methods never came into common usage.

Generation and Enforcement of Codes

Many enforcement codes utilize discrete measures such as A-weighted sound levels to monitor noise output of individual vehicles and to objectively determine excessive noise output. To accomplish this, four elements are necessary: (1) a well-written noise code, (2) requisite equipment to use in enforcement, (3) specific techniques for measuring the noise source(s) stated in the code (e.g., distance from the source), and (4) adequately trained personnel to operate the equipment and to enforce the code. In

1975, the U.S. Environmental Protection Agency (EPA) Office of Noise Abatement and Control published a model community noise control ordinance (U.S. EPA, 1975). The structure of the ordinance is provided, but the dB levels are to be inserted by each community according to the needs of that locale. A case history of the application of that model noise ordinance to a city has been prepared to describe experiences in Colorado Springs, Colorado (U.S. EPA, 1979). The model noise ordinance and case history can be of benefit to those interested in updating or initiating local noise regulations.

The code will normally specify maximum permissible sound levels. Further, it is useful to break down these permissible levels by type of sound source. Otherwise, a single-number code could constitute a "license to steal" for operators of vehicles that are usually less noisy, for example. Thus, stationary sound sources should meet sould level restrictions appropriate to them, trucks should be made to conform to specific allowable levels, motorcyles should have special consideration of their unique sound production, and passenger cars should meet more stringent permissible levels. Speed of the vehicle is a further consideration. "Under 35 mph" and "over 35 mph" are commonly used speeds in setting allowable noise output of various vehicle types. When a city code is introduced, it is often "staged." For the initial period the code is in effect, somewhat more lenient permissible levels are enacted into the law. After that period has elapsed, the code automatically becomes more stringent. In some cases, more than two "stages" have been utilized.

Setting forth the levels permissible is valueless unless specific procedures for measuring noise emitted by the vehicles are also given. It is unfortunate that some community noise codes reflect considerable thought regarding the vehicle noise levels to be permitted and the time periods for the "staging" but no instructions are given on how to enforce the code. One method that has found favor is to establish that sound level of passing vehicles should be monitored with the sound measuring equipment located 50 feet to the side of the center of the lane of travel. Such specifics for measurement of regulated sound sources *must* be given, or the code will be unenforceable.

Enforcement of noise level codes in a manner comparable to setting up radar speed traps has proven to be effective in some communities. This style of enforcement is time-consuming, and therefore, costly. It often places a strain on the staff of an enforcement agency assigned the task of monitoring noise output of individual vehicles. Noise control is a relatively low priority item for most enforcement agencies. Therefore, if adequate staff is not available, noise control enforcement is left wanting.

LAY ENFORCEMENT EFFORTS. Effecting adequate enforcement by assigned agencies is often not possible, thus, alternative approaches to decreasing community noise have been developed. One method that has been applied in a few communities is to encourage one or more civic organizations to form a group called the "racket squad" (Lipscomb, 1974). Volunteers from the organization work to encourage quieter noise output of commercial vehicles. With an inexpensive sound level meter, the racket squad sets up in various locations of the community and takes notes of those commercial vehicles that "sound off" in excess of appropriate sound levels. On its letterhead, the civic club sends a reminder to the owner of the commercial firm operating the vehicle to point out that the company is sending "X" dB of bad publicity through the streets of the city. Then, an appeal for efforts to quiet the vehicles is made. This approach is particularly effective when the news media provides coverage. The hearing conservationist is well suited to assist in this endeavor because all of the sound measurement procedures utilized are common to industrial hearing conservation work.

INDUSTRIAL ZONE NOISE. Many communities have industrial districts that border directly on residential neighborhoods. With encroachment of residences or because of expansion of the industrial part, noise becomes a source of friction between the two elements. Since about 1965, there has been widespread coverage of noise as a problem. One theme of numerous articles is that unnecessary noise should not be tolerated. As a result, many residents who at one time found industrial noise output to be acceptable, or at least tolerable, have now changed their opinions and have complained against the noise. In some cases, legal action has been threatened or initiated. Often, this problem can be avoided if a public relations effort is undertaken by the offending industries, or if plans for extended working hours, new construction, or expansion include consideration of the impact such activities have on adjacent residences.

A small community had been successful in attracting several sizable manufacturing facilities to its industrial park. At about the same time, a suburban housing development was constructed bordering on that industrial district. There was a buffer of woods between the homes nearest the industrial district and existing industrial plants. Then, a new manufacturing facility was proposed for location close to the homes. At a meeting between the company proposing the building and residents opposing the construction, promises were made that the new facility would make no impact upon the residential neighborhood. As construction progressed, it became obvious that the promises would not be kept. The woods were cut down and the main building was located at the closest point to the residences legally possible. Two or three small outbuildings housing compressors and other noisy machinery were placed between the plant and the homes. Exhaust fans were located on the side of the roof facing the homes, so fan noise and sounds originating within the plant were efficiently broadcast toward the neighborhood. The result of this "planning" was that the residents could no longer enjoy use of their patios and yards during hours of plant operation. To add insult to injury, working hours were expanded to include night-time hours. Thus, neighbor-hood residents could no longer sleep with open windows because the sound originating at the plant entered the homes at disturbingly high levels. Attempts to seek cooperation from the company failed. Thus, a law suit was initiated.

The community did not have an environmental noise code, so the boundary noise limits imposed by some cities did not have the force of law. However, boundary noise measures (as described in Chapter 4) were taken. It was found that those realistic criterion levels were exceeded dramatically. During the time those measures were being taken, the plant manager noted the activity and shut down the plant to prohibit further noise measures and to frustrate the sound assessment effort—or, so he thought. By quieting the plant, the opportunity was provided to take measures of the ambient sound environment comparable to that which was present before the facility was constructed. Thus, a comparison of sound environments ''before and after'' was possible.

At a later time, sound measures were taken inside the homes in the vicinity of the plant during the hours of evening operation. Of particular interest were the ''open window'' and ''closed window'' measures in bedrooms of the residences. In all cases, the sound levels measured with open windows exceeded the appropriate Noise Criterion Curve for a sleeping environment (see Chapter 5 for information on NC Curves).

All of these measures and their interpretations were presented during the trial. The judge rendered a decision favorable to the residents. In doing so, he gave strict instructions to the manufacturing concern to stop the noise problem within 90 days or be closed down. Although this was considered to be a victory by the residents, the sad truth is that much agony, time and cost could have been avoided if the community had an enforceable noise regulation that could have been put to use. Or, if the designer of the manufacturing plant had given consideration to the noise emissions

they were designing into the facility, the problems could have been avoided. Retrofitting, redesign, and reconstruction are universally more costly than controlling the problem by the initial design.

LAND-USE PLANNING

Noise is only one of several factors to be considered by planners. Unfortunately, as the foregoing example shows, potential for excess noise is often overlooked until complaints are lodged. Stung repeatedly by complaints after the facts, highway developers have become cognizant of potential noise problems. An example of the value of pre-planning, the following story relates how an extensive program for highway development was permitted in the vicinity of a home that was listed in the Registry of National Heritage Homes.

A plan was initiated to locate an interstate loop within 250 feet of one side of this lovely old country home. Then, to the other side, a new four-lane ''spur'' road was contemplated. The small street running in front of the home was to be widened to a four-lane thoroughfare. Hence, the plans meant the house would be surrounded by 14 lanes of road, 12 of which would be new. Learning of this, the owners of the home moved to have all construction blocked. Apparently, such a threat from owners of a National Heritage homesite carries considerable weight. The planners had to deal with each of the anticipated problems enumerated by the owners. Chief among these were visible traffic movement and disturbing noise. A landscape architect was contracted to preserve some of the stately old trees on the property and to develop other landscaping modifications that would obliterate any view from the house of the new highways. An audiologist was engaged to assist in the planning of barriers between the home and the interstate loop. An earth berm of appropriate height was constructed and landscaped so as to baffle sound originating from the highway. The crest of the berm was continued in each direction by a concrete wall to further buffer the home from highway noise. Sound level measures were taken prior to the beginning of highway construction to provide a baseline against which post-construction measures could be taken, should the owners complain. The intial plans worked satisfactorily so that later sound measures were not called for.

By concentrating noise control efforts in design stages, it was possible to avert a costly renovation of existing highways or having road construction delayed by lengthy legal battles. Unfortunately, this happy scenario occurs far too seldom. In this case, it was necessary for the audiologist to be capable of taking sound measures and utilizing some predictive data to determine how high and wide the earth berm and wall had to be in order to serve as an effective sound barrier. All of this information is readily available and can be put to use to develop an effective noise control strategy by advanced planning.

CONCLUSION

This chapter is brief because the topic is one somewhat outside the normal daily pursuits of hearing conservationists. However, as personal experience suggests, it is not outside the realm of possibility that conservationists will be pressed into service for community noise assessment, abatement, and/or planning. It is not unusual that the hearing conservationist is the only person available who has requisite equipment and background sufficient to undertake some forms of community noise work. The caution should be stated that one should never engage in this activity if he or she does not possess adequate competency to accomplish the tasks.

REFERENCES

Galloway, W. J., & Bishop, D. E. (1970, August). *Noise exposure forecasts: evolution, evaluation, extensions and land use interpretations.* (Federal Aviation Authority publication: FAA-NO-70-9). Washington, DC.

Goldstein, J. (1978). Fundamental concepts in sound measurement. In D. M. Lipscomb (Ed.). *Noise and audiology* (pp. 3–58). Baltimore: University Park Press.

Kryter, K. D. (1970). *The effects of noise on man.* New York: Academic Press.

Lipscomb, D. M. (1974). *Noise, the unwanted sounds.* Chicago: Nelson-Hall Company.

Ratering, E. G. (1978). Highway and rail traffic noise. In D. M. Lipscomb & A. C. Taylor, Jr. (Eds.), *Noise Control: Handbook of principles and practices* (pp. 248–278). New York: Van Nostrand Reinhold Company.

Sperry, W. C. (1978). Aircraft and airport noise. In D. M. Lipscomb & A. C. Taylor, Jr. (Eds.), *Noise control: Handbook of principles and practices* (pp. 206–247). New York: Van Nostrand Reinhold Company.

Taylor, A. C., Jr., & Lipscomb, D. M. (1978). The use and measurement of equivalent sound level. In D. M. Lipscomb & A. C. Taylor, Jr. (Eds.), *Noise control: Handbook of principles and practices* (pp. 61–82). New York: Van Nostrand Reinhold Company.

U.S. Environmental Protection Agency. (1975). *Model community noise control ordinance.* (EPA document 550/9-76-003). Washington, DC.

U.S. Environmental Protection Agency. (1979). *Colorado Springs, Colorado: Case history of a municipal noise control program.* Washington, DC.

Williams, K. (1978). An introduction to the assessment and measurement of sound. In D. M. Lipscomb & A. C. Taylor, Jr. (Eds.), *Noise control: Handbook of principles and practices* (pp. 33–60). New York: Van Nostrand Reinhold Company.

SUGGESTED READINGS

Beranek, L. L. (Ed.) (1960). *Noise Reduction.* New York: McGraw-Hill.

Beranek, L. L. (Ed.) (1971). *Noise and vibration control.* New York: McGraw-Hill.

Burns, W. (1973). *Noise and man* (2nd ed.). Philadelphia: J. B. Lippincott Co.

Cunniff, P. C. (1977). *Environmental noise polution.* New York: John Wiley & Sons.

Chalupnik, J. D. (Ed.). (1970). *Transportation noises.* Seattle: University of Washington Press.

Harris, C. M. (Ed.). (1979). *Handbook of noise control* (2nd ed.). New York: McGraw-Hill.

Harrison, R. (1978). Control of noise from

recreational activities. In D. M. Lipscomb & A. C. Taylor, Jr. (Eds.), *Noise control: Handbook of principles and practices* (pp. 341–360). New York: Van Nostrand Reinhold Company.

Jensen, P. (1978). Control of noise in the home. In D. M. Lipscomb & A. C. Taylor, Jr. (Eds.), *Noise control: Handbook of principles and practices* (pp. 325–340). New York: Van Nostrand Reinhold Company.

Peterson, A. P. G. (1980). *Handbook of noise measurement* (9th ed.) Concord, MA: Genrad.

U.S. Environmental Protection Agency (1971). *Community noise.* (NTID300.3). Washington, DC.

The Business of Hearing Conservation

SECTION

4

Holding sufficient knowledge of hearing conservation techniques and maintaining the skills to put that knowledge to work is desirable. However, this application within our profession has certain business implications that are not always known to persons preparing for a career in hearing conservation. Seasoned professionals have agreed to share their knowledge with students so that business applications can be part of the composite training effort for which this textbook is being developed.

Managing a Hearing Conservation Service

Wayne G. Bodenhemier

The audiologist who chooses to begin or manage a hearing conservation service will likely find himself or herself facing challenges and decisions for which a traditional education in the field of audiology has probably not prepared him or her. Regardless of whether the service is nationwide in scope or more regional or local in nature, the manager must be well informed and confident in areas that reach far beyond the basic audiological considerations associated with hearing conservation. These areas include some knowledge of basic business principles and philosophy, staffing, marketing and sales, policy development and coordination of activities, and liability/risk assessment. In this chapter, I will offer some observations and thoughts and, in the process, likely raise additional questions in each of these areas based on my experience in managing a nationwide industrial hearing conservation service since 1981.

There are several "reasons" why one might want to get into the business of hearing conservation. First, from the strictly humanitarian viewpoint, it is generally recognized that of the more than 20 million Americans who are exposed to hazardous noise on and off the job (*Hearing conservation: A guide to preventing hearing loss*, 1983), more than 16 million have some degree of sensorineural hearing loss caused by *occupational* exposure (Glorig, 1980). In fact, "noise in the workplace is undoubtedly the most serious contributor to the hearing impairment of the American worker because it is the noise source to which he is most continually exposed" (Miller, 1978). Second, federal and many state regulations governing noise exposure *require* employers to provide hearing conservation programs for employees at risk of occupationally induced noise related hearing loss (Bodenhemier, 1984). Third, employers can protect themselves from the continuously escalating costs of state worker's compensation claims by implementing effective hearing conservation programs. Fourth, the business of hearing conservation *can* be profitable. Regardless of the reason(s) for getting into the hearing conservation business, it must be remembered that, in fact, it *is* a business and therefore, the same principles and considerations will apply to it as to any other type of business.

THE BUSINESS OF
HEARING CONSERVATION

Most businesses in the United States today, whether they are large or small, belong to the private enterprise system in which success or failure depends on how well the offerings of competitors are matched and countered. Competition is the continuing struggle among businesses for acceptance by the consumer, and sales and profits are generally the yardsticks by which acceptance is measured. The private enterprise system demands that companies continually adjust their service standards, product offerings, operating procedures and strategies, etc. Otherwise, the competition will gain a larger share of an industry's sales and profits. Competition is a critical mechanism for assuring that the private enterprise system will continue to offer the products and services that provide a high standard of living. Few organizations can escape the influence of competition, and a hearing conservation service is no exception. The manager or owner of such a service must decide how to deal with the competition within a framework of ground rules that govern competitive activity. This framework includes the fact that the U.S. government has passed laws to prohibit practices designed to eventually eliminate the competition. It has also established rules that prohibit price discrimination, fraudulent dealings in financial markets, and deceptive practices in advertising and packaging.

In addition to earning profits, successful businesses must also seek to meet social responsibilities. This concept refers to management's consideration of the social as well as economic effects of its decisions. Businesses must be responsible in their dealings with employees, consumers, suppliers, competitors, government, and the general public if they expect to be successful in the long run.

The importance of effective management to organizational success cannot be overestimated. One of the leading causes of small business failure is quite frequently identified as "poor management." Management can be defined as the achievement of organizational objectives through people and other resources (Duncan, 1983). It is the manager's job to combine human and technical resources in the best possible way to achieve these objectives. In a larger hearing conservation service, the manager would probably not typically be involved in "production," that is, the provision of services to the customer. More often than not, the manager in this instance directs the efforts of others toward the company goals. In the smaller, more typical hearing conservation service, however, the program manager may be just as involved with direct service provision as he or she is with making critical decisions concerning the goals and objectives of the business. Management is the critical ingredient in the basic resources (management, man/woman power, materials, money, and machinery) of any firm, and the principles and concepts of effective management apply to profit-oriented businesses as well as to non-profit, service-oriented organizations such as hospitals, city governments, and charitable agencies.

Every manager, regardless of his or her level in the organization, must possess three basic managerial skills: technical skills, human relation skills, and conceptual skills. Although the importance and relevance of each skill varies at different levels, managers of hearing conservation services will find that they need to use all three types. Table 16–1 presents a summary of these three important managerial skills as they relate to a hearing conservation service.

Technical skills refer to the manager's ability to understand and use techniques, knowledge, and tools of a specific discipline or department. To be effective, the manager of a hearing conservation service *must* be

TABLE 16-1.

Managerial Skills in Hearing Conservation

Technical Skills

Thorough knowledge and understanding of federal and state regulations including state worker's compensation statutes

Familiarity with noise monitoring techniques and instrumentation

Understanding of audiometric testing procedures, audiometers, and calibration

Knowledge of personal hearing protection devices

Grasp of training requirements, educational programs

Knowledge of record keeping requirements

Human Relations Skills

"Personality" skills that allow the manager to extract the maximum effort from all employees in order to achieve short-term goals and long-term objectives

Leadership skills

Personnel policy development skills

Ability to gain respect of staff and create environment for productivity

Conceptual Skills

Ability to understand the contributions of all facets of the hearing conservation service (sales, scheduling, marketing, data processing, etc.) so that decisions concerning future direction can be made (especially important for top management)

Ability to project future activities, staff needs, etc.

Ability to conceptualize the value and impact of a hearing conservation program on workers and industry and project that conceptualization to all associated parties

---

completely familiar with all of the aspects of hearing conservation. As a minimum, these include: a thorough knowledge of the federal and state regulation of occupational noise exposure, including relevant state worker's compensation laws; noise monitoring techniques and associated instrumentation and calibration considerations; audiometric testing procedures, audiometers, and calibration requirements; personal hearing protection devices and methods of determining effectiveness; training requirements for employees included in a hearing conservation program; and record keeping regulations. The demand for such

a wide range of technical skill on the part of a hearing conservation manager seems to indicate that, as a minimum, a master's degree in audiology with more than superficial background and experience in industrial hearing conservation is required.

The second type of skills, *human relations skills*, are "people" oriented. They involve the manager's ability to work effectively with and through people. In a hearing conservation service, these skills are particularly important because the manager will have to deal directly with people at all levels of involvement; from the employee whose hearing is "being conserved," to the audiometric technician administering baseline and annual hearing tests, to the president of the company that is paying for the hearing conservation program. These human relations skills involve communicating, leading, and motivating employees to accomplish assigned activities. The ability to interact with superiors and others outside the immediate work environment is crucial to the growth and success of any hearing conservation service.

Conceptual skills refer to the ability of the manager to see the organization (a hearing conservation service) as a unified whole while also understanding how each part of the overall organization relates to other parts. By acquiring, analyzing, and interpreting information using these skills, a manager is able to "see the big picture." The future direction of the organization is dependent on the manager's ability to develop imaginative and analytical plans. This skill is greatly affected by one's ability to understand the contributions of *all* aspects of the organization such as finance, marketing, human resources, public relations, and data processing.

STAFFING

Staffing a hearing conservation service will depend largely on the size, scope, and financial resources of the business. Many beginning hearing conservation services are one-person operations, with the owner actually performing all of the necessary functions. Many services are part of larger organizations such as hospitals, private clinics, or universities, and can take advantage of the financial resources and other ancillary resources that are available, such as public relations, marketing and advertising, and data processing.

This discussion of staffing will focus on what I view as an ideal hearing conservation service that is conducting business with relatively large numbers of customers over a rather wide geographic area. Obviously, as the size of the program is reduced relative to the number of clients served and the physical area, fewer staff positions are necessary.

Consultant Positions

It is likely that three positions can actually be maintained on a consultant or retainer basis rather than as paid, full-time staff. A hearing conservation business that retains the services of a physician (preferably an otologist) as "medical director" or "medical consultant" of the program increases, justifiably or not, the credibility of that program. Many clients are more confident and comfortable with the overall service because "a doctor is in charge," and others feel that they are getting more for their money when a physician is involved in the program. In addition, a physician who is in fact an active and interested part of the hearing conservation service *is* a valuable attribute. Questionable audiograms or those with possible medical implications can be reviewed and discussed with staff or directly with the medical personnel at the client company. Sales will almost surely increase if the physician is willing to talk to potential clients concerning the need for conserving hearing and the advisability of a comprehensive hearing conservation program.

The second consultant position that I see as necessary to a hearing conservation service is that of an attorney serving as legal counsel who has the expertise and practical experience in the area of industrial hearing loss and who clearly understands the liabilities of a hearing conservation service in providing industrial hearing protection. Not only should the attorney be succinctly familiar with the OSHA Hearing Conservation Amendment but with the requirements and nuances of state worker's compensation regulations as well.

The third position that I have found to be valuable to maintain on a consultant basis is that of an acoustical engineer. Often, a client company will encounter noise problems that can be reduced or eliminated through the implementation of appropriate engineering and/or administrative controls. The hearing conservation company that can offer this type of service to its customers gains an advantage by the comprehensive nature of the help it has available.

Staff Positions

The hierarchy of paid staff positions begins with the owner, the president, director, executive vice-president, or whatever other descriptive title the top administrative official assumes. This is the "front person" for the hearing conservation service who makes all final decisions concerning policy and operations and who also will likely be heavily involved in sales. A solid background in audiology and particularly in noise and hearing is a definite asset. Business acumen is a requirement. A master's degree or a doctorate in audiology would often provide an advantage, much as having a physician as medical director does. Also, it is appropriate that this person have certification as a Council for Accreditation in Occupational Hearing Conservation (CAOHC) Course Director in order that the hearing conservation service can conduct training courses for employees of potential client companies. The importance of being able to offer this type of training course will be discussed in the next section dealing with marketing and sales.

The next staff position necessary is that of "manager of services." This person is responsible for the day-to-day operation of the hearing conservation service. As a minimum, a master's degree in audiology and the Certificate of Clinical Competence in Audiology as issued by the American Speech-Language-Hearing Association (ASHA) are required. This is the person who reviews and evaluates all industrial audiograms for completeness and accuracy, and makes comments and appropriate recommendations when indicated. Maintenance and calibration checks of all equipment, including audiometers, sound level meters, frequency analyzers, and dosimeters are usually the responsibility of the manager of services. This person might also be responsible for recruiting and interviewing other potential staff members and for making recommendations relative to hiring and/or release of personnel. In addition, the manager of services is responsible for seeing that the entire hearing testing/data processing/client reporting process is occurring in an efficient, effective manner. Contact with client companies relative to questions or problems in any of these areas occurs through the manager of services as well as ongoing liaison with corporate and plant management personnel and assistance when necessary in the education and training of plant management in all aspects of hearing conservation. Finally, this person may or may not also have certification as a CAOHC Course Director, but nonetheless plays an integral part in the training and certification of the audiometric technicians and in the constant review and supervision of their performance in the field.

The production supervisor is the staff person whose main responsibility is to plan, coordinate, and assign testing schedules to the audiometric technicians. This must be done in a cost-effective and efficient manner so as to minimize expenses and maximize the revenues of each trip. Consideration must be given to such details as distance to be traveled, number of client companies within a certain geographic area, number of hearing tests to be administered, and the number of work shifts operated by each company. The production supervisor serves as the liaison between the hearing conservation service and the contact person responsible for the hearing conservation program at each client company. Problems encountered by the audiometric technicians, such as broken or malfunctioning audiometers, disagreeable or uncooperative contact persons, technician illness or tardiness, incapacitated mobile van, and the like, are the responsibility of the production supervisor. In the case of a hearing conservation service that utilizes mobile vans, the production supervisor would also be responsible for tracking the use, care, and maintenance of the mobile units. Finally, complete and accurate statistics and reports of expenses and revenues generated during each testing trip are the responsibility of this very important staff member.

Another staff position that I believe is necessary to a hearing conservation service operating on a full-time basis is that of data processing coordinator. This person is in charge of the coordination and assembly of audiometric test results for all client companies and testing trips, data entry, automated processing, and the maintenance of files for immediate information retrieval. Timely reporting of test results to the client is a prime responsibility of the data processing coordinator. Explanation of computer print-outs (summary reports and result slips, for instance), processing delays, and general information relative to any aspect of data management are also important parts of this person's job.

The audiometric technician is a critical part (some would argue the *most* essential part) of any hearing conservation service. The minimum requirement for this position is certification as an Occupational Hearing Conservationist issued by CAOHC. The responsibilities of this position include all hearing testing for client companies on a local, regional, and national basis as required; conducting sound level surveys and employee training programs as requested by the client; and answering any and all questions in the field concerning hearing conservation in general and the hearing conservation service's programs in particular. The technician is often faced with long hours of administering accurate pure tone air conduction threshold hearing tests, and must deal with a multitude of different people and varying personalities on a one-to-one basis every day. Therefore, the technician must be a "people person" who possesses a pleasant personality and the ability to deal and communicate with people effectively. In my view, technicians should actually be considered, at least partially, to be sales staff for the hearing conservation service since they are the ones who make personal contact with the client companies (and potential client companies) on at least an annual basis. The customer, therefore, equates the performance of the audiometric technician with the service itself and his or her opinion of the technician contributes heavily to the judgment of a hearing conservation service. As a consequence, the selection of audiometric technicians must be a careful task that is not to be taken lightly.

The staff positions and consultants discussed to this point are viewed as essential to any hearing conservation service that is trying to operate on a fairly large-scale basis. There are other positions that should

be mentioned that may or may not be staffed, depending again on the size, scope, financial resources, and short- and long-term growth objectives of the hearing conservation service. For instance, can a sales staff be justified or are sales to be the responsibility of all existing staff from the owner through the audiometric technicians? Is the volume of sales of hearing protection devices and educational/motivational materials large enough to require an inventory clerk or can this responsibility be assumed or shared by existing staff? Should a marketing/public relations/advertising expert be hired or retained on a consultant basis? How many secretaries/receptionists are required to handle the workload of existing staff? And finally, what are the staff requirements for bookkeeping and accounting—how many people will be required to handle the business accounts receivable and payable, payroll, tax returns, financial reporting, and so on? Figure 16-1 is an organizational chart that summarizes what I consider to be a workable hierarchy of staff and consultant positions in a hearing conservation service.

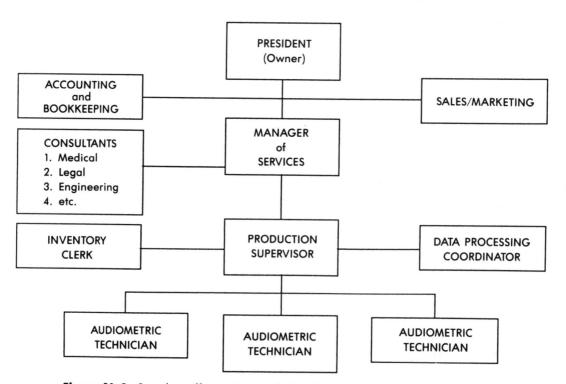

Figure 16-1. Sample staff organizational chart for a hearing conservation service.

In addition to the previous considerations concerning which positions should be staffed, several other issues related to staffing must be addressed. For instance, what will company policy be concerning staff hiring, evaluation, and termination? Will there be paid vacations and/or holidays? What employee expenses, if any, will be reimbursed by the company? What will salaries be and how often will employees be paid? What standards will apply to personal appearance? What guidelines will be set governing acceptable leave due to employee illness? What type (if any) of group health and life insurance policy will be available for employees and will there be a pension plan, profit-sharing plan, or some other type of retirement program for long-time employees? These are examples of a few of the more important questions that the manager of a hearing conservation program must answer when the decision to begin adding staff has been made. As any business begins to develop and grow, addition of staff becomes unavoidable. However, the manager must be cognizant of the fact that this step, although geared to meeting the demands and requirements of the business, will add another dimension to the already-existing problems of that business—those relating to employee relationships and policies.

MARKETING AND SALES

Marketing

Marketing is the primary link between any business and the consumer. All businesses must serve consumer needs if they are to succeed. For purposes of this discussion, marketing can be defined as the development and efficient distribution of goods, services, ideas, issues, and concepts for chosen consumer groups. Marketing is much more than selling! It is a very complex activity that reaches into many aspects of the organization and its interactions with its consumers such as buying, selling, transporting, storing, grading, financing, risk taking, and marketing information. In order to effectively reach consumers, all businesses need to develop a marketing strategy. This is a two-step process that involves selecting the market target and developing a marketing mix designed to satisfy that market target.

A *market target* is a group of consumers toward which a firm decides to direct its marketing efforts. This market consists of people with purchasing power and the authority to buy, whether they are individual consumers, purchasing departments of a nearby plant, or the president of a company with a significant noise problem. Market targets are selected as the result of marketing research and analysis. Some of the tools used in market research include surveys, direct observation, personal interviews, and analysis of published data. The target of any hearing conservation service will depend on a number of factors, including the size and scope of the service, geographic location, financial resources, and competition. An example of a market target for this type of service would be "any company with a noise problem within a 250 mile radius." This is obviously a very general target and perhaps one much less easy to track and sell to than, for instance, "any heavy iron and steel company with more than 50 employees within a 250 mile radius." The results of good market research will help answer the question as to which is the best market target for a particular hearing conservation service.

The second step in the development of a marketing strategy is formulation of a *marketing mix* to reach the market target. This is the mechanism that allows companies to match consumer needs with product and service offerings. A marketing mix is the combination of the firm's product-service offerings, pricing, distribution, and promotional strategies. *Product strategy* is the part of the marketing decision making that deals with specific products and services, package design, trademarks, new product and service development, and so on. *Pricing strategy*, one of the most difficult parts of marketing decision making, deals with the methods used to set profitable and justifiable prices. In making these decisions, government regulations, public opinion, and competitive pricing must be considered. Obviously, the potential buyer of hearing conservation services, as is the case with the buyer of any service or product, is looking for the best possible service at the best possible price. Pricing strategy must take this fact into consideration. *Distribution strategy* involves the actual physical distribution of goods and services to the consumer. Finally, *promotional strategy* is the part of marketing decision making that involves blending personal selling, advertising, and sales promotion tools to produce effective communication between the firm and the market target.

Sales

One of the first suggestions relating to successful sales of hearing conservation services (or *any* services or products for that matter) is to determine exactly which person in a firm has the authority to make a purchase decision of this nature. I am aware of many instances where the manager or sales staff of a hearing conservation service wasted countless hours convincing the plant physician or nurse about the merits of his or her program, only to discover that the ultimate buying decision actually rests with the safety director. Had this fact been discovered early on, much time (and money!) could have been saved. In an editorial written in 1983, the responses given by buyers of hearing conservation services to the question "What do you *really* want?" were presented (Radcliffe, 1983). These responses are quite instructive to someone on a sales call and range from "Why are you as well or better qualified than another vendor to provide services to us?" through "What resources do you bring to the job?" to "Tell me about your rate structure."

Other important considerations relative to the success of sales revolve around the company image of the hearing conservation service and the packaging of that image. For instance, are professional-looking brochures available which clearly explain the services and products being offered as well as price structure? Are business cards used? Are clear, concise summary reports and management reports used so that personnel who are not familiar with hearing conservation terminology can make sense of them? What educational and/or motivational materials does the service provide such as newsletters, training manuals, films, slide/tape presentations, etc.? It should be recognized that, all other things such as pricing, qualifications, service, and so forth being pretty much equal, the hearing conservation service that projects a more professional and business-like image has the better chance of closing the sale. In addition, there are some important messages that should be conveyed to the potential customer during any sales call. First, it should be emphasized that the primary purpose of an industrial hearing conservation program is to protect the hearing of the employees, thus helping to maintain overall health and well-being. As a result, productivity will likely improve. Second, the initiation of a hearing conservation program will help the company to comply with OSHA regulations that specifically deal with employee exposure to potentially hazardous noise (see Chapter 13). Finally, it should be emphasized that having a hearing conservation program can assist the potential customer in limiting liability as it relates to state worker's compensation claims. This is usually quite a powerful and meaningful message for a prospective client, because ignoring this fact has the likely end result of hitting where it hurts the most—in the pocketbook—as a result of increased compensation claims and compensation costs!

Generating Sales Leads

If a new hearing conservation business expects to survive and be successful or if an already-existing, established business expects to grow, sales leads must be generated. There are many different ways of generating these leads, some of which are probably quite obvious and others not so obvious.

One common method of generating sales is to employ a sales person or a sales staff who is "knocking on doors" and calling prospective clients on the phone as well as making cold calls. For a small hearing conservation service, this may not be an economically feasible or practical approach to finding new customers, but for the medium to large service it often makes good business sense. Another fairly obvious method of generating sales leads is through advertising. Depending on budgetary allowances, this advertising can run the gamut from a display ad in the yellow pages of the phone directory to the placement of neat and informative ads in local, regional, and national publications such as professional journals read by safety and industrial hygiene personnel, physicians, and occupational nurses. It should be common practice that the effectiveness and impact of different advertising approaches be carefully monitored and tracked on at least an annual basis to determine which are most productive and cost effective and which might need to be replaced or eliminated. Related to this direct advertising approach is the fact that most managers of hearing conservation services are, or should be, members of professional organizations such as the American Industrial Hygiene Association and the American Society of Safety Engineers. What better way to advertise services than by getting to know personally the safety directors, occupational nurses, and industrial hygienists in your marketing area? Programs dealing with hearing loss, occupational noise exposure and industrial hearing conservation might be offered to the local and regional chapters of these organizationns. Finally, many hearing conservation services have begun to use direct-mail literature to attract potential clients. As with all other types of direct-mail pieces, the expected response should probably not exceed 1 to 2 percent of the total mailing and it is necessary that careful research and planning be conducted prior to initiating this type of a sales approach. If a "scatter-gun" method of direct mail is used, very little substantive response can be expected.

Some of the not-so-obvious ways of generating sales leads include: informational seminars, CAOHC training courses, referrals from current clients, and referrals by insurance companies. It is sometimes surprising how much interest can be generated by offering a free informational seminar to current customers as well as to representatives of companies who *might* be shopping for hearing conservation services. This seminar could cover topics ranging from current federal and state regulations governing occupational noise exposure, through a discussion of the audiometric testing program, to a review of the current status of personal hearing protection devices. Again, careful planning of this type of seminar and deciding who to invite are critical issues. Many hearing conservation services offer CAOHC training courses if the manager or other personnel are certified as a CAOHC Course Director. These courses are often attended by employees of companies who merely want to learn about industrial hearing conservation and do not necessarily want to conduct their own baseline and annual hearing tests as part of an in-house program. They are basically uninformed concerning the entire area and are simply trying to find out what their responsibilities are and what kind of help is available to them. These course participants become prime prospects as potential buyers of a hearing conservation service. An often overlooked source of sales leads is through existing customers. If the hearing conservation service is providing a competitively priced service that meets the needs and requirements of the customer, that customer is often more than happy to talk to other businesses in the area about their program and to make referrals to the hearing conservation service. However, they usually have to be *asked* to do this, and the manager of the hearing conservation service must be comfortable and confident enough about the program to be willing to ask customers to "spread the good word" about the business. Finally, a good source of sales leads can be established through insurance companies. Because they make inspection visits to client companies, they often are the first to recognize a potential noise problem. If they have been made aware of the scope of services provided by a particular hearing conservation service and are confident that their client will receive cost-effective, reliable, and comprehensive help, they are usually eager to make a referral. When this happens, the possibility of a sale increases because the client company is usually anxious to solve a problem when it is brought to its attention by its insurance carrier.

POLICY DEVELOPMENT AND COORDINATION OF ACTIVITIES

The key to policy development and coordination of activities of a hearing conservation service is active communication at *all* levels. Within the hearing conservation business itself, it is important that ongoing communication occur among and between all employees, but it is especially vital that open lines of discourse exist between the audiometric technicians, the production supervisor, the manager of services, and the top administrative official of the business. If customers are to be retained, a good rapport and relationship must be established with *every* client, especially the key individual who has been made personally responsible for the hearing conservation program at each testing location. This type of relationship can only be established and maintained with ongoing contact throughout the year. This contact can occur through correspondence, newsletters, phone calls, and personal visits whenever possible. If the person in charge at the client company is comfortable with the fact that the hearing conservation service is sincerely interested in helping on an *ongoing* basis, a firm business relationship can be established. However, if the person in charge has the notion that the only contact from the service is once a year when it is time to schedule annual testing, he or she may begin to feel ignored and somewhat used. The client must be made to feel comfortable with the fact that he or she can personally contact anyone at

the hearing conservation service for answers to questions or solutions to problems—from the top administrative official through the data processing coordinator to the audiometric technician. This type of communication goes a long way toward establishing a solid, longstanding relationship as well as assuring that the coordination of activities between the hearing conservation service and the client company occurs smoothly.

Policy development is, in large part, guided by a combination of economic factors and the various state and federal regulations regarding occupational noise exposure. It is necessary that any legitimate hearing conservation service be able to either directly offer assistance, or recommend sources of assistance, in six distinct areas: noise exposure monitoring, audiometric testing, selection of hearing protection, employee training/education, appropriate engineering and administrative controls, and assistance in record keeping. These areas have been covered in detail elsewhere (Bodenhemier, 1984) as well as in this volume. Suffice to say, however, that the policies of any hearing conservation service will be directed by the local, state, and/or federal regulations governing each of these areas, as well as the needs and requirements of each individual client company. Because most hearing conservation services deal with clients who are involved in a wide variety of manufacturing/noise-producing processes, it is important to realize that policy development and enforcement, of necessity, needs to be flexible. Finally, all staff should have input relative to policy development *before* a final decision is made. Each staff position mentioned previously plays an integral part in the overall effectiveness and efficiency of a hearing conservation service and the information received from each of these areas is important and necessary to making an informed decision.

LIABILITY/RISK ASSESSMENT

The issues of liability and risk may not be among the first priorities of the manager of a hearing conservation service, but the following review will suggest that it is an area that demands careful thought and consideration in view of the recent and ever-increasing growth of liability claims.

The term *liability* will be used here to mean legal responsibility and the term *risk*, the uncertainty about potential loss or injury. Both are unavoidable parts of any business and the manager of a hearing conservation service must find ways of addressing and dealing with each. The important first step is to recognize that both do exist, and that ignoring this fact will not make them disappear.

Because the OSHA Hearing Conservation Amendment *seems* to have created a legislated windfall for audiologists and their practices, many hearing conservation managers are new to the field. Consequently, they have a tendency to underestimate or they do not fully understand their professional and personal liability in providing hearing conservation services to industry. At the sixth annual meeting of the Hearing Conservation Association in 1982, Attorney John E. Turner identified several of the major issues surrounding professional liability for the hearing conservation manager (Radcliffe, 1982). Among the more thought-provoking of these issues are the following:

- Incorporation of a business does not protect the individual from liability in all instances! If the individual owners are personally involved in delivering the services, the employee of a client company can sue the corporation *and* the individual owners.
- Compliance with OSHA is no defense. Merely complying with minimum OSHA regulations can, in fact, be cause for suit. Any hearing conservation program should do more than simply comply with OSHA.
- The employer can, in most cases, use the availability of worker's compensation as a defense. However, a third party such as a hearing conservation professional cannot use the same defense. The employee can accept worker's compensation and *still* sue the professional as a third party.

- The worker's compensation carrier may sue the hearing conservation program manager to recover any amount of compensation awarded to an employee.
- Juries do not have the expertise to know when an expert has or has not acted properly. The manager of a hearing conservation service may be found guilty and liable even when all of the correct procedures were followed and followed well.

Turner summarized his presentation by advising that the hearing conservation service manager must be able to show that he or she *knows* about hearing conservation, that he or she *does* everything reasonable to enhance hearing conservation, and that he or she *cares* about hearing conservation. Based on this summary of some of the main pratfalls of professional liability for the hearing conservation program manager, it would seem incumbent upon him or her (as suggested in the previous discussion on staffing) to consult a knowledgeable attorney who specializes in the area *before* deciding whether or not to assume these rather extensive legal responsibilities.

As with legal liability, risk is unavoidable in any business. The list of risk-filled decisions in a hearing conservation service is a long one. For instance, the main office, which might contain audiometers, sound level meters, dosimeters, computer records, etc., faces the risk of fire, burglary, water damage, and physical deterioration. Accidents, judgments due to lawsuits, and non-payment of bills by customers are other risks. In general, two major types of risk exist and must be dealt with: *speculative risk* and *pure risk*.

As an example of speculative risk, expansion of hearing conservation operations into a new market or in a new region may mean higher profits or could mean the loss of invested funds. On the other hand, pure risk involves only the chance of loss. Automobile drivers (or the drivers of industrial hearing conservation mobile vans) always face the risk of accidents. If they occur, the drivers (as well as others) may suffer financial and physical loss. If they do not occur, however, there is no gain.

Once the presence of risk is recognized, the manager has essentially four methods available for dealing with it as summarized below:

- Avoid it
- Reduce frequency and/or severity
- Self-insure
- Shift risk to insurance companies

Companies unwilling to assume risk by avoiding it are content to produce and market products and services that have a stable demand and offer an adequate profit margin. This strategy usually ensures some degree of profitability, but it stifles innovation. Many types of risk can be reduced or even eliminated by eliminating hazards. Safety programs are often begun to educate employees about potential hazards (driving a mobile van, for instance) and the proper methods of performing specific tasks. Installation of fire-retardant building materials and/or an automatic sprinkler system can help protect equipment, records, inventory, etc. from fire. Adequate credit checks allow managers to make careful decisions concerning which customers should be extended credit. All of these actions can reduce the risk involved in business operations, but they cannot eliminate all risk.

A self-insurance fund is a special fund created by setting aside cash reserves on a periodic basis to be drawn upon only in the event of a financial loss resulting from the assumption of a pure risk. The regular payments to the fund are invested in interest-bearing securities, and losses are charged to it. The alternative of self-insurance can be a realistic choice for large multilocation companies because the likelihood of several fires in different locations is small, and the likelihood of a single fire can be calculated. For the single-location company, however, one fire can prove disastrous, and contributions to a reserve fund for large potential fire damage can be prohibitively high.

Although steps can be taken to avoid or reduce risk, the most common method of dealing with it is to shift it to others in the form of insurance. Although literally hundreds of different types of insurance policies are available for purchase by individuals and businesses, they can be conveniently divided into three broad categories: (1) property and liability insurance, (2) health insurance, and (3) life insurance. This discussion will focus on property and liability insurance only, even though, as discussed previously, the manager of a hearing conservation service would have to consider the pros and cons of health insurance and life insurance for the employees.

Property and liability insurance provides protection against a number of perils. Property losses are financial losses resulting from interruption of business operations or physical damage to property as a result of fires, accidents, windstorms, theft, or other destructive occurrences. Liability losses are financial losses suffered by a business or individual should the business or individual be held responsible for property damage or injuries suffered by others. Table 16–2 identifies the major types of property and liability insurance that the manager of a hearing conservation service must be familiar with, as well as a brief explanation of the purpose of each. These types of insurance will not be discussed further except to make several comments concerning public liability insurance. Since many of these policies limit such claims to $100,000, many businesses purchase *umbrella liability insurance* to extend the amount of coverage limits. Product liability insurance, a form of public liability insurance, is designed to protect businesses against claims for damages resulting from the use of a company's products. It is extremely important for the manager of a hearing conservation service to be aware that he or she or the company *could* be liable for damages claimed by customer's employees due to the use of personal hearing protection devices sold, recommended, or fitted by the hearing conservation service. Realizing that the numbers of product liability cases have increased by 500 percent during a 5-year period in the late 1970s, and that the average claim rose from $476,000 to $1.7 million over this period (*Insurance Facts*

TABLE 16–2.
Types of Property and Liability Insurance and Their Functions

Type of Insurance	Protection
Fire insurance	Losses due to fire and (with extended coverage) damage caused by wind, hail, water, riot, and smoke
Business interruption insurance	Losses occurring when one of the above causes a business to close temporarily
Automobile (van) insurance	Losses due to theft, fire, or collision; claims resulting from damage to other property due to collision or injury or death of another person
Burglary, robbery, and theft insurance	Losses due to the unlawful taking of the insured's property by force or by burglaries
Marine insurance	Losses to property that is being shipped from one location to another
Fidelity, surety, and credit insurance	Misappropriation of funds and employee dishonesty (fidelity bond); failure to perform a job (surety bond); failure to repay loans (credit insurance)
Public liability insurance	Claims against property owner for injuries or damage to individuals or the property of others as a result of falls, malpractice, negligence, or faulty products

1982-83, 1983), public/product liability insurance must be considered a necessity for any hearing conservation service.

The manager of a hearing conservation service must carefully study the exact circumstances of his or her company and employees with the aid of a qualified insurance agent. Since most agents work on a commission basis, it is important for the consumer to understand the various types of insurance and the specific needs for which insurance is being considered. At this point, then, a determination of the proper balance between savings and protection that is best for the company and the individual(s) involved can be rationally made.

SUMMARY

Although most audiologists today have at least some training and, perhaps, some experience related to the effects of noise on hearing and industrial hearing conservation, little, if any, of that background relates to the business considerations of starting and managing a hearing conservation service. In this chapter, I have tried to highlight some

of the more important of these considerations, and I hope they will be informative and enlightening to anyone who wishes to become involved in a hearing conservation business.

REFERENCES

Bodenhemier, W. G. (1984). Recent considerations in industrial hearing conservation. *Seminars in hearing, 5*(2). New York, NY: Thieme-Stratton, Inc.

Duncan, W. J. (1983). *Management* (pp. 13–14). New York: Random House.

Glorig, A. (1980). Noise: Past, present and future. *Ear and Hearing, 1*(1).

Hearing conservation: A guide to preventing hearing loss. (1983). Injury Prevention Library. Daly City, CA: PAS Publishing.

Insurance Facts 1982–83 (1983). (p. 53). New York: Insurance Information Institute.

Miller, M. (1978). Hearing conservation in noise. In *Council for Accreditation in Occupational Hearing Conservation Manual* (Chapter 1). Cherry Hill, NJ: Fischler's Printing.

Radcliffe, D. (1982, July/August). Current issues in hearing conservation. *Hearing Aid Journal.*

Radcliffe, D. (1983, August). Pointed Questions. *The Hearing Journal, 36*(8).

SUGGESTED READINGS

Kravity, W. (1983). *Bookkeeping the easy way.* Woodbury, NY: Barron's Educational Series, Inc.

Peters, T. J., & Waterman, R. H., Jr. (1983). *In search of excellence.* New York: Warner Books.

Simini, J. P. (1967). *Accounting made simple.* Garden City, NY: Doubleday & Company, Inc.

Taylor, C. (1985). *The entrepreneurial workbook: A step-by-step guide to starting and operating your own small business.* New York: New American Library.

Worthington, E. R., & Worthington, A. E. (1985). *Staffing a small business: Hiring, compensation and evaluating.* Sunnyvale, CA: Oasis Press.

Equipping a Hearing Conservation Program

John Balko

I f one is contemplating entering the area of hearing conservation, getting started includes obtaining the right equipment. This chapter is intended to give a brief insight into some of the choices one has concerning the setting and equipment in order to help one become a more knowledgeable consumer of this equipment. The first choice must be to determine whether hearing conservation will be practiced in-house or on a mobile basis. The advantages and disadvantages of each will be explained.

IN-HOUSE OPERATION

"In-house" may be an individual's private practice office, school system, or industrial plant complex.

Office

One may wish to incorporate a hearing conservation program with a clinical practice or hearing aid dispensing operation.

This can work very well for small companies in the local area. It is also common to provide retests and/or complete diagnostic follow-up test services to large companies in the area who test with mobile services. Usually, a mobile service would rather find a local in-house professional to provide these services instead of bringing their test unit back for a small number of retests.

Industrial Plant Complex

Convenience is the key word for an in-plant hearing testing operation. Management can have a test completed whenever they desire. It can be incorporated into the plant pre-employment physical examination of the new employee, and is convenient to test an employee who is retiring or terminating employment. Management may have an employee tested after an extended sick leave to determine if the illness or any of the medications the employee took during his or her sick leave affected hearing acuity.

The disadvantage of an in-house plant hearing testing operation is the time involvement for personnel during the testing. Time is especially important for an employee with other duties, for example, a plant nurse. If a company has 2,000 employees, a plant nurse could possibly test five employees per shift. This a common number considering other duties the nurse has to perform on a daily basis. This would mean 10 employees could be tested per day, five on each shift. Taking into consideration 20 working days per month, it would take 10 months to test the hearing of all 2,000 employees in the plant. This assumes there are no injuries or problems during the day to take precedence over the nurse's testing time. In this example, by the time all hearing tests are completed, along with the retests, it is time to start over. This example also does not allow for time to consult with employees regarding their hearing acuity, to fit and evaluate hearing protectors, or to compare annual tests to baseline tests, a very cumbersome operation when doing it manually. One can easily see how this plan cannot assure the continuity of a hearing conservation program. Further, production can be influenced in that employees might come to the nurse's station for their hearing tests, only to find the nurse does not have time to test them. Thus, they would either wait in the nurse's office, which would be unproductive time, or go back to their work station only to be called back for a second time.

Disadvantages of an In-House Operation—Private Practice

In most instances, an in-house operation cannot compete in the area of marketing hearing conservation services to industry. Companies do not want to send their employees to an in-house location, as management loses valuable production time when employees are in transit. Also, once the employee leaves the plant property, management loses control of the employee in that he or she may elect to stop on the way to the testing site or on the way from the

testing site for a variety of reasons. However, one must realize that management is still liable for this employee should he or she be involved in a traffic accident while on company time and on company business. Sometimes, management will elect to bus or taxi employees to an in-house operation, but this is usually the case when there are other medical diagnostic tests which management wants to complete on this employee, such as x-ray, a general physical, or perhaps blood and urine analysis. Obviously, the additional cost involved in the bus or taxi is another factor that makes this approach less attractive to industry.

Advantages of an In-House Operation—Private Practice

There are obvious cost savings if one needs a stationary location for other segments of the practice: by incorporating industrial hearing conservation testing services to the practice, audiologists are not adding a great deal to their overhead structure. All that is basically needed is a test environment and an audiometer which, if diagnostic hearing tests are part of the activity schedule, the facility would already have the necessary equipment to complete industrial hearing tests.

MOBILE SERVICES

Disadvantages

The biggest disadvantage of a mobile service is the initial cost of purchasing the mobile unit and equipping it. After this procedure is completed, there are maintenance costs (gas, tires, oil, engine repairs), insurance, licenses, and often, parking and storage fees. These are all fixed monthly costs, regardless of the number of hearing tests being performed each month.

Many individuals or organizations decide to purchase a mobile vehicle before they determine if they have or can acquire enough work to support it on an annual basis. Month-to-month fixed expenses must be factored into the decision process. In today's industrial sector, approximately 80 percent of the companies that need hearing conservation services are already receiving these services from someone. New arrivals will be competing for the 20 percent of the marketplace that is not being serviced and/or trying to lure clients away from someone else who is already servicing them. This is extremely difficult in most situations because of the computerized comparison to baseline data. If a new firm is successful in taking a client from someone else, it will be necessary to extrapolate former baseline data into the computer system in order to determine if there is a threshold shift from the baseline or the base reference audiogram. This is an additional procedure which involves additional cost. Unless new entries into the marketplace are willing to absorb this cost in order to get the business, they will essentially be overpricing themselves by adding the cost of including the baseline audiogram to their computerization expenses.

There are some manufacturers of mobile hearing test units who will rent or lease mobile hearing testing vehicles. This is a good way to find out if mobile testing is appropriate for one's organization.

Advantages

A mobile hearing test unit will often have the capability of servicing a variety of geographical locations without the overhead of rent and equipment at each of these locations. It also gives the capability of servicing locations on a spontaneous and randomly scheduled basis. Obviously, the closer these locations are to one's home base, the more cost effective the program will be because of the lower mileage involvement and time on the road for staff.

If possible, one should try to develop ways for the mobile facility to complement other areas of the practice. This may include hearing screenings at the local shopping mall, or health fairs to refer patients to the office for hearing aid sales or diagnostic audiological evaluations. It should be noted

here that industrial hearing testing has approximately a 7 percent primary medical referral rate. For example, if an individual tests 100 new employees at a company, seven of them should be referred to an ear, nose, and throat specialist for diagnosis and treatment of a medical ear or hearing problem. From that new 7 percent primary referral group, a secondary referral group of 14 percent (14 new patients) will be sent to that same medical practice. These will be children, spouses, parents, or other relatives or friends of the original group. One can easily see this type of industrial referral is an extremely benefical factor for a medical practice. Obviously the type of care, distance from the medical practice, and the cost of services are all factors in the real world situation.

Mobile hearing conservation testing vehicles may also be used as satellite hearing aid dispensing offices. In some cases, the size of the test environment must be increased and the test capability of the audiometer must be enhanced. Instead of only pure tone capability, the audiometer must be equipped to provide complete diagnostic tests, and the test environment must be large enough to provide freefield testing along with patient talkback capability. This increase in equipment cost is minimal compared to the increased revenue capability of the mobile unit. This additional equipment will also give one the ability to offer on-site clinical audiology services which may be marketed to residential geriatric facilities, physicians' offices that do not offer these services in-house, and/or small hospitals and clinics that do not offer these services.

TEST ENVIRONMENT: MINI-BOOTH VS. SINGLE ROOM

All the major manufacturers of audiometric test environments make a single-person test room that costs approximately $2,000. More than one of these closet-type structures may be placed in a mobile unit or in a single area so that more than one person may be tested at a time. The single-person test booth isolates each individual testee both visually and acoustically. Most professionals feel this is a distinct advantage over a single multi-station room for reliability of testing. Also, since each person is in his or her own booth, as soon as the hearing test is finished he or she can be removed and the next person can be tested, thus speeding up the testing activity. Whereas, in a single multi-station room, no one can be removed until everyone is finished without disrupting the test for everyone in the room. Also, there is the possibility of many contaminants inside a single room multi-station environment, for example, talking, coughing, or movement of feet on the floor.

The main advantage of a single multi-station room in mobile testing is that it is less expensive and takes up less space than the equivalent number of single person mini-booths. (The approximate cost of a six-person single sound room is $8,000 installed, compared to eight single rooms at approximately $16,000).

There are certain factors that should be considered regardless of the type of acoustic environment chosen. First, a background sound level measurement should be taken at each and every testing location in each room. For example, if one has three mini-booths in the unit and it is moved to a different location at a plant site, the client should get six copies of background noise measurement documentation with the hearing test report. Second, the testee should never be able to see the tester or audiometer in his or her visual field. Do not rely on seating alone to achieve this; the testee will move his or her seat or sit sideways on a fixed seat or bench. The most cost-effective way to achieve adequate isolation is to frost the bottom section of the booth window to block the testee's view of the audiometer or tester. One-way glass is only effective in some situations and, unfortunately, unless the inside of the test environment is extremely bright, one-way glass does not work at all in a mobile test environment. Bright lights create heat and one should try to keep

the test environment cool and comfortable. Remember to keep the headsets and response buttons on wall hooks inside the booths, preferably behind the testee in the sitting position, so that the tester can quickly and efficiently position the headphones and hand the testee the response button. Remember, time is money in hearing conservation testing.

AUDIOMETERS

If one wants to be able to provide more comprehensive examinations than pure tone thresholds, the choice for an audiometer has already been made. A clinical diagnostic manual audiometer will be required. The cost for such an instrument will range from $1,400 to $7,500, depending upon quality and options.

If plans include pure tone threshold testing only, any of the following will suffice:

Manual Audiometers

Advantages: These devices are less expensive—$400 to $1,000. They are fastest of all the pure tone threshold audiometers in the hands of a trained technician or audiologist. Manual units are durable. If the testee is having problems, the tester can slow down or change technique without the need for adjustments to equipment or complete retesting.

Disadvantages: Only one person can be tested at a time.

Self-Recording Audiometers

Advantages: One tester can conduct up to eight hearing tests simultaneously.

Disadvantages: These units are more expensive than manual audiometers, costing about $3,500 each. Approximately 6 percent to 9 percent of testee population cannot take a reliable self-recorded test. The total test time required is longer and repeat testing is necessary more often than other hearing test methods.

Micro-Processors

Advantages: One tester can test up to eight testees simultaneously. In some cases, the micro-processor audiometer can be used as a standard manual audiometer if the testee is having problems with the micro-processor sequence. Some manufacturers have computerized interpretation programs and printers so that test results can be interpreted and printed on hard copy automatically.

Disadvantages: Automatic audiometers are expensive—$1,600 to $2,800 for each audiometer, and in some cases, a main terminal is required to print out hard copy results, which costs an additional $2,500 to $4,000. A certain segment of the population has problems taking a reliable micro-processor test quickly. If the testing has to be interrupted to feed identification data into this system, the time involved is not a cost-effective time trade off for the time saved.

SOUND LEVEL MEASURING EQUIPMENT

It should be noted here that I do not suggest that a neophyte in hearing conservation services should be completing employee noise exposure measurements. This is an extremely sophisticated area of hearing conservation. An inexperienced person, even using the new state-of-the-art equipment, can make some serious errors in the measurement of noise exposure in a high-intensity noise environment (see Chapter 4).

Requisite equipment for sound measures includes at least a Type 2 sound level meter (SLM) equipped with an octave filter set. This piece of equipment can be justified for most applications because it is needed to measure the background noise level in the hearing test environment. The SLM will only be needed periodically in an in-house operation to determine if the ambient noise in the test environment is constant. It will be needed on a daily basis in a mobile situation to determine and document background levels at the different frequency ranges for each testing location.

SPECIAL CONSIDERATIONS: TRAILER MOBILE UNITS VERSUS SELF-PROPELLED MOBILE UNITS

If mobile on-site hearing conservation services are planned, a second choice is required: the acquisition of a trailer mobile unit versus a self-propelled mobile unit.

Trailer Mobile Unit

Advantages: A trailer unit will be less expensive if the organization already has a proper towing vehicle. The towing vehicle must not only have the power to pull the size and weight trailer in the terrain encountered, but the towing vehicle must also have the weight to stop and guide the trailer safely. A general rule of thumb in the trailer industry is that the towing vehicle should never weigh less than one-third as much as the trailer being towed.

A trailer unit will offer more flexibility to move around. The trailer can be unhooked at the testing location allowing staff personnel freedom to use the towing vehicle for purposes of transportation. They can leave the test site and go to a restaurant for lunch or dinner if they have time; they may also leave the trailer at a remote location and drive their towing vehicle home at night or home for the weekend. Another advantage enjoyed by larger providers of industrial hearing conservation services, and especially by the military, is that several trailers can be moved around individually by one towing vehicle, a feature that is economically attractive. Finally, interior space is available at approximately 70 percent of the square foot cost of equivalent space in a self-propelled vehicle, not taking into consideration the cost of the towing vehicle.

Disadvantages: In general, trailer mobile units are more difficult to operate than self-propelled units. They are more difficult to maneuver in city traffic. Simple things like pulling into a gas station, a motel, or restaurant can become a major problem for an inexperienced driver. Backing up is almost impossible without assistance behind the towing vehicle. The same is true for hitching or hooking up the towing vehicle to a trailer by an inexperienced operator.

Overhead costs are doubled in many ways. Two license plates, two insurance policies, two sets of tires are required. In many states, the operator will need a special commercial license to drive the towing vehicle while it is hooked to the mobile unit. Finally, from personal experience, the mobile hearing conservation trailer unit will have less resale value than will a self-propelled unit.

Self-Propelled Mobile Unit

Advantages: Self-propelled mobile units are generally easy to maneuver and operate. They are more efficient to use when there is a requirement for daily operation in that there is not constant hitching and unhitching. Self-propelled units have a better resale value than do trailers. Only one person is required to drive or operate these vehicles in-transit.

Disadvantages: These units are more expensive to acquire than are trailer-oriented facilities. If technicians must leave the test site, they must stow and secure all of the test equipment to drive the unit. This must be done whenever the unit is moved. If the unit is rather large and this is a cumbersome operation, then one of the technicians must drive another vehicle so that they can leave the test unit at the test site. This is an added expense to the operation. Since they are larger than the vehicle needed to tow a trailer unit, selfpropelled units are more expensive to maintain, e.g. tires, engine maintenance and fuel.

RECOGNIZING DIFFERENT TYPES OF MOBILE UNITS

In Figure 17-1 (pp. 294–297), line drawings showing different types of vehicles and trailers that may be used for on-site mobile audiometric testing are provided. With each drawing, a short caption gives the name of the type of mobile unit, a brief description, the approximate interior dimensions, and the approximate cost of such a vehicle.

CONSTRUCTION

This section discusses features to look for in the actual construction of a quality mobile unit.

Tires

Be extremely careful of purchasing a unit with radial tires. Radial tires have a tendency to roll because of the sidewall flex that is inherent in the radial construction. There are recorded experiences where the sidewall of the tire rolled away from the rim, and the operator of the vehicle had an extremely difficult time maintaining control. Also be sure to install winter snow tires in the rear of the mobile unit in snow country during the snow season or on the rear tires of the towing vehicle for a trailer unit. One should not depend on the weight of the vehicle for traction to get by with summer tread tires during the snow season.

Suspension System

The suspension system of a trailer or self-propelled unit is the mechanism between the tires and axles and the frame of the unit. The better the suspension system, the less vibration audiometric equipment and the sound room will sustain in-transit. Excessive vibration will create acoustic leaks in the sound room. Most good manufacturers of mobile medical units utilize quality suspension systems in their units. In self-propelled units, air ride or air bag suspensions along with helper springs are usually sufficient. The suspension system is more important in a trailer unit than in a self-propelled unit. Most quality medical trailer manufacturers utilize a rubber-ride independent suspension, which allows for three-tire towing in emergencies. The rubber eliminates shock and gives a smoother ride to the trailer and its equipment. There are no springs to break; no shackles to grease and maintain. The rubber-ride axle is virtually a lifetime maintenance-free mechanism. All wheels on the trailer should have electric or hydraulic brakes. All axles should be sized in weight in specific accordance with the weight of the trailer, and of the equipment to be installed in the mobile unit.

Framing

The internal frame of the body of the trailer or self-propelled unit is usually steel. However, most of the better custom body builders have changed to an all-aluminum roll cage construction for framing of their medical bodies and their trailer units. Aluminum has the strength-to-weight ratio of 0.1, which is higher than structural steel. Therefore, strength can be built into the trailer using significantly lower weight. A stronger trailer or truck body weighing 20 percent to 50 percent less than a comparable steel unit gives better traveling characteristics, lower engine displacements are required on the vehicle, and better gas mileage is achieved. The obvious advantage of this construction is the strength-to-weight ratio. Not only is the thickness of the framing material important, but also the positions at center-to-center distances of the framework of the vehicle must be considered. The closer the center-to-center distances of the main upright supports of the frame of the body, the more structural integrity the unit will have. Also, it is important that there is enough side wall strength around the perimeter of the interior body so that the mounting brackets for equipment can be adhered to a side wall strong enough to support the weight of the equipment being attached to them.

Exterior Coverings

I am partial to an aluminum skin as I have had excellent experience with aluminum exterior vehicles. There is no place in the country that has worse weather conditions than western Pennsylvania. The highway department in Pennsylvania uses a great deal of salt in the winter to de-ice the roads and create better traction for the vehicles. My experience with aluminum exterior truck bodies and trailers illustrate that aluminum resists the corrosive attacks from road salts and other industrial corrosion far better than steel. Where aluminum surfaces are exposed to the atmosphere, an aluminum oxide skin forms, which protects the metal from further oxidation. What all this means to the end purchaser of an aluminum exterior trailer is that there will be no rust and very little exterior skin corrosion on the vehicles.

The next choice for exterior is fiberglass. Fiberglass is an excellent body covering in mobile medical units. However, there can be problems with stress cracks, especially with extremely cold weather exposure. If the fiberglass exterior becomes pitted from industrial corrosion, it becomes difficult to maintain.

A new substance that is just beginning to become available in mobile medical units is a laminate construction of plywood and fiberglass called *fiberglass reinforced plywood* (FRP). It provides a smooth interior and exterior surface. It is cost effective to use; however, with the typical construction seen today, the long flat walls do not lend themselves to an aerodynamic body design. Some manufacturers have discussed the problems they are having with side wall ossilation, which certainly would not be advantageous in an industrial audiometric testing unit. At this writing, FRP has not been in use long enough in medical unit construction to make a definitive decision on it. However, it will be utilized more extensively in the future.

Insulation

Insulation is extremely important in an audiometric test environment. Not only does it provide thermal insulation from exterior to interior, but it will also add to the sound attenuation characteristics of the vehicle. The best insulation for this purpose is foamed-in-place urethane which should be sandwiched between the exterior skin of the vehicle and the interior side wall panels.

VANS: 10′ Long—5½′ Wide
These are the least expensive. They are easy to drive; however, they have very limited space. Usually they have a single mini-booth inside. Approximate cost: $20,000.

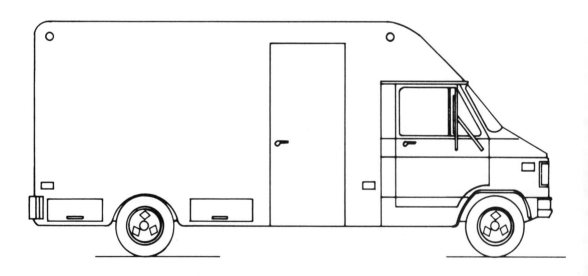

CUT-AWAY VANS: 10′ to 14′ Long—7½′ Wide
These units are very cost effective. They have access from driving area to body area. They may have the capability of testing up to three people simultaneously. Approximate cost: $27,000.

FIGURE 17-1

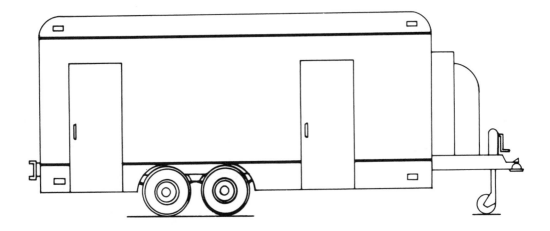

BUMPER PULL TRAILER: 12' to 20' Long—7½' Wide
These are less expensive; smaller so that a ¾-ton pickup truck or large station wagon can be used as a safe towing vehicle. Approximate cost: $18,000.

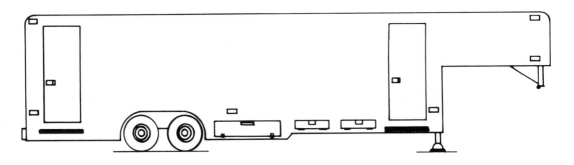

GOOSE NECK/FIFTH WHEEL TRAILER: 17' to 34' Long—8' Wide
Very roomy; must be pulled by 1-ton dual wheel truck. This type of hitch assembly is actually safer to use for large trailers than a bumper pull style hitch. Approximate cost: $30,000.

FIGURE 17-1 (*continued*)

CAB AND CHASSIS (LIGHT DUTY) - 12,000 LB. GVW: 12' to 16' Long—7½' Wide
The author's favorite; easiest to drive, custom designed body to purchaser's needs. All-aluminum bodies, extremely long life. Can be transferred to new cab and chassis in 45 minutes. Approximate cost: $35,000.

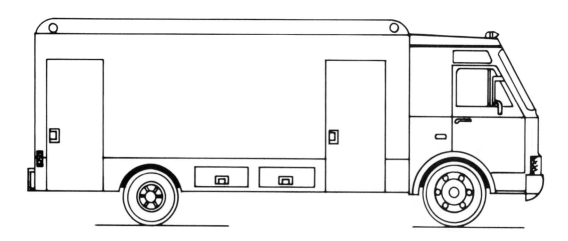

CAB AND CHASSIS (MEDIUM TO HEAVY DUTY) - 1 to 2½ TON GVW: 16' to 30' Long—8' Wide
Diesel motors, heavy duty frame and suspension systems. All-aluminum or aluminum-fiberglass composite bodies lend this vehicle to a 20-plus year usable life. The tradeoff is that these units are difficult to drive, expensive, and require special operators license. Approximate cost: $60,000.

FIGURE 17–1 (continued)

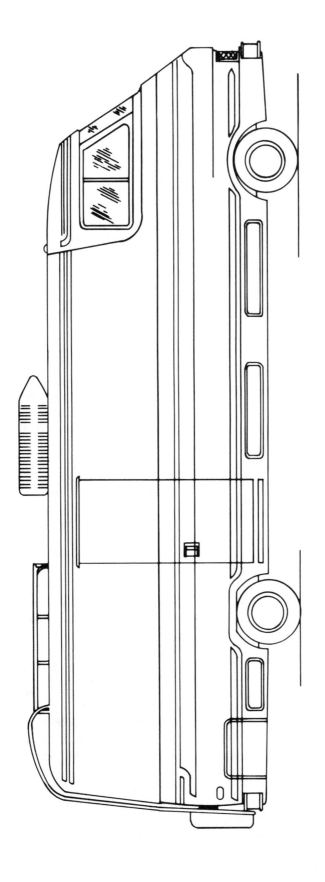

MOTOR HOME CONVERSION: 18' to 38' Long—7½' Wide

Asthetically appealing, expensive, not very efficient use of space. (Walls normally belly out; difficult to install sound rooms.) Most RV manufacturers do not intend their vehicles for commercial use and many will not honor their warranties if they are used commercially. Approximate cost: $45,000.

FIGURE 17-1 (*continued*)

This application will discourage development of thermal or acoustical leaks. Foamed-in-place insulation is difficult to work with, it is labor-intensive, and it is often expensive. However, it does provide an excellent thermal and acoustical barrier, and will add to the structural integrity of the vehicle.

The other type of insulation is styrofoam panels of various thicknesses. Styrofoam in itself is an excellent insulator. However, at the seams where the panels meet, air leaks are created and cold spots occur in the vehicle. This also creates flanking paths for sound.

Another type of insulation often used in mobile vehicles is batten fiberglass similar to that used in residential construction. If the fiberglass can be held in place and stays dry, it is satisfactory. However, if the fiberglass becomes wet from leakage or condensation, it offers very little insulation quality.

Electrical Systems

The electrical system of the vehicle will either be an on-board generator, a towed-behind generator, or a lead-in shoreline power cable. The majority of mobile audiometric test units in use today utilize lead-in shoreline power cables. These units will either have a conventional 3-wire 120-volt 60-cycle single-phase 20-amp electrical system or a 220-volt system. The advantage of the 110-volt system is that no special wiring hook-up needs to be obtained where the vehicle is located. If the vehicle is planned for other forms of industrial hearing testing such as hearing aid dispensing, audiometric screening, or diagnosis, it is easier to utilize this type of vehicle rather than one that has a 220-volt lead-in power source.

All wiring should be installed inside conduits, and commerical quality breaker box circuit systems should be designed into the mobile unit electrical system. All air conditioning, heating, audiometric equipment, and interior/exterior lighting circuits should have different breakers so that if there is a problem with equipment, it is easy to identify which circuit is involved.

Air Conditioning and Heating

Many of the units utilized for mobile audiometry have roof-top mounted air conditioners. These systems are the most efficient. As the cooled air settles to the floor, the entire unit has even temperature. However, if these air conditioners are not properly vibration isolated, a great deal of noise contamination in the testing environment will derive from the vibration of the air conditioner. Many manufacturers of mobile audiometric units utilize an exterior mounted heating/air conditioning unit which is either mounted outside of the body itself, or, in some situations, underneath the body in its own compartment suspended from the bottom of the frame of the vehicle by vibration mounts. This exterior mounting of the heating and air conditioning system is then connected to an air duct which should also be isolated for vibration purposes. Then the air duct is directed into the body of the test environment, preferably via a flexible air conditioning assembly that does not transmit the noise or vibration into the test environment.

Before making a decision on these units, be sure to check the air conditioner noise in the "high" fan setting. Then obtain a sound level reading to determine the ambient noise in the test environment when the air conditioner is on and the compressor is working. The air conditioner can be the most significant single noise contaminant in the acoustical environment of the mobile unit.

Heating Systems

In some cases, heating can be a problem, especially in the extreme northern areas of the country. Propane heating systems on mobile medical units are not recommended. The propane tanks are normally stored outside for safety purposes. If the propane is exposed to extreme cold weather it will have a tendency to jell, reducing tank pressure, and the propane heater will not work properly.

Self-propelled units should have an auxiliary heater in the test area. While in-transit to the test location, initial heating of the test area will begin with warmth from the engine block. Upon arrival at the test area, the unit will warm to a comfortable temperature. A radiant electric fan-forced heater will then keep the temperature at a comfortable level. In most instances, with the insulation that a quality medical body will have for acoustical and thermal purposes, once it is warmed up to a comfortable level, it is extremely easy to keep the unit warm to this level.

Interior Lighting

Interior lighting of the test unit is usually provided by fluorescent light fixtures surface mounted to the interior body ceiling with plastic diffuser panels. A full spectrum fluorescent light is advised. This not only enables one to see better, but studies have demonstrated less eye fatigue and better worker productivity under this type of natural lighting. All work surfaces should have a minimum light density of 75 candles at working distances in all parts of the interior test unit. If diffusers are used, they should be equipped with a latching device to prevent them from falling off in-transit. A 12-volt lighting source should be installed within the test unit with a switch at the door entrance. This enables one to enter for brief periods without hooking the unit up to a 110- or 220-volt shoreline or starting the generator.

Interior Wall and Floor Furnishings

Most quality audiometric mobile units will utilize some type of acoustical wall covering or even a short nap carpeting on the walls. This is very attractive, it is easy to maintain, and it also has an acoustical absorption quality. There should be a bright but not glossy finish on the ceilings. Acoustical ceiling tile is attractive, and can be utilized very effectively for ceiling coverings.

Floor coverings can vary. I feel that carpeting is probably the worst choice for an industrial test vehicle, as it will require replacing on an annual basis. Commerical quality Congoleum is a good choice; however, the best choice for flooring in any type of a medical mobile unit is the newer style, more expensive, industrial raised vinyl disc flooring. It is good looking and wears well. Most importantly, it is easy on a person's feet and legs. There is much less fatigue when standing on this type of flooring and moving around on it all day.

Miscellaneous Items

There are various options that may either be overlooked when purchasing a vehicle or perhaps viewed as an unnecessary expense. The following are several to be incorporated in all mobile audiometric test units.

FIRE EXTINGUISHERS. A Halon-type fire extinguisher should be installed in the interior body area of the vehicle in the event of an electrical fire. This type of extinguisher emits material that will be less likely to ruin audiometric equipment. Furthermore, there should be a dry chemical Type-A fire extinguisher in the cab of the vehicle or in the cab of the towing vehicle to combat an engine fire.

BACKUP ALARM. A backup alarm belongs on the vehicle, whether it be a trailer or a self-propelled unit. It should be an audible high/low frequency alarm that will automatically activate whenever the vehicle transmission is placed into reverse gear. In some industrial environments, these backup alarms are mandatory.

RADIO. A good quality radio is a necessity in a mobile audiometric test unit. Technicians will be spending long hours driving this vehicle, and a radio can make this time more enjoyable. The radio reduces driver fatigue, and it may even keep one from falling asleep at the wheel during long drives.

CB RADIO/MOBILE PHONE. If technicians operating the vehicle are not familiar with possible mechanical problems, a CB radio is a must. Better yet, a cellular mobile phone is recommended if they are available in the area being serviced. Another advantage of a cellular mobile phone is that the liasion person at the testing site can be conveniently alerted as to the arrival time at the plant.

Size and Floor Plan

When buying a unit, the person selling the units deals on a profit margin based on total cost. Thus, in many instances, people are oversold on the size of the mobile audiometric test vehicles they need for what they are going to do with these vehicles. One should take into consideration not only what the monthly purchase or lease cost is going to be, but also maintenance, license, and insurance.

It's great to have a large test facility to test 1,000 plus employees at one location; but unless one has the luxury of having a small test facility as well, that same unit will go down the road to the plant that has 35 people to test. Calculate how cost effective this is going to be. Although there is a place for large multi-person test units, it is important to take into consideration the current needs in relation to the size of the vehicle under consideration. Be realistic. Everyone hopes to have large mobile test organizations at some time, however, it is best to buy for today—not next year.

Regarding floor plans and design layouts, everyone works a little differently. I am an audiologist who has spent countless hours in mobile vehicles engaged in all types of audiometric activities—industrial, hearing aid dispensing, and clinical audiometry. I have established a definite feeling for the type of design layout that is most efficient and most functional for me. However, another person may hate it. When making a design layout of a proposed hearing test facility, take the perimeters of space that are available in the vehicle and lay it out full scale on a floor somewhere. This could be a basement floor at home or a large room in an office. Measure each specific wall and door, then using masking tape, lay out the inside fixtures: desk, stools, benches, sound enclosures, etc. Then walk through it. Take friends or whoever will be working with this unit through it. Let them walk through with mock-up employees or patients to see, if, in fact, this particular floor plan is functional and efficient for the type of testing envisioned. Is there a bottleneck? Is there something about this floor plan that may be upsetting if one has to work in it over a long period of time? Remember, mobile units are cramped for space to start with and if there is a problem with the design at the onset, it can only get worse.

The most common problem or bottleneck regarding patient or employee flow are door openings that are too small. Exterior entrance doors should not be less than 30'' wide, preferably 32''. Interior passageways, for example, between the corner of the audiometer console and the wall, should be at least 28'' wide. Normally, a mobile hearing conservation unit is utilized to test the hearing acuity of fully grown men. They must be able to move through the unit quickly, comfortably, and efficiently.

CONCLUSION

When becoming involved with mobile hearing testing, it will be extremely beneficial to consult with a professional already active in this area, preferably, one in the same specific area of mobile testing. One may have to do some inquiring to find the right person and then incur the traveling expense to visit them. It is certainly not advisable to consult the person closest to you regarding this matter, as in all probability, that will be the competition.

If approached in the proper perspective, mobile audiology services can be a cost-effective asset to any type of audiology practice.

18

CHAPTER

Shaping Managerial Attitudes for Hearing Conservation

David M. Lipscomb

Preceding sections have offered background information and techniques for developing and maintaining a quality hearing conservation program. The emphasis in this section of the book has been on "marketing" the services one hopes to provide to industry, the military, or to school systems. Marketing techniques range widely in style and content. Attitude shaping, attitude manipulation, and attitude alteration are part of everyday life. When two people set about to accomplish a task, they both have somewhat different ideas about how to undertake the job. Each works on the attitude of the other until the project becomes a joint venture. Otherwise, the task will remain undone or the time required to finish is lengthened.

There are numerous approaches to promoting hearing conservation. At some point, management must be included because funding for the program comes under the purview of managers at several levels. It is not uncommon for operatives in a

hearing conservation program to encounter resistance from management, particularly middle management. These managers are challenged to see that products are produced with a profit or to see that an operation runs smoothly with minimal expense. The superficial view they may take of the hearing conservation program is that it is costly and cuts into the infamous "bottom line." Further, the hearing conservation program may have been forced on them by upper management due to enforcement agency pressure. Therefore, the introduction these people have to hearing conservation may be negative. Somehow, an evolutionary process must be initiated to develop a positive attitude for hearing conservation in the minds of management at all levels.

Pockets of managerial resistance to the hearing conservation program can severely limit the success of that program. Opposition develops for several reasons:

First, budgetary concerns cause managers to oppose any new program that will

301

either require additional staff or take time from duties of existing staff. In the case of hearing conservation, management must be made to realize that two and three decades into the future, reduced compensation costs (including legal fees) will more than pay for the costs and efforts extended in the present time.

Second, the OSHA guidelines had more than a 10 year gestation period. During that time, the specter of hearing conservation was on the horizon, but many managers got into the habit of saying "We don't know what will be required, so we will wait until the directives come down." Now, the guidelines have been promulgated; have met a court challenge; have been rescinded; have been reinstated; and, at this writing, they have the force of law. Most salient features of the program are outlined in the OSHA guidelines (Appendix A) for those locations that come under the regulatory authority of the Department of Labor. Those outside that regulatory control can adopt the OSHA guidelines for their operation as well. Managers of operations outside OSHA regulatory authority are acutely aware that they do not have to meet regulations. However, that does not absolve their operation from future legal actions by employees seeking compensation for noise-induced hearing impairment.

Third, because they do not appreciate the reasons behind the hearing conservation program, some members of management feel that the program as proposed to them makes little or no sense. It is a "spur road" that does not contribute directly to the manufacturing of a product or to providing a service. Therefore, managers may wonder: "Why is it necessary to undertake this project?" People understand why safety hats and glasses are worn. They comprehend the value of gloves and steel-toed shoes. But earplugs too?

POSSIBLE APPROACHES

There are a number of points one can make when speaking with management about the merits of a hearing conservation program:

It's the law! The OSHA guidelines have the force of law for most employers in the United States. In fact, hearing conservation has been mandated for at least some segments of the manufacturing arena since 1969. Further, it can be stated that failure to develop a hearing conservation program deemed to be acceptable by enforcement authorities can result in citation and fines. This is the ultimate inducement for some to accept the hearing conservation activity.

It can reduce legal claims! An effective hearing conservation program will forestall any hearing impairment attributable to occupational noise exposure. The only remaining segment of the work force that may have some claim would be those individuals in the employ of the company prior to the installation of the hearing conservation program. The additional point can be made that combating legal claims will significantly increase costs due to attorney and court costs. This provides a valid economic reason for management to smile upon the hearing conservation program.

It can get the union off our back! Hearing conservation programs have become a negotiable item between management and labor in some segments of the industrial world. Shortly after the initial legislative action in the revision to the Walsh-Healy Act in 1969, organized labor took note of the fact that improvements in work place conditions should include acoustic considerations and hearing conservation programs. During the negotiation stages, agreement to develop or expand hearing conservation activities may be a useful pawn in the discussion process.

The costs can be amortized over the years! Expenditures for this year will bear fruit literally decades into the future. Thus, the program can appropriately be touted as a meaningful hedge against future financial outlay.

The competitors are doing it! In those areas where competition between entities is not only for a share of the market but also for the workforce personnel themselves, initiation of a hearing conservation program by one entity may tip the scales in their favor with regard to obtaining qualified and quality employees.

It will preserve hearing! We who are engaged in the business and services involving hearing conservation should never lose sight of the fact that hearing is our most precious social and learning sense. This observation must be made repeatedly when speaking with management. It may seem naïve to emphasize the human qualities of hearing conservation in light of the other compelling reasons cited. Yet, preservation of that feature of human existence which largely distinguishes humans from animals cannot be deemed "hokey" or "unmentionable." Regardless of the facade presented by management, they are still human beings with feelings for the human condition. It is appropriately within the context of hearing conservation to speak in praise of the ear in order to foster a new and deeper appreciation of the human qualities provided by hearing. Further, we should attempt to illustrate the depth of loss when someone is deprived of some portion of the hearing sense due to thoughtless disregard by persons who have some responsibility for that person's welfare.

The remainder of this chapter will be devoted to listing some of the dazzling features of the ear and hearing. It is my intent to provide an additional list of reasons for hearing conservation to hearing conservationists. At the same time, some ammunition for discussions with management will be offered.

General Observations

The ear is primarily a physiological device which operates in the mechanical domain. It is a passive receiver moving in direct response to acoustic driving signals. The inner ear is an analyzer which mechanically separates complex acoustic signals into their pure tone component parts, taking heed of the relative amplitude of all frequencies represented in a sound. Thus, this tiny, fluid-filled compartment that is smaller than a pencil eraser performs Fourier analysis on acoustic messages *in real time*. The ear is a complex mechanism because it is designed to handle information presented in complex acoustical stimuli. Ultimately, the inner ear encodes the "intelligence" and initiates nerve impulses that are then transmitted along intricate neurological pathways to the central nervous system.

Sensitivity

The ear is capable of sensory response to sound whose pressure at the ear drum is no greater than two ten-thousandths of a millionth of barometric pressure (0.0002 microbar). In his presentations, Dr. Wayne Rudmose equated that force to "about the push of a healthy mosquito." This pressure moves the ear drum about one one-hundred-millionth of an inch (Davis & Silverman, 1970, p. 20). That dimension is approximately one one-hundredth the width of a hydrogen molecule, the tiniest of all known molecules. Therefore, throughout a significant portion of the ear's dynamic range, it is moving in *sub-molecular dimensions*. That's sensitivity!

Dynamic Range

One would estimate that a device with the impressive sensitivity of the ear would also be delicate and fragile. That is true. However, the ear still provides us with a surprisingly wide dynamic range. Between the softest sound audible (0 dB) to the highest level sound tolerable (approximately 140 dB—the pain threshold for sound) are sound energy increments of 10^{14} power (Hirsh, 1952, p. 107). Therefore, the dynamic range of the human ear encompasses a sound energy increase ratio of one hundred trillion to one.

Importance of the Ear and Hearing

There are numerous indications of the vital character of the ear and hearing:

FETAL DEVELOPMENT. The inner ear is one of the first complex organisms to be completed in the developing fetus. The cochlea is of nearly adult size and complexity by the end of the third month of pregnancy.

BACKUP SYSTEM. The ear has an independent and complete backup system—there are two ears. Diseases or injuries that impair or destroy one ear does not leave the victim without hearing. In this context, it might be surmised that hearing is more important to the survival of the species than is the case with other vital systems. This point comes clear when one reviews the phylogenic development of the ear (Stevens & Warshofsky, 1971; van Bergeijk, 1967). Each body is equipped with only one heart, and there is no backup. Of course, the heart is far more vital to sustaining an *individual's* life than are the ears, but for perpetuating the species, binaural hearing is vital. It provides our only distant 360° warning sense. (Vision is good for slightly over 180° of the visual field. Taction also provides 360° sensitivity, but it does not offer the distant warning.)

AUDITORY LATERALIZATION AND LOCALIZATION. Two ears provide distinct advantages over one in that two points are provided allowing for *auditory triangulation* (Durrant & Lovrinic, 1984, p. 251; Shaw, 1974, p. 1848; Yost & Nielsen, 1977, p. 156). In addition to the previously mentioned backup system in the event one ear becomes abnormal, there is the *localization* capability, which is possible only with two points with which to *aud* a given signal. Differences in time of arrival (phase differences) and variations in level of a sound reaching the two ears provides the ability to *lateralize* sound—to tell whether a sound is arriving from a point midline to the head or displaced to one side or the other. The convoluted configuration of the concha (the opening portion of the ear canal) causes a variety of reflections of sound depending upon the source of the sound. Therefore, the ability is given to determine whether a sound originated from front, back, up or down—*sound localization* (Shaw, 1974). In a work environment, it is often necessary to quickly determine the source of some sound

that is indicating impending danger. The loss of part of one ear can impair that ability and increase the danger quotient for workers.

REDUNDANCY. It has long been known that the most important social communicative stimulus we hear, speech, is redundant by nature. Significant portions of the speech signal can be degraded or removed (e.g., by frequency filtering or by rapidly switching the signal on and off) with little or no loss of intelligibility of the signal. In essence, we have more speech signal than is absolutely necessary for the communication process. In like manner, the encoding of neural signals by the ear is more redundant than is absolutely necessary to process complex signals like speech. This is fortunate for those who have lost some hearing. Their residual hearing allows them to continue in a communicating society without total loss of the contribution of hearing. Naturally, the loss of some of the sensory cells in the organ of Corti reduces the redundant features of the hearing sense (physiologically degrades the signal). Therefore, difficult listening conditions are made more difficult, and in some cases, impossible.

Complexity of Auditory Function

There are approximately 16,000 sensory hair cells in the cochlea of each ear (roughly 12,000 outer hair cells and 4,000 inner hair cells) (Zemlin, 1968, p. 411). These sensory cells are stimulated in a predictable manner by various sounds according to the frequency and amplitude of each tonal component of the acoustic signal. Serving these hair cells are approximately 31,000 afferent (sensory) nerve fibers (Zemlin, 1968, p. 411). There is not a simple 2:1 relationship between the nerves and the cells, however. About 95 percent of the nerve fibers serve the 4,000 inner hair cells, leaving only 5 percent of the fibers to innervate the 12,000 outer hair cells (Spoendlin, 1976). This must be accomplished by having several outer hair cells served by a single nerve fiber in a complex pattern which is not yet fully understood. But each inner hair cell has a cluster of afferent (sensory) nerve fibers. The pathway to the auditory centers of the central nervous system is also complex, including several way-stations en route to the temporal lobe. By putting together the research findings of several experimental neurophysiologists, one can estimate that a single sensory nerve fiber leaving the cochlea has the potential to stimulate as many as 618,000 cells in the auditory cortex. The fiber next to it has the same potential, but with a slightly different pattern, resulting in a discrete "auditory picture" presented to the cortex by each reporting nerve fiber. With approximately 31,000 fibers involved, it is possible to know, but perhaps not to grasp, the complexity involved in auditory neural function. This complexity baffles the research community to this day. But, without such complex function, humans would still be communicating using a primitive series of squawks, squeaks, grunts, and whistles. The loss of sensory components leaves blank some portions of those "auditory pictures," impairing the ability to receive and act upon speech and other forms of communication input.

Discrimination Ability

The "normal" human ear can detect up to 280 changes in amplitude from threshold to the highest level sound tolerable (difference limens for intensity). Further, the "normal" ear is able to distinguish about 1,400 discrete pitch changes throughout the human audible spectrum (difference limens for frequency). Combining those values, it is possible to state that the ear can make approximately 400,000 discrete amplitude/frequency discriminations (Denes & Pinson, 1973, p. 115). Again, the loss of sensory elements due to noise exposure significantly decreases that impressive discrimination ability.

Dealing with Noise

Psychoacoustical studies of human response to speech in the presence of noise have shown that a person with normal hearing can discriminate speech even when the level of the noise is greater (in dB) than the level of speech. This indicates that one can expect to communicate with at least some degree of adequacy in a negative signal-to-noise (S/N) condition. However, when the sensory hair cell population of the cochlea is depleted in numbers, one requires a more and more favorable (positive) S/N to understand speech communication in the presence of noise. That concept forms the basis for differences between normal and nonnormal ears in performing psychophysical tuning curves (Durrant & Lovrinic, 1984, pp. 239–242).

Learning and Communication

Hearing provides us with the most valuable sense of all with respect to communication and learning. A congenitally deaf child is born with a serious "learning problem" in that hearing, of all the senses, is most capable of helping the developing child to understand concepts such as colors, time, locations, sizes, names, words, speech, and grammar. Adventitiously deafened adults begin to lose the fine points of voice control and speech production accuracy quite soon following severe hearing impairment. Although adults will have acquired most of the basic concepts that are so difficult for deaf children, progressive learning through the auditory channels will be seriously impacted, which can increase the time necessary for learning new materials or can decrease one's ultimate achievement.

Slight hearing impairments are also damaging to learning and speech communication. It is not uncommon for persons with slight sensory hearing impairments to have speech discrimination problems, particularly in difficult listening situations. That can be a handicap in a noisy classroom or work place. The advent of a hearing problem can result in social withdrawal due to the self-consciousness that some persons develop. They misunderstand questions and comments, then feel ridiculous upon discovering how inappropriate their response was judged to be. To avoid such ignominy, they decide to avoid those situations, thus, the social withdrawing process begins. It is sad to note that many of these social problems could have been avoided by earlier hearing conservation efforts. The most distinguishing aspect of noise-induced hearing impairment is that, in most cases, it is completely preventable.

CONCLUSION

All of the impressive features mentioned previously are found in a single sensory mechanism, the receptor portion of which is contained in a tiny capsule deep within the most dense bone in the body (the petrous bone of the skull). Nature has signaled to us the significance of this structure, but many have not heeded those signals intelligently. It remains for the hearing conservationist, regardless of the environment in which he or she works, to strive to preserve this precious sensory mechanism and its function for those whom we serve.

Anything to be conserved must be worthy of the effort required. It is abundantly evident that the ear and hearing meets that criterion. By keeping the remarkable functions of the ear in mind, one is restimulated to continue and to increase conservation endeavors.

The development of the hearing mechanism; its value to us today, and the need to preserve this sense is summarized in what might be called *the economy of the creation* (Lipscomb, 1982, pp. 178–179):

- The ear takes up very little space, an economy that keeps the head no larger than necessary
- Throughout the progressive stages of development, ear forms were provided which adequately served but did not outperform the central nervous system, an economy that precluded expanding ability beyond the point of diminishing returns

- Redundancy in auditory signals has been matched by redundancy inherent in auditory function to economize on transfer of the information content from each stimulus
- The developing inner ear took on an increasing role as an analyzer, leaving the central nervous system free to develop greater perceptual skills, an economy in neural organization
- Providing the simplest structure that could accomplish a given task has been the policy in design of the auditory mechanism throughout the range of structure variations, an economy in tissue acquisition and usage
- No special anatomic, biochemical, mechanical, or architectural principles were invented to serve audition. The ear was designed utilizing existing principles according to the most economical method to employ each principle or set of principles
- In sum, the design and structure of the ear are so complete and so intelligent that it is an impossibility to conceive or to manufacture a system of the size and complexity of the ear which even comes close to the quality of performance we take for granted in our hearing sense.

It makes imminent sense to extend our efforts to preserve a sensory function that has such a dignified history and impressive design.

REFERENCES

Davis, H., and Silverman, S. R. (1970) *Hearing and Deafness.* (3rd ed.). New York: Holt, Rinehart and Winston.

Denes, P. B. & Pinson, E. N. (1973). *The speech chain: The physics and biology of spoken language.* Garden City, NY: Doubleday.

Durrant, J. D. & Lovrinic, J. H. (1984). *Bases of hearing science* (2nd. ed.). Baltimore: Williams & Wilkins.

Hirsh, I. J. (1952). *The measurement of hearing.* New York: McGraw-Hill Book Company, Inc.

Lipscomb, D. M. (1982). Anatomy and physiology of the hearing mechanism. In N. J. Lass, L.

V. McReynolds, J. L. Northern, and D. E. Yoder (Eds.), Speech, langauge and hearing: Vol. 1. Normal Processes (pp. 156–179). Philadelphia: W. B. Saunders Company.

Shaw, E. A. G. (1974). Transformation of sound pressure level from the free field to the eardrum in the horizontal plane. *J. Acoust. Soc. Am.*, *56*,(6), 1848–1861.

Spoendlin, H. (1976). Neural connections of the outer hair cell system. *Acta otolaryngol*, *87*, 381–387.

Stevens, S. S. & Warshofsky, F. (1971). *Sound and hearing*. New York: Time-Life Books.

van Bergeijk, W. A. (1967). The evolution of vertebrate hearing. In W. D. Neff, (Ed.), *Contributions to sensory physiology*. (Vol. 2). New York: Academic Press.

Yost, W. A. and Nielsen, D. W. (1977). *Fundamentals of hearing: An introduction*. New York: Holt, Rinehart and Winston.

Zemlin, W. R. (1968). *Speech and hearing science: Anatomy and physiology*. Englewood Cliffs, NJ: Prentice-Hall, Inc.

Epilogue

David M. Lipscomb ■

Hearing conservation is a remarkably descriptive term for that aspect of our profession which offers assistance to those who experience threats to their hearing in one or more facets of their life. It offers a challenge to persons willing to undertake the work required to become a capable hearing conservation professional. Information contained in this and other books on the subject will give readers ideas, facts, techniques, and reference bases. The one feature that no book can instill is the desire to serve and to contribute to the value of individual lives by conserving one of the valuable senses provided to us. That characteristic must come from within each hearing conservationist.

Approximately half of this text was prepared by outstanding professionals who took time from overcrowded schedules to share their knowledge and insight with professionals in training. I am deeply grateful to each one of these friends who responded to my invitation to participate in the preparation of the book and who put up with my reminders about schedules and quality control. What we have prepared is a compilation of information that transcends that which any single person could provide to the students in training and those professionals who utilize this book as reference material. The topic has been covered just about as completely as I know to do it—yet, there is more.

Space and time would not allow the insertion of consideration of legal matters, which is another of the important topical areas professionally adjacent to the hearing conservation movement. Doubtless, hearing conservationists will find themselves being called to court as an expert witness. The legal arena presents emerging professional opportunities that are sufficiently numerous and information-packed that a separate text is being planned for that topic alone.

APPENDIX

Department of Labor Occupational Noise Exposure Standard

(Code of Federal Regulations, Title 29, Chapter XVII, Part 1910, Subpart G, 36 FR 10466, May 29, 1971; Amended by 46 FR 4161, January 16, 1981; Amended by 48 FR 9776, March 8, 1983)

§ 1910.95 Occupational noise exposure.

(a) Protection against the effects of noise exposure shall be provided when the sound levels exceed those shown in Table G–16a when measured on the A scale of a standard sound level meter at slow response. When noise levels are determined by octave band analysis, the equivalent A-weighted sound level may be determined as follows:

(b)(1) When employees are subjected to sound exceeding those listed in Table G–16a, feasible administrative or engineering controls shall be utilized. If such controls fail to reduce sound levels within the levels of Table G–16a, personal protective equipment shall be provided and used to reduce sound levels within the levels of the table.

(2) If the variations in noise level involve maxima at intervals of 1 second or less, it is to be considered continuous.

(c) *Hearing conservation program.* (1) The employer shall administer a continuing, effective hearing conservation program, as described in paragraphs (c) through (o) of this section, whenever employee noise exposures equal or exceed an 8-hour time-weighted average sound level (TWA) of 85 dB measured on the A scale (slow response) or, equivalently, a dose of fifty percent. For purposes of the hearing conservation program, employee noise exposures shall be computed in

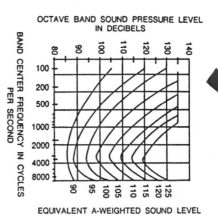

Figure G.9. Equivalent sound level contours. Octave band sound pressure levels may be converted to the equivalent A-weighted sound level by plotting them on this graph and noting the A-weighted sound level corresponding to the point of highest penetration into the sound level contours. This equivalent A-weighted sound level, which may differ from the actual A-weighted sound level of the noise, is used to determine exposure limits from Table G–16a.

accordance with Appendix A and Table G–16a, and without regard to any attenuation provided by the use of personal protective equipment.

(2) For purposes of paragraph (c) through (n) of this section, an 8 hour time-weighted average of 85 dB or a dose of fifty percent shall also be referred to as the action level.

(d) *Monitoring.* (1) When information indicates that any employee's exposure may equal or exceed an 8 hour time-weighted average of 85 dB, the employer shall develop and implement a monitoring program. (i) The sampling strategy shall be designed to identify employees for inclusion in the hearing conservation program and to enable the proper selection of hearing protectors.

(ii) Where circumstances such as high worker mobility, significant variations in sound level, or a significant component of impulse noise make area monitoring generally inappropriate, the employer shall use representative personal sampling to comply with the monitoring requirements of this paragraph unless the employer can show that area sampling produces equivalent results.

(2)(i) All continuous, intermittent and impulsive sound levels from 80 to 130 dB shall be integrated into the noise measurements.

(ii) Instruments used to measure employee noise exposure shall be calibrated to ensure measurement accuracy.

(3) Monitoring shall be repeated whenever a change in production, process, equipment or controls increases noise exposures to the extent that:

(i) Additional employees may be exposed at or above the action level; or

(ii) The attenuation provided by hearing protectors being used by employees may be rendered inadequate to meet the requirements of paragraph (j) of this section.

(e) *Employee notification.* The employer shall notify each employee exposed at or above an 8 hour time-weighted average of 85 dB of the results of the monitoring.

(f) *Observation of monitoring.* The employer shall provide affected employees or their representatives with an opportunity to observe any noise measurements conducted pursuant to this section.

(g) *Audiometric testing program.* (1) The employer shall establish and maintain an audiometric testing program as provided in this paragraph by making audiometric testing available to all employees whose exposures equal or exceed an 8-hour time-weighted average of 85 dB.

(2) The program shall be provided at no cost to employees.

(3) Audiometric tests shall be performed by a licensed or certified audiologist, otolaryngologist, or other physician, or by a technician who is certified by the Council of Accreditation in Occupational Hearing Conservation, or who has satisfactorily demonstrated competence in administering audiometric examinations, obtaining valid audiograms, and properly using, maintaining and checking calibration and proper functioning of the audiometers being used. A technician who operates microprocessor audiometers does not need to be certified. A technician who performs audiometric test must be responsible to an audiologist, otolaryngologist or physician.

(4) All audiograms obtained pursuant to this section shall meet the requirements of Appendix C: *Audiometric Measuring Instruments.*

(5) *Baseline audiogram.* (i) Within 6 months of an employee's first exposure at or above the action level, the employer shall establish a valid baseline audiogram against which subsequent audiograms can be compared.

(ii) *Mobile test van exception.* Where mobile test vans are used to meet the audiometric testing obligation, the employer shall obtain a valid baseline audiogram within 1 year of an employee's first exposure at or above the action level. Where baseline audiograms are obtained more than 6 months after the employee's first exposure at or above the action level, employees shall wear hearing protectors for any period exceeding 6 months after first exposure until the baseline audiogram is obtained.

(iii) Testing to establish a baseline audiogram shall be preceded by at least 14 hours without exposure to workplace noise. Hearing protectors may be used as a substitute for the requirement that baseline audiograms be preceded by 14 hours without exposure to workplace noise.

(iv) The employer shall notify employees of the need to avoid high levels of non-occupational noise exposure during the 14-hour period immediately preceding the audiometric examination.

(6) *Annual audiogram.* At least annually after obtaining the baseline audiogram, the employer shall obtain a new audiogram for each employee exposed at or above an 8 hour time-weighted average of 85 dB.

(7) *Evaluation of audiogram.* (i) Each employee's annual audiogram shall be compared to that employee's baseline audiogram to determine if the audiogram is valid and if a standard threshold shift as defined in paragraph (g)(10) of this section has occurred. This comparison may be done by a technician.

(ii) If the annual audiogram shows that an employee has suffered a standard threshold shift, the employer may obtain a retest within 30 days and consider the results of the retest and the annual audiogram.

(iii) The audiologist, otolaryngologist, or physician shall review problem audiograms and shall determine whether there is a need for further evaluation. The employer shall provide to the person performing this evaluation the following information:

(A) A copy of the requirements for hearing conservation as set forth in paragraphs (c) through (n) of this section;

(B) The baseline audiogram and most recent audiogram of the employee to be evaluated;

(C) Measurements of background sound pressure levels in the audiometric test room as required in Appendix D: Audiometric Test Rooms.

(D) Records of audiometer calibrations required by paragraph (h)(5) of this section.

(8) *Follow-up procedures.* (i) If a comparison of the annual audiogram to the baseline audiogram indicates a standard threshold shift as defined in paragraph (g)(10) of this section has occurred, the employee shall be informed of this fact in writing, within 21 days of the determination.

(ii) Unless a physician determines that the standard threshold shift is not work related or aggravated by occupational noise exposure, the employer shall ensure that the following steps are taken when a standard threshold shift occurs:

(A) Employees not using hearing protectors shall be fitted with hearing protectors, trained in their use and case, and required to use them.

(B) Employees already using hearing protectors shall be refitted and retained in the use of hearing protectors and provided with hearing protectors offering a greater attenuation if necessary.

(C) The employee shall be referred for a clinical audiological evaluation or an otological examination, as appropriate, if additional testing is necessary or if the employer suspects that a medical pathology of the ear is caused or aggravated by the wearing of hearing protectors.

(D) The employee is informed of the need for an otological examination if a medical pathology of the ear that is unrelated to the use of hearing protectors is suspected.

(iii) If subsequent audiometric testing of an employee whose exposure to noise is less than an 8 hour TWA of 90 dB indicates that a standard threshold shift is not persistent, the employer:

(A) Shall inform the employee of the new audiometric interpretation; and

(B) May discontinue the required use of hearing protectors for that employee.

(9) *Revised baseline.* An annual audiogram may be substituted for the baseline audiogram when, in the judgment of the audiologist, otolaryngologist or physician who is evaluating the audiogram:

(i) The standard threshold shift revealed by the audiogram is persistent; or

(ii) The hearing threshold shown in the annual audiogram indicates significant improvement over the baseline audiogram.

(10) *Standard threshold shift.* (1) As used in this section, a standard threshold shift is a change in hearing threshold relative to the baseline audiogram of an average of 10 dB or more at 2000, 3000, and 4000 Hz in either ear.

(ii) In determining whether a standard threshold shift has occurred, allowance may be made for the contribution of aging (presbycusis) to the change in hearing level by correcting the annual audiogram according to the procedure described in Appendix F: *Calculation and Application of Age Correction to Audiograms.*

(h) *Audiometric test requirements.* (1) Audiometric tests shall be pure tone, air conduction, hearing threshold examinations, with test frequencies including as a minimum 500, 1000, 2000, 3000, 4000, and 6000 Hz. Tests at each frequency shall be taken separately for each ear.

(2) Audiometric tests shall be conducted with audiometers (including microprocessor audiometers) that meet the specifications of, and are maintained and used in accordance with, American National Standard Specification for Audiometers, S3.6–1969.

(3) Pulsed-tone and self-recording audiometers, if used, shall meet the requirements specified in Appendix C: *Audiometric Measuring Instruments.*

(4) Audiometric examinations shall be administered in a room meeting the requirements listed in Appendix D: *Audiometric Test Rooms.*

(5) *Audiometer calibration.* (i) The functional operation of the audiometer shall be checked before each day's use by testing a person with known, stable hearing thresholds and by listening to the audiometer's output to make sure that the output is free from distorted or unwanted sounds. Deviations of 10 dB or greater require an acoustic calibration.

(ii) Audiometer calibration shall be checked acoustically at least annually in accordance with Appendix E: *Acoustic Calibration of Audiometers.* Test frequencies below 500 Hz and above 6000 Hz may be omitted from this check. Deviations of 15 dB or greater require an exhaustive calibration.

(iii) An exhaustive calibration shall be performed at least every two years in accordance with sections 4.1.2; 4.1.3; 4.1.4.3; 4.2; 4.4.1; 4.4.2; 4.4.3; and 4.5 of the American National Standard Specification for Audiometers, S3.6–1969. Test frequencies below 500 Hz and above 6000 Hz may be omitted from this calibration.

(i) *Hearing protectors.* (1) Employers shall make hearing protectors available to all employees exposed to an 8 hour time-weighted average of 85 dB or greater at no cost to the employees. Hearing protectors shall be replaced as necessary.

(2) Employers shall ensure that hearing protectors are worn:

(i) By an employee who is required by paragraph (b)(1) of this section to wear personal protective equipment; and

(ii) By any employee who is exposed to an 8 hour time-weighted average of 85 dB or greater, and who:

(A) Has not yet had a baseline audiogram established pursuant to paragraph (g)(5)(ii); or

(B) Has experienced a standard threshold shift.

(3) Employees shall be given the opportunity to select their hearing protectors from a variety of suitable protectors provided by the employer.

(4) The employer shall provide training in the use and care of all hearing protectors provided to employees.

(5) The employer shall ensure proper initial fitting and supervise the correct use of all hearing protectors.

(j) *Hearing protector attenuation.* (1) The employer shall evaluate hearing protector attenuation for the specific noise environments in which the protector will be used. The employer shall use one of the evaluation methods described in Appendix B: *Methods for Estimating the Adequacy of Hearing Protection Attenuation.*

(2) Hearing protectors must attenuate employee exposure at least to an 8 hour time-weighted average

of 90 dB as required by paragraph (b) of this section.

(3) For employees who have experienced a standard threshold shift, hearing protectors must attenuate employee exposure to an 8 hour time-weighted average of 85 dB or below.

(4) The adequacy of hearing protector attenuation shall be re-evaluated whenever employee noise exposures increase to the extent that the hearing protectors provided may no longer provide adequate attenuation. The employee shall provide more effective hearing protectors where necessary.

(k) *Training program.* (1) The employer shall institute a training program for all employees who are exposed to noise at or above an 8 hour time-weighted average of 85 dB, and shall ensure employee participation in such a program.

(2) The training program shall be repeated annually for each employee included in the hearing conservation program. Information provided in the training program shall be updated to be consistent with changes in protective equipment and work processes.

(3) The employer shall ensure that each employee is informed of the following:

(i) The effects of noise on hearing;

(ii) The purpose of hearing protectors, the advantages, disadvantages, and attenuation of various types, and instruction on selection, fitting, use, and care; and

(iii) The purpose of audiometric testing, and an explanation of the test procedures.

(l) *Access to information and training materials.* (1) The employer shall make available to affected employees or their representatives copies of this standard and shall also post a copy in the workplace.

(2) The employer shall provide to affected employees any informational materials pertaining to the standard that are supplied to the employer by the Assistant Secretary.

(3) The employer shall provide, upon request, all materials related to the employer's training and education program pertaining to this standard to the Assistant Secretary and the Director.

(m) *Record keeping.*—(1) *Exposure measurements.* The employer shall maintain an accurate record of all employee exposure measurements required by paragraph (d) of this section.

(2) *Audiometric tests.* (i) The employer shall retain all employee audiometric test records obtained pursuant to paragraph (g) of this section:

(ii) This record shall include:

(A) Name and job classification of the employee;

(B) Date of the audiogram;

(C) The examiner's name;

(D) Date of the last acoustic or exhaustive calibration of the audiometer; and

(E) Employee's most recent noise exposure assessment.

(F) The employer shall maintain accurate records of the measurements of the background sound pressure levels in audiometric test rooms.

(3) *Record retention.* The employer shall retain

records required in this paragraph (m) for at least the following periods.

(i) Noise exposure measurement records shall be retained for two years.

(ii) Audiometric test records shall be retained for the duration of the affected employee's employment.

(4) *Access to records.* All records required by this section shall be provided upon request to employees, former employees, representatives designated by the individual employee, and the Assistant Secretary. The provisions of 29 CFR 1910.20 (a)-(e) and (g)-(i) apply to access to records under this section.

(5) *Transfer of records.* If the employer ceases to do business, the employer shall transfer to the successor employer all records required to be maintained by this section, and the successor employer shall retain them for the remainder of the period prescribed in paragraph (m)(3) of this section.

(n) *Appendices.* (1) Appendices A, B, C, D, and E, to this section are incorporated as part of this section and the contents of these Appendices are mandatory.

(2) Appendices F and G to this section are informational and are not intended to create any additional obligations not otherwise imposed or to detract from any existing obligations.

(o) *Exemptions.* Paragraphs (c) through (n) of this section shall not apply to employers engaged in oil and gas well drilling and servicing operations.

(p) *Startup date.* Baseline audiograms required by paragraph (g) of this section shall be completed by March 1, 1984.

*   *   *   *   *

Appendix A: Noise Exposure Computation

This Appendix is Mandatory

I. Computation of Employee Noise Exposure

(1) Noise dose is computed using Table G–16a as follows:

(i) When the sound level, L, is constant over the entire work shift, the noise dose, D, in percent, is given by: D = 100 C/T where C is the total length of the work day, in hours, and T is the reference duration corresponding to the measured sound level, as given in Table G–16a or by the formula shown as a footnote to that table.

(ii) When the workshift noise exposure is composed of two or more periods of noise at different levels, the total noise dose over the work day is given by:

$$D = 100 \ (C_1/T_1 + C_2/T_2 + \ldots + C_n/T_n),$$

where C_n, indicates the total time of exposure at a specific noise level, and T_n indicates the reference duration for that level as given by Table G–16a.

(2) The 8 hour, time-weighted average sound level (TWA), in dB, may be computed from the dose, in percent, by means of the formula: TWA = 16.61 $\log_{10}$ (D/100) + 90. For an eight hour workshift with the noise level constant over the entire shift, the TWA is equal to the measured sound level.

TABLE G–16a.

A-weighted sound level, L (dB)	Reference duration, T (hour)
80	32
81	27.9
82	24.3
83	21.1
84	16.4
85	16
86	13.9
87	12.1
88	10.6
89	9.2
90	8
91	7.0
92	6.1
93	5.3
94	4.6
95	4
96	3.5
97	3.0
98	2.6
99	2.3
100	2
101	1.7
102	1.5
103	1.3
104	1.1
105	1
106	0.87
107	0.76
108	0.66
109	0.57
110	0.5
111	0.44
112	0.38
113	0.33
114	0.29
115	0.25
116	0.22
117	0.19
118	0.16
119	0.14
120	0.125
121	0.11
122	0.095
123	0.082
124	0.072
125	0.063
126	0.054
127	0.047
128	0.041
129	0.036
130	0.031

(3) A table relating dose and TWA is given in Section II.

In the above table the reference duration, T, is computed by

$$T = \frac{8}{2^{(L-90)/5}}$$

where L is the measured A-weighted sound level.

II. Conversion Between "Dose" and "8 Hour Time Weighted Average" Sound Level

Compliance with paragraphs (c)–(r) of this regulation is determined by the amount of exposure to noise in the workplace. The amount of exposure is usually measured with an audiodosimeter which gives a read-out in terms of "dose." In order to better understand the requirements of the amendment, dosimeter readings can be converted to an "8 hour time-weighted average sound level" (TWA).

In order to convert the reading of a dosimeter into TWA, see Table A–1, below. This table applies to dosimeters that are set by the manufacturer to calculate dose or percent exposure according to the relationships in Table G–16a. So, for example, a dose of 91 percent over an eight hour day results in a TWA of 89.3 dB, and, a dose of 50 percent corresponds to a TWA of 85 dB.

If the dose as read on the dosimeter is less than or greater than the values found in Table A–1, the TWA may be calculated by using the formula: TWA = 16.61 $\log_{10}$ (D/100) + 90 where TWA = 8 hour time-weighted average sound level and D = accumulated dose in percent exposure.

Appendix B: Methods for Estimating the Adequacy of Hearing Protector Attenuation

This Appendix is Mandatory

For employees who have experienced a significant threshold shift, hearing protector attenuation must be sufficient to reduce employee exposure to a TWA of 85 dB. Employers must select one of the following methods by which to estimate the adequacy of hearing protector attenuation.

The most convenient method is the Noise Reduction Rating (NRR) developed by the Environmental Protection Agency (EPA). According to EPA regulation, the NRR must be shown on the hearing protector package. The NRR is then related to an individual worker's noise environment in order to assess the adequacy of the attenuation of a given hearing protector. This Appendix describes four methods of using the NRR to determine whether a particular hearing protector provides adequate protection within a given exposure environment. Selection among the four procedures is dependent upon the employer's noise measuring instruments.

Instead of using the NRR, employers may evaluate the adequacy of hearing protector attenuation by using one of the three methods developed by the National Institute for Occupational Safety and Health (NIOSH), which are described in the "List of Personal Hearing Protectors and Attenuation Data," HEW Publication No. 76–120, 1975, pages 21–37. These methods are known as NIOSH methods #1, #2 and #3. The NRR described below is a simplification of NIOSH method #2. The most complex method is NIOSH method #1, which is probably the most accurate method since it uses the largest amount of spectral information from the individual employee's noise environment. As in the case of the NRR method described below, if one of the

TABLE A–1.
Conversion from "Percent Noise Exposure" or "Dose" to "8 Hour Time-Weighted Average Sound Level" (TWA)

Dose or percent noise exposure	TWA
10	73.4
15	76.3
20	78.4
25	80.0
30	81.3
35	82.4
40	83.4
45	84.2
50	85.0
55	85.7
60	86.3
65	86.9
70	87.4
75	87.9
80	88.4
81	88.5
82	88.6
83	88.7
84	88.7
85	88.8
86	88.9
87	89.0
88	89.1
89	89.2
90	89.2
91	89.3
92	89.4
93	89.5
94	89.6
95	89.6
96	89.7
97	89.8
98	89.9
99	89.9
100	90.0
101	90.1
102	90.1
103	90.2
104	90.3
105	90.4
106	90.4
107	90.5
108	90.6
109	90.6
110	90.7
111	90.8
112	90.8
113	90.9
114	90.9
115	91.1
116	91.1
117	91.1
118	91.2
119	91.3
120	91.3
125	91.6
130	91.9
135	92.2
140	92.4
145	92.7
150	92.9

Dose or percent noise exposure	TWA
155	93.2
160	93.4
165	93.6
170	93.8
175	94.0
180	94.2
185	94.4
190	94.6
195	94.8
200	95.0
210	95.4
220	95.7
230	96.0
240	96.3
250	96.6
260	96.9
270	97.2
280	97.4
290	97.7
300	97.9
310	98.2
320	98.4
330	98.6
340	98.8
350	99.0
360	99.2
370	99.4
380	99.6
390	99.8
400	100.0
410	100.2
420	100.4
430	100.5
440	100.7
450	100.8
460	101.0
470	101.2
480	101.3
490	101.5
500	101.6
510	101.8
520	101.9
530	102.0
540	102.2
550	102.3
560	102.4
570	102.6
580	102.7
590	102.8
600	102.9
610	103.0
620	103.2
630	103.3
640	103.4
650	103.5
660	103.6
670	103.7
680	103.8
690	103.9
700	104.0
710	104.1
720	104.2
730	104.3

(continued)

TABLE A-1. (continued)

Dose or percent noise exposure	TWA
740	104.4
750	104.5
760	104.6
770	104.7
780	104.8
790	104.9
800	105.0
810	105.1
820	105.2
830	105.3
840	105.4
850	105.4
860	105.5
870	105.6
880	105.7
890	105.8
900	105.8
910	105.9
920	106.0
930	106.1
940	106.2
950	106.2
960	106.3
970	106.4
980	106.5
990	106.5
999	106.6

NIOSH methods is used, the selected method must be applied to an individual's noise environment to assess the adequacy of the attenuation. Employers should be careful to take a sufficient number of measurements in order to achieve a representative sample for each time segment.

Note.—The employer must remember that calculated attenuation values reflect realistic values only to the extent that the protectors are properly fitted and worn.

When using the NRR to assess hearing protector adequacy, one of the following methods must be used:

(i) When using a dosimeter that is capable of C-weighted measurements:

(A) Obtain the employee's C-weighted dose for the entire workshift, and convert to TWA (see Appendix A, II).

(B) Subtract the NRR from the C-weighted TWA to obtain the estimated A-weighted TWA under the ear protector.

(ii) When using a dosimeter that is not capable of C-weighted measurements, the following method may be used:

(A) Convert the A-weighted dose to TWA (see Appendix A.

(B) Subtract 7 dB from the NRR.

(C) Subtract the remainder from the A-weighted TWA to obtain the estimated A-weighted TWA under the ear protector.

(iii) When using a sound level meter set to the A-weighting network:

(A) Obtain the employee's A-weighted TWA.

(B) Subtract 7 dB from the NRR, and subtract the remainder from the A-weighted TWA to obtain the estimated A-weighted TWA under the ear protector.

(iv) When using a sound level meter set on the C-weighting network:

(A) Obtain a representative sample of the C-weighted sound levels in the employee's environment.

(B) Subtract the NRR from the C-weighted average sound level to obtain the estimated A-weighted TWA under the ear protector.

(v) When using area monitoring procedures and a sound level meter set to the A-weighting network.

(A) Obtain a representative sound level for the area in question.

(B) Subtract 7 dB from the NRR and subtract the remainder from the A-weighted sound level for that area.

(vi) When using area monitoring procedures and a sound level meter set to the C-weighting network:

(A) Obtain a representative sound level for the area in question.

(B) Subtract the NRR from the C-weighted sound level for that area.

Appendix C: Audiometric Measuring Instruments

This Appendix is Mandatory

1. In the event that pulsed-tone audiometers are used,, they shall have a tone on-time of at least 200 milliseconds.

2. Self-recording audiometers shall comply with the following requirements:

(A) The chart upon which the audiogram is traced shall have lines at positions corresponding to all multiples of 10 dB hearing level within the intensity range spanned by the audiometer. The lines shall be equally spaced and shall be separated by at least ¼ inch. Additional increments are optional. The audiogram pen tracings shall not exceed 2 dB in width.

(B) It shall be possible to set the stylus manually at the 10-dB increment lines for calibration purposes.

(C) The slewing rate for the audiometer attenuator shall not be more than 6 dB/sec except that an initial slewing rate greater than 6 dB/sec is permitted at the beginning of each new test frequency, but only until the second subject response.

(D) The audiometer shall remain at each required test frequency for 30 seconds (±3 seconds). The audiogram shall be clearly marked at each change of frequency and the actual frequency change of the audiometer shall not deviate from the frequency boundaries marked on the audiogram by more than ±3 seconds.

(E) It must be possible at each test frequency to place a horizontal line segment parallel to the time axis on the audiogram, such that the audiometric tracing crosses the line segment at least six times at that test frequency. At each test frequency the threshold shall be the average of the midpoints of the tracing excursions.

Appendix D: Audiometric Test Rooms

This Appendix is Mandatory

Rooms used for audiometric testing shall not have background sound pressure levels exceeding those in Table D–1 when measured by equipment conforming at least to the Type 2 requirements of American National Standard Specification for Sound Level Meters, S1.4–1971 (R1976), and to the Class II requirements of American National Standard Specification for Octave, Half-Octave, and Third-Octave Band Filter Sets, S1.11–1971 (R1976).

TABLE D–1.
Maximum Allowable Octave-Band Sound Pressure Levels for Audiometric Test Rooms

Octave-band center frequency (Hz)	500	1000	2000	4000	5000
Sound pressure level (dB)	40	40	47	57	62

Appendix E: Acoustic Calibration of the Audiometer

This Appendix is Mandatory

Audiometer calibration shall be checked acoustically, at least annually, according to the procedures described in this Appendix. The equipment necessary to perform these measurements is a sound level meter, octave-band filter set, and a National Bureau of Standards 9A coupler. In making these measurements, the accuracy of the calibrating equipment shall be sufficient to determine that the audiometer is within the tolerances permitted by American Standard Specification for Audiometers, S3.6–1969.

(1) Sound Pressure Output Check

A. Place the earphone coupler over the microphone of the sound level meter and place the earphone on the coupler.

B. Set the audiometer's hearing threshold level (HTL) dial to 70 dB.

C. Measure the sound pressure level of the tones that each test frequency from 500 Hz through 6000 Hz for each earphone.

TABLE E–1.
Reference Threshold Levels for Telephonics—TDH–39 Earphones

Frequency, Hz	Reference threshold level for TDH–39 earphones, dB	Sound level meter reading, dB
500	11.5	81.5
1000	7	77
2000	9	79
3000	10	80
4000	9.5	79.5
6000	15.5	85.5

TABLE E–2.
Reference Threshold Levels for Telephonics—TDH–49 Earphones

Frequency, Hz	Reference threshold level for TDH–49 earphones, dB	Sound level meter reading, dB
500	13.5	83.5
1000	7.5	77.5
2000	11	81.0
3000	9.5	79.5
4000	10.5	80.5
5000	13.5	83.5

D. At each frequency the readout on the sound level meter should correspond to the levels in Table E–1 or Table E–2, as appropriate, for the type of earphone, in the column entitled "sound level meter reading."

(2) Linearity Check

A. With the earphone in place, set the frequency to 1000 Hz and the HTL dial on the audiometer to 70 dB.

B. Measure the sound levels in the coupler at each 10-dB decrement from 70 dB to 10 dB, noting the sound level meter reading at each setting.

C. For each 10-dB decrement on the audiometer the sound level meter should indicate a corresponding 10 dB decrease.

D. This measurement may be made electrically with a voltmeter connected to the earphone terminals.

(3) Tolerances

When any of the measured sound levels deviate from the levels in Table E–1 or Table E–2 by ±3 dB at any test frequency between 500 and 3000 Hz, 4 dB at 4000 Hz, or 5 dB at 6000 Hz, an exhaustive calibration is advised. An exhaustive calibration is required if the deviations are greater than 10 dB at any test frequency.

Appendix F: Calculations and Application of Age Corrections to Audiograms

This Appendix is Mandatory

In determining whether a standard threshold shift has occurred, allowance may be made for the contribution of aging to the change in hearing level by adjusting the most recent audiogram. If the employer chooses to adjust the audiogram, the employer shall follow the procedure described below. This procedure and the age correction tables were developed by the National Institute for Occupational Safety and Health in the criteria document entitled "Criteria for a Recommended Standard...Occupational Exposure to Noise," [(HSM)-11001].

For each audiometric test frequency:

(i) Determine from Tables F–1 or F–2 the age correction values for the employee by:

(A) Finding the age at which the most recent audiogram was taken and recording the corresponding values of age corrections at 1000 Hz through 6000 Hz;

(B) Finding the age at which the baseline audiogram was taken and recording the corresponding values

TABLE F–1.
Age Correction Values in Decibels for Males

	Audiometric test frequency (Hz)				
Years	1000	2000	3000	4000	6000
20 or younger	5	3	4	5	8
21	5	3	4	5	8
22	5	3	4	5	8
23	5	3	4	6	9
24	5	3	5	6	9
25	5	3	5	7	10
26	5	4	5	7	10
27	5	4	6	7	11
28	6	4	6	8	11
29	6	4	6	8	12
30	6	4	6	9	12
31	6	4	7	9	13
32	6	5	7	10	14
33	6	5	7	10	14
34	6	5	8	11	15
35	7	5	8	11	15
36	7	5	9	12	16
37	7	6	9	12	17
38	7	6	9	13	17
39	7	6	10	14	18
40	7	6	10	14	19
41	7	6	10	14	20
42	6	7	11	16	20
43	8	7	12	16	21
44	8	7	12	17	22
45	8	7	13	18	23
46	8	8	13	19	24
47	8	8	14	19	24
48	9	8	14	20	25
49	9	9	15	21	26
50	9	9	16	22	27
51	9	9	16	23	28
52	9	10	17	24	29
53	9	10	18	25	30
54	10	10	18	26	31
55	10	11	19	27	32
56	10	11	20	28	34
57	10	11	21	29	35
58	10	12	22	31	36
59	11	12	22	32	37
60 or older	11	13	23	33	38

TABLE F–2.
Age Correction Values in Decibels for Females

	Audiometric test frequency (Hz)				
Years	1000	2000	3000	4000	6000
20 or younger	7	4	3	3	6
21	7	4	4	3	6
22	7	4	4	4	6
23	7	5	4	4	7
24	7	5	4	4	7
25	8	5	4	4	7
26	8	5	5	4	8
27	8	5	5	5	8
28	8	5	5	5	8
29	8	5	5	5	9
30	8	6	5	5	9
31	8	6	6	5	9
32	9	6	6	6	10
33	9	6	6	6	10
34	9	6	6	6	10
35	9	6	7	7	11
36	9	7	7	7	11
37	9	7	7	7	12
38	10	7	7	7	12
39	10	7	8	8	12
40	10	7	8	8	13
41	10	8	8	8	13
42	10	8	9	9	13
43	11	8	9	9	14
44	11	8	9	9	14
45	11	8	10	10	15
46	11	9	10	10	15
47	11	9	10	11	16
48	12	9	11	11	16
49	12	9	11	11	16
50	12	10	11	12	17
51	12	10	12	12	17
52	12	10	12	13	18
53	13	10	13	13	18
54	13	11	13	14	19
55	13	11	14	14	19
56	13	11	14	15	20
57	13	11	15	15	20
58	14	12	15	16	21
59	14	12	16	16	21
60 or older	14	12	16	17	22

of age corrections at 1000 Hz through 6000 Hz.

(ii) Subtract the values found in step (i)(A) from the value found in step (i)(B).

(iii) The difference calculated in step (ii) represented that portion of the change in hearing that may be due to aging.

Example: Employee is a 32-year-old male. The audiometric history for his right ear is shown in decibels below.

	Audiometric test frequency (Hz)				
Employee's age	1000	2000	3000	4000	6000
26	10	5	5	10	5
*27	0	0	0	5	5
28	0	0	0	10	5
29	5	0	5	15	5
30	0	5	10	20	10
31	5	10	20	15	15
*32	5	10	10	25	20

The audiogram at age 27 is considered the baseline since it shows the best hearing threshold levels. Asterisks have been used to identify the baseline and most recent audiogram. A threshold shift of 20 dB exists at 4000 Hz between the audiograms taken at ages 27 and 32.

(The threshold shift is computed by subtracting the hearing threshold at age 27, which was 5, from the hearing threshold at age 32, which is 25). A retest audiogram has confirmed this shift. The contribution of aging to this change in hearing may be estimated in the following manner:

Go to Table F–1 and find the age correction values (in dB) for 4000 Hz at age 27 and age 32.

The difference represents the amount of hearing loss that may be attributed to aging in the time period between the baseline audiogram and the most recent audiogram. In this example, the difference at 4000 Hz is 3 dB. This value is subtracted from the hearing level at 4000 Hz, which in the most recent audiogram is 25,

	Frequency (Hz)				
	1000	**2000**	**3000**	**4000**	**6000**
Age 32	6	5	7	10	14
Age 27	5	4	6	7	11
Difference	1	1	1	3	3

yielding 22 after adjustment. Then the hearing threshold in the baseline audiogram at 4000 Hz (5) is subtracted from the adjusted annual audiogram hearing threshold at 4000 Hz (22). Thus the age-corrected threshold shift would be 17 dB (as opposed to a threshold shift of 20 dB without age correction).

Appendix G: Monitoring Noise Levels
Non-Mandatory Informational Appendix

This appendix provides information to help employers comply with the noise monitoring obligations that are part of the hearing conservation amendment.

What is the purpose of noise monitoring?

The revised amendment requires that employees be placed in a hearing conservation program if they are exposed to average noise levels of 85 dB or greater during an 8 hour workday. In order to determine if exposures are at or above this level, it may be necessary to measure or monitor the actual noise levels in the workplace and to estimate the noise exposure or "dose" received by employees during the workday.

When is it necessary to implement a noise monitoring program?

It is not necessary for every employer to measure workplace noise. Noise monitoring or measuring must be conducted only when exposures are at or above 85 dB. Factors which suggest that noise exposures in the workplace may be at this level include employee complaints about the loudness of noise, indications that employees are losing their hearing, or noisy conditions which make normal conversation difficult. The employer should also consider any information available regarding noise emitted from specific machines. In addition, actual workplace noise measurements can suggest whether or not a monitoring program should be initiated.

How is noise measured?

Basically, there are two different instruments to measure noise exposures: the sound level meter and the dosimeter. A sound level meter is a device that measures the intensity of sound at a given moment. Since sound level meters provide a measure of sound intensity at only one point in time, it is generally necessary to take a number of measurements at different times during the day to estimate noise exposure over a workday. If noise levels fluctuate, the amount of time noise remains at each of the various measured levels must be determined.

To estimate employee noise exposures with a sound level meter it is also generally necessary to take several measurements at different locations within the workplace. After appropriate sound level meter readings are obtained, people sometimes draw "maps" of the sound levels within different areas of the workplace. By using a sound level "map" and information on employee locations throughout the day, estimates of individual exposure levels can be developed. This measurement method is generally referred to as area noise monitoring.

A dosimeter is like a sound level meter except that it stores sound level measurements and integrates these measurements over time, providing an average noise exposure reading for a given period of time, such as an 8 hour workday. With a dosimeter, a microphone is attached to the employee's clothing and the exposure measurement is simply read at the end of the desired time period. A reader may be used to read-out the dosimeter's measurements. Since the dosimeter is worn by the employee, it measures noise levels in the locations in which the employee travels. A sound level meter can also be positioned within the immediate vicinity of the exposed worker to obtain an individual exposure estimate. Such procedures are generally referred to as *personal* noise monitoring.

Area monitoring can be used to estimate noise exposure when the noise levels are relatively constant and employees are not mobile. In workplaces where employees move about in different areas or where the noise intensity tends to fluctuate over time, noise exposure is generally more accurately estimated by the personal monitoring approach.

In situations where personal monitoring is appropriate, proper positioning of the microphone is necessary to obtain accurate measurements. With a dosimeter, the microphone is generally located on the shoulder and remains in that position for the entire workday. With a sound level meter, the microphone is stationed near the employee's head, and the instrument is usually held by an individual who follows the employee as he or she moves about.

Manufacturer's instructions, contained in dosimeter and sound level meter operating manuals, should be followed for calibration and maintenance. To ensure accurate results, it is considered good professional practice to calibrate instruments before and after each use.

How often is it necessary to monitor sound levels?

The amendment requires that when there are significant changes in machinery or production processes that may result in increased noise levels, remonitoring must be conducted to determine whether additional employees need to be included in the hearing conservation program. Many companies choose to remonitor periodically (once every year or two) to ensure that all exposed employees are included in their hearing programs.

Walsh-Healey Noise Standard

APPENDIX

The hearing conservation portion of the 1971 Occupational Health and Safety Act (Section 1910.95) was taken directly from the 1969 Revision of the Walsh-Healey Public Contracts Act.

§ 1910.95 Occupational Noise Exposure (Walsh-Healey Noise Standard)

(a) Protection against the effects of noise exposure shall be provided when the sound levels exceed those shown in Table G-16 when measured on the A scale of a standard sound level meter at slow response. When noise levels are determined by octave band analysis,the equivalent A-weighted sound level may be determined as follows:

(b)(1) When employees are subjected to sound exceeding those listed in Table G-16, feasible administrative or engineering controls shall be utilized. If such controls fail to reduce sound levels within the levels of Table G-16, personal protective equipment shall be provided and used to reduce sound levels within the levels of the table.

(2) If the variations in noise level involve maxima at intervals of 1 second or less, it is to be considered continuous.

(3) In all cases where the sound levels exceed the values shown herein, a continuing, effective hearing conservation program shall be administered.

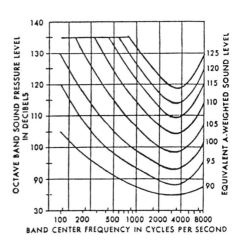

Figure C-9. Rules and Regulations—Equivalent sound level contours. Octave band sound pressure levels may be converted to the equivalent A-weighted sound level by plotting them on this graph and noting the A-weighted sound level corresponding to the point of highest penetration into the sound level contours. This equivalent A-weighted sound level, which may differ from the actual A-weighted sound level of the noise, is used to determine exposure limits from Table I.G-16.

TABLE G–16.

Permissible Noise Exposures[1]

Duration per day, hours	Sound level dBA slow response
8	90
6	92
4	95
3	97
2	100
1½	102
1	105
½	110
¼ or less	115

[1]When the daily noise exposure is composed of two or more periods of noise exposure of different levels, their combined effect should be considered, rather than the individual effect of each. If the sum of the following fraction: $C1/T1 + C2/T2$ Cn/Tn exceeds unity, then, the mixed exposure should be considered to exceed the limit value. Cn indicates the total time of exposure at a specified noise level, and Tn indicates the total time of exposure permitted at that level.

Exposure to impulsive or impact noise should not exceed 140 dB peak sound pressure level.

■ *Author Index*

▪ *Subject Index*

Italic page numbers refer to tables and figures.

Notes

Notes